SUPERCONDUCTOR TECHNOLOGY: APPLI(
ELECTRICAL MACHINES, AND PROPULSION

OPTICAL COMPUTING: AN INTRODUCTIO!

MW00844956

INTRODUCTION TO ELECTROMAGNETIC A!
Gabriel D. Colef, and Raymond L. Camisa

MILLIMETER WAVE OPTICAL DIELECTRIC INTEGRATED GUIDES AND CIRCUITS •
Shiban K. Koul

MICROWAVE DEVICES, CIRCUITS AND THEIR INTERACTION • Charles A. Lee and
G. Conrad Dalman

ADVANCES IN MICROSTRIP AND PRINTED ANTENNAS • Kai-Fong Lee and Wei Chen (eds.)

SPHEROIDAL WAVE FUNCTIONS IN ELECTROMAGNETIC THEORY • Le-Wei Li, Xiao-Kang Kang,
and Mook-Seng Leong

ARITHMETIC AND LOGIC IN COMPUTER SYSTEMS • Mi Lu

OPTICAL FILTER DESIGN AND ANALYSIS: A SIGNAL PROCESSING APPROACH •
Christi K. Madsen and Jian H. Zhao

THEORY AND PRACTICE OF INFRARED TECHNOLOGY FOR NONDESTRUCTIVE TESTING •
Xavier P. V. Maldague

OPTOELECTRONIC PACKAGING • A. R. Mickelson, N. R. Basavanhally, and Y. C. Lee (eds.)

OPTICAL CHARACTER RECOGNITION • Shunji Mori, Hirobumi Nishida, and Hiromitsu Yamada

ANTENNAS FOR RADAR AND COMMUNICATIONS: A POLARIMETRIC APPROACH •
Harold Mott

INTEGRATED ACTIVE ANTENNAS AND SPATIAL POWER COMBINING • Julio A. Navarro and
Kai Chang

ANALYSIS METHODS FOR RF, MICROWAVE, AND MILLIMETER-WAVE PLANAR TRANSMIS-
SION LINE STRUCTURES • Cam Nguyen

FREQUENCY CONTROL OF SEMICONDUCTOR LASERS • Motoichi Ohtsu (ed.)

WAVELETS IN ELECTROMAGNETICS AND DEVICE MODELING • George W. Pan

SOLAR CELLS AND THEIR APPLICATIONS • Larry D. Partain (ed.)

ANALYSIS OF MULTICONDUCTOR TRANSMISSION LINES • Clayton R. Paul

INTRODUCTION TO ELECTROMAGNETIC COMPATIBILITY • Clayton R. Paul

ELECTROMAGNETIC OPTIMIZATION BY GENETIC ALGORITHMS • Yahya Rahmat-Samii and Eric
Michielssen (eds.)

INTRODUCTION TO HIGH-SPEED ELECTRONICS AND OPTOELECTRONICS •
Leonard M. Riaziat

NEW FRONTIERS IN MEDICAL DEVICE TECHNOLOGY • Arye Rosen and Harel Rosen (eds.)

ELECTROMAGNETIC PROPAGATION IN MULTI-MODE RANDOM MEDIA • Harrison E. Rowe

ELECTROMAGNETIC PROPAGATION IN ONE-DIMENSIONAL RANDOM MEDIA •
Harrison E. Rowe

NONLINEAR OPTICS • E. G. Sauter

COPLANAR WAVEGUIDE CIRCUITS, COMPONENTS, AND SYSTEMS • Rainee N. Simons

ELECTROMAGNETIC FIELDS IN UNCONVENTIONAL MATERIALS AND STRUCTURES •
Onkar N. Singh and Akhlesh Lakhtakia (eds.)

FUNDAMENTALS OF GLOBAL POSITIONING SYSTEM RECEIVERS: A SOFTWARE
APPROACH, Second Edition • James Bao-yen Tsui

InP-BASED MATERIALS AND DEVICES: PHYSICS AND TECHNOLOGY • Osamu Wada
and Hideki Hasegawa (eds.)

COMPACT AND BROADBAND MICROSTRIP ANTENNAS • Kin-Lu Wong

Fundamentals of Global Positioning System Receivers

WILEY SERIES IN MICROWAVE AND OPTICAL ENGINEERING

KAI CHANG, Editor
Texas A&M University

A complete list of the titles in this series appears at the end of this volume.

Fundamentals of Global Positioning System Receivers
A Software Approach

SECOND EDITION

JAMES BAO-YEN TSUI

A JOHN WILEY & SONS, INC., PUBLICATION

Library of Congress Cataloging-in-Publication Data

Tsui, James Bao-yen.
   Fundamentals of global positioning system receivers : a software approach / James Bao-yen Tsui. – 2nd ed.
      p. cm.
   Includes bibliographical references (p.).
   ISBN 0-471-70647-7 (cloth)
   1. Global Positioning System. I. Title.
G109.5.T85 2005
910′.285 – dc22

                                                              2004053458

Printed in the United States of America.

10 9 8 7 6 5 4 3 2

Contents

Preface

In this new edition of the book, only minor changes were made to the original nine chapters but three new chapters treat topics of increasing interest to GPS users and equipment developers. One topic, improving the GPS receiver sensitivity may extend their operations into buildings, which is becoming important for emergency rescue and urban warfare. Thus, Chapters 10 and 11 are devoted to the processing of weak signals, as well as the limitations of autonomous GPS receivers. These same approaches are also applicable to GPS receivers in noisy environments and under interference conditions. Other subjects new to this edition, such as using the almanac data to simplify signal acquisition; determining the number of analog-to-digital converter bits required for the GPS receiver to work under strong interference; and, using GPS signals reflected from the ground as an altimeter are covered in Chapter 12.

I constantly discuss technical subjects with Mr. D. Lin and Dr. L. L. Liou, my colleagues at AFRL, and Dr. Y. T. Morton of Miami University. They worked closely with me and made tremendous contributions in this edition. I very much appreciate their help. I would especially like to thank Drs. J. Morton and T. Y. Morton of Miami University and Dr. J. Garrison of Purdue University for reviewing my manuscripts.

The management in AFRL/SNR as usual provided excellent guidance and support. Special thanks to W. Moore, K. Loree, M. Longbrake, B. Holsapple, and Dr. S. Hary. I also would like to thank my new colleagues, M. Berarducci, J. Buck, J. Coker, J. C. Ha, Dr. M. Miller, S. Moore, T. Nguyen, H. Noffke, N. Wilkins, J. McCartney, T. Niedzwiecki, M. Thompson, and C. Tolle for their help.

Preface to the First Edition

The purpose of this book is to present detailed fundamental information on a global positioning system (GPS) receiver. Although GPS receivers are popularly used in every-day life, their operation principles cannot be easily found in one book. Most other types of receivers process the input signals to obtain the necessary information easily, such as in amplitude modulation (AM) and frequency modulation (FM) radios. In a GPS receiver the signal is processed to obtain the required information, which in turn is used to calculate the user position. Therefore, at least two areas of discipline, receiver technology and navigation scheme, are employed in a GPS receiver. This book covers both areas.

In the case of GPS signals, there are two sets of information: the civilian code, referred to as the coarse/acquisition (C/A) code, and the classified military code, referred to as the P(Y) code. This book concentrates only on the civilian C/A code. This is the information used by commercial GPS receivers to obtain the user position.

The material in this book is presented from the software receiver viewpoint for two reasons. First, it is likely that narrow band receivers, such as the GPS receiver, will be implemented in software in the future. Second, a software receiver approach may explain the operation better. A few key computer programs can be used to further illustrate some points.

This book is written for engineers and scientists who intend to study and understand the detailed operation principles of GPS receivers. The book is at the senior or graduate school level. A few computer programs written in Matlab are listed at the end of several chapters to help the reader understand some of the ideas presented.

I would like to acknowledge the following persons. My sincere appreciation to three engineers: Dr. D. M. Akos from Stanford University, M. Stockmaster from Rockwell Collins, and J. Schamus from Veridian. They worked with me at the Air Force Research Laboratory, Wright Patterson Air Force Base on the xvi PREFACE TO THE FIRST EDITION

design of a software GPS receiver. This work made this book possible. Dr. Akos also reviewed my manuscripts. I used information from several courses on GPS receivers given at the Air Force Institute of Technology by Lt. Col. B. Riggins, Ph.D. and Capt. J. Requet, Ph.D. Valuable discussion with Drs. F. VanGraas and M. Braasch from Ohio University helped me as well. I am constantly discussing GPS subjects with my coworkers, D. M. Lin and V. D. Chakravarthy.

The management in the Sensor Division of the Air Force Research Laboratory provided excellent guidance and support in GPS receiver research. Special thanks are extended to Dr. P. S. Hadorn, E. R. Martinsek, A. W. White, and N. A. Pequignot. I would also like to thank my colleagues, R. L. Davis, S. M. Rodrigue, K. M. Graves, J. R. McCall, J. A. Tenbarge, Dr. S. W. Schneider, J. N. Hedge Jr., J. Caschera, J. Mudd, J. P. Stephens, Capt. R. S. Parks, P. G. Howe, D. L. Howell, Dr. L. L. Liou, D. R. Meeks, and D. Jones, for their consultation and assistance.

Last, but not least, I would like to thank my wife, Susan, for her encouragement and understanding.

CHAPTER ONE

Introduction

1.1 INTRODUCTION[1-13]

This book presents detailed information in a compact form about the global positioning system (GPS) coarse/acquisition (C/A) code receiver. Using the C/A code to find the user location is referred to as the standard position service (SPS). Most of the information can be found in references 1 through 13. However, there is much more information in the references than the basics required to understand a GPS receiver. Therefore, one must study the proper subjects and put them together. This is a tedious and cumbersome task. This book does this job for the reader.

This book not only introduces the information available from the references, it emphasizes its applications. Software programs are provided to help understand some of the concepts. These programs are also useful in designing GPS receivers. In addition, various techniques to perform acquisition and tracking on the GPS signals are included.

This book concentrates only on the very basic concepts of the C/A code GPS receiver. Any subject not directly related to the basic receiver (even if it is of general interest, i.e., differential GPS receiver and GPS receiver with carrier-aided tracking capacity) will not be included in this book. These other subjects can be found in reference 1.

1.2 HISTORY OF GPS DEVELOPMENT[1,5,12]

The discovery of navigation seems to have occurred early in human history. According to Chinese storytelling, the compass was discovered and used in wars

Fundamentals of Global Positioning System Receivers: A Software Approach, Second Edition, by James Bao-Yen Tsui
ISBN 0-471-70647-7   Copyright © 2005 John Wiley & Sons, Inc.

during foggy weather before recorded history. There have been many different navigation techniques to support ocean and air transportation. Satellite-based navigation started in the early 1970s. Three satellite systems were explored before the GPS programs: the U.S. Navy Navigation Satellite System (also referred to as the Transit), the U.S. Navy's Timation (TIMe navigATION), and U.S. Air Force project 621B. The Transit project used a continuous wave (cw) signal. The closest approach of the satellite can be found by measuring the maximum rate of Doppler shift. The Timation program used an atomic clock that improves the prediction of satellite orbits and reduces the ground control update rate. The Air Force 621B project used the pseudorandom noise (PRN) signal to modulate the carrier frequency.

The GPS program was approved in December 1973. The first satellite was launched in 1978. In August 1993, GPS had 24 satellites in orbit and in December of the same year the initial operational capability was established. In February 1994, the Federal Aviation Agency (FAA) declared GPS ready for aviation use.

1.3 A BASIC GPS RECEIVER

The basic GPS receiver discussed in this book is shown in Figure 1.1. The signals transmitted from the GPS satellites are received from the antenna. Through the radio frequency (RF) chain the input signal is amplified to a proper amplitude and the frequency is converted to a desired output frequency. An analog-to-digital converter (ADC) is used to digitize the output signal. The antenna, RF chain, and ADC are the hardware used in the receiver.

After the signal is digitized, software is used to process it, and that is why this book has taken a software approach. Acquisition means to find the signal of a certain satellite. The tracking program is used to find the phase transition of the navigation data. In a conventional receiver, the acquisition and tracking are

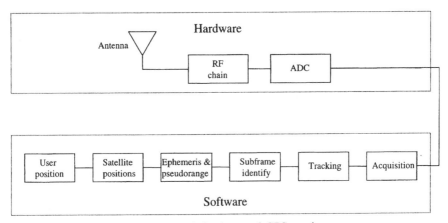

FIGURE 1.1 A fundamental GPS receiver.

performed by hardware. From the navigation data phase transition the subframes and navigation data can be obtained. Ephemeris data and pseudoranges can be obtained from the navigation data. The ephemeris data are used to obtain the satellite positions. Finally, the user position can be calculated for the satellite positions and the pseudoranges. Both the hardware used to collect digitized data and the software used to find the user position will be discussed in this book.

1.4   APPROACHES OF PRESENTATION

There are two possible approaches to writing this book. One is a straightforward way to follow the signal flow shown in Figure 1.1. In this approach the book would start with the signal structure of the GPS system and the methods to process the signal to obtain the necessary the information. This information would be used to calculate the positions of the satellites and the pseudoranges. By using the positions of the satellites and the pseudoranges the user position can be found. In this approach, the flow of discussion would be smooth, from one subject to another. However, the disadvantage of this approach is that readers might not have a clear idea why these steps are needed. They could understand the concept of the GPS operation only after reading the entire book.

The other approach is to start with the basic concept of the GPS from a system designers' point of view. This approach would start with the basic concept of finding the user position from the satellite positions. The description of the satellite constellation would be presented. The detailed information of the satellite orbit is contained in the GPS data. In order to obtain these data, the GPS signal must be tracked. The C/A code of the GPS signals would then be presented. Each satellite has an unique C/A code. A receiver can perform acquisition on the C/A code to find the signal. Once the C/A code of a certain satellite is found, the signal can be tracked. The tracking program can produce the navigation data. From these data, the position of the satellite can be found. The relative pseudorange can be obtained by comparing the time a certain data point arrived at the receiver. The user position can be calculated from the satellite positions and pseudoranges of several satellites.

This book takes this second approach to present the material because it should give a clearer idea of the GPS function from the very beginning. The final chapter describes the overall functions of the GPS receiver and can be considered as taking the first approach for digitizing the signal, performing acquisition and tracking, extracting the navigation data, and calculating the user position.

1.5   SOFTWARE APPROACH

This book uses the concept of software radio to present the subject. The software radio idea is to use an analog-to-digital converter (ADC) to change the input signal into digital data at the earliest possible stage in the receiver. In other words, the input signal is digitized as close to the antenna as possible. Once the signal is digitized, digital signal processing will be used to obtain the necessary information. The primary goal of the software radio is minimum hardware use in a radio. Conceptually, one can tune the radio through software or even change the function of the radio such as from amplitude modulation (AM) to frequency modulation (FM) by changing the software; therefore great flexibility can be achieved.

The main purpose of using the software radio concept to present this subject is to illustrate the idea of signal acquisition and tracking. Although using hardware to perform signal acquisition and tracking can also describe GPS receiver function, it appears that using software may provide a clearer idea of the signal acquisition and tracking. In addition, a software approach should provide a better understanding of the receiver function because some of the calculations can be illustrated with programs. Once the software concept is well understood, the readers should be able to introduce new solutions to problems such as various acquisition and tracking methods to improve efficiency and performance. At the time (December 1997) this chapter was being written, a software GPS receiver using a 200 MHz personal computer (PC) could not track one satellite in real time. When this chapter was revised in December 1998, the software had been modified to track two satellites in real time with a new PC operating at 400 MHz. Although it is still impossible to implement a software GPS receiver operating in real time, with the improvement in PC operating speed and software modification it is likely that by the time this book is published a software GPS receiver will be a reality. Of course, using a digital signal processing (DSP) chip is another viable way to build the receiver. When this second edition was prepared, software receivers could already operate at real time. In Section 12.10, some of the results will be presented.

Only the fundamentals of a GPS receiver are presented in this book. In order to improve the performance of a receiver, fine tuning of some of the operations might be necessary. Once readers understand the basic operation principles of the receiver, they can make the necessary improvement.

1.6 POTENTIAL ADVANTAGES OF THE SOFTWARE APPROACH

An important aspect of using the software approach to build a GPS receiver is that the approach can drastically deviate from the conventional hardware approach. For example, the user may take a snapshot of data and process them to find the location rather than continuously tracking the signal. Theoretically, 30 seconds of data are enough to find the user location. This is especially useful when data cannot be collected in a continuous manner. Since the software approach is in the infant stage, one can explore many potential methods.

The software approach is very flexible. It can process data collected from various types of hardware. For example, one system may collect complex data referred to as the inphase and quadrature-phase (I and Q) channels. Another system may collect real data from one channel. The data can easily be changed from one form to another. One can also generate programs to process complex signals from programs processing real signals or vice versa with some simple modifications. A program can be used to process signals digitized with various sampling frequencies. Therefore, a software approach can almost be considered as hardware independent.

New algorithms can easily be developed without changing the design of the hardware. This is especially useful for studying some new problems. For example, in order to study the antijamming problem one can collect a set of digitized signals with jamming signals present and use different algorithms to analyze it.

1.7 ORGANIZATION OF THE BOOK

This book contains twelve chapters. Chapter 2 introduces the user position requirements, which lead to the GPS parameters. Also included in Chapter 2 is the basic concept of how to find the user position if the satellite positions are known. Chapter 3 discusses the satellite constellation and its impact on the GPS signals, which in turn affects the design of the GPS receiver. Chapter 4 discusses the earth-centered, earth-fixed system. Using this coordinate system, the user position can be calculated to match the position on every-day maps. The GPS signal structure is discussed in detail in Chapter 5. Chapter 6 discusses the hardware to collect data, which is equivalent to the front end of a conventional GPS receiver. Changing the format of data is also presented. Chapter 7 presents several acquisition methods. Some of them can be used in hardware design and others are suitable for software applications. Chapter 8 discusses two tracking methods. One uses the conventional phase-locked loop approach and the other one is more suitable for the software radio approach. Chapter 9 is a summary of the previous chapters. It takes all the information in the first eight chapters and presents in it an order following the signal flow in a GPS receiver. Chapters 10 and 11 are devoted to weak GPS signal processing. Not only the processing but the limitation of an autonomous GPS receiver is also defined. Chapter 12 includes various subjects related to GPS receivers.

Computer programs written in Matlab are listed at the end of several chapters. Some of the programs are used only to illustrate ideas. Others can be used in the receiver design. In the final chapter all of the programs related to designing a receiver will listed. These programs are by no means optimized and they are used only for demonstration purposes.

REFERENCES

1. Parkinson, B. W., Spilker, J. J. Jr., *Global Positioning System: Theory and Applications*, vols. 1 and 2, American Institute of Aeronautics and Astronautics, 370 L'Enfant Promenade, SW, Washington, DC, 1996.
2. "System specification for the navstar global positioning system," SS-GPS-300B code ident 07868, March 3, 1980.
3. Spilker, J. J., "GPS signal structure and performance characteristics," *Navigation*, Institute of Navigation, vol. 25, no. 2, pp. 121–146, Summer 1978.

4. Milliken, R. J., Zoller, C. J., "Principle of operation of NAVSTAR and system characteristics," Advisory Group for Aerospace Research and Development (AGARD) Ag-245, pp. 4–1–4.12, July 1979.

5. Misra, P. N., "Integrated use of GPS and GLONASS in civil aviation," *Lincoln Laboratory Journal*, Massachusetts Institute of Technology, vol. 6, no. 2, pp. 231–247, Summer/Fall, 1993.

6. "Reference data for radio engineers," 5th ed., Howard W. Sams & Co. (subsidiary of ITT), Indianapolis, 1972.

7. Bate, R. R., Mueller, D. D., White, J. E., *Fundamentals of Astrodynamics*, pp. 182–188, Dover Publications, New York, 1971.

8. Wells, D. E., Beck, N., Delikaraoglou, D., Kleusbery, A., Krakiwsky, E. J., Lachapelle, G., Langley, R. B., Nakiboglu, M., Schwarz, K. P., Tranquilla, J. M., Vanicek, P., *Guide to GPS Positioning*, Canadian GPS Associates, Frederiction, N.B., Canada, 1987.

9. "Department of Defense world geodetic system, 1984 (WGS-84), its definition and relationships with local geodetic systems," DMA-TR-8350.2, Defense Mapping Agency, September 1987.

10. *"Global Positioning System Standard Positioning Service Signal Specification"*, 2nd ed., GPS Joint Program Office, June 1995.

11. Bate, R. R., Mueller, D. D., White, J. E., *Fundamentals of Astrodynamics*, Dover Publications, New York, 1971.

12. Riggins, B., "Satellite navigation using the global positioning system," manuscript used in Air Force Institute of Technology, Dayton OH, 1996.

13. Kaplan, E. D., ed., *Understanding GPS Principles and Applications*, Artech House, Norwood, MA, 1996.

CHAPTER TWO

Basic GPS Concept

2.1 INTRODUCTION

This chapter will introduce the basic concept of how a GPS receiver determines its position. In order to better understand the concept, GPS performance requirements will be discussed first. These requirements determine the arrangement of the satellite constellation. From the satellite constellation, the user position can be solved. However, the equations required for solving the user position turn out to be nonlinear simultaneous equations. In addition, some practical considerations (i.e., the inaccuracy of the user clock) will be included in these equations. These equations are solved through a linearization and iteration method. The solution is in a Cartesian coordinate system and the result will be converted into a spherical coordinate system. However, the earth is not a perfect sphere; therefore, once the user position is found, the shape of the earth must be taken into consideration. The user position is then translated into the earth-based coordinate system. Finally, the selection of satellites to obtain better user position accuracy and the dilution of precision will be discussed.

2.2 GPS PERFORMANCE REQUIREMENTS[1]

Some of the performance requirements are listed below:

1. The user position root mean square (rms) error should be 10–30 m.
2. It should be applicable to real-time navigation for all users including the high-dynamics user, such as in high-speed aircraft with flexible maneuverability.

Fundamentals of Global Positioning System Receivers: A Software Approach, Second Edition, by James Bao-Yen Tsui
ISBN 0-471-70647-7 Copyright © 2005 John Wiley & Sons, Inc.

3. It should have worldwide coverage. Thus, in order to cover the polar regions the satellites must be in inclined orbits.

4. The transmitted signals should tolerate, to some degree, intentional and unintentional interference. For example, the harmonics from some narrow-band signals should not disturb its operation. Intentional jamming of GPS signals is a serious concern for military applications.

5. It cannot require that every GPS receiver utilize a highly accurate clock such as those based on atomic standards.

6. When the receiver is first turned on, it should take minutes rather than hours to find the user position.

7. The size of the receiving antenna should be small. The signal attenuation through space should be kept reasonably small.

These requirements combining with the availability of the frequency band allocation determines the carrier frequency of the GPS to be in the L band (1–2 GHz) of the microwave range.

2.3 BASIC GPS CONCEPT

The position of a certain point in space can be found from distance measured from this point to some known positions in space. Let us use some examples to illustrate this point. In Figure 2.1, the user position is on the x-axis; this is a one-dimensional case. If the satellite position S_1 and the distance to the satellite x_1 are both known, the user position can be at two places, either to the left or right of S_1. In order to determine the user position, the distance to another satellite with known position must be measured. In this figure, the positions of S_2 and x_2 uniquely determine the user position U.

Figure 2.2 shows a two-dimensional case. In order to determine the user position, three satellites and three distances are required. The trace of a point with constant distance to a fixed point is a circle in the two-dimensional case. Two satellites and two distances give two possible solutions because two circles intersect at two points. A third circle is needed to uniquely determine the user position.

For similar reasons one might decide that in a three-dimensional case four satellites and four distances are needed. The equal-distance trace to a fixed point is a sphere in a three-dimensional case. Two spheres intersect to make a circle. This circle intersects another sphere to produce two points. In order to determine which point is the user position, one more satellite is needed.

FIGURE 2.1 One-dimensional user position.

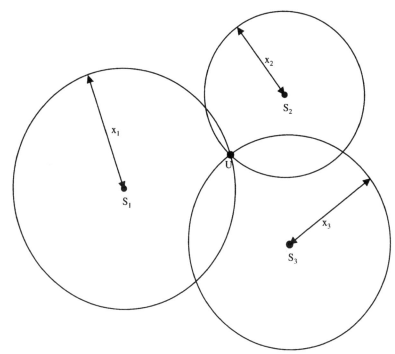

FIGURE 2.2 Two-dimensional user position.

In GPS the position of the satellite is known from the ephemeris data transmitted by the satellite. One can measure the distance from the receiver to the satellite. Therefore, the position of the receiver can be determined.

In the above discussion, the distance measured from the user to the satellite is assumed to be very accurate and there is no bias error. However, the distance measured between the receiver and the satellite has a constant unknown bias, because the user clock usually is different from the GPS clock. In order to resolve this bias error one more satellite is required. Therefore, in order to find the user position five satellites are needed.

If one uses four satellites and the measured distance with bias error to measure a user position, two possible solutions can be obtained. Theoretically, one cannot determine the user position. However, one of the solutions is close to the earth's surface and the other one is in space. Since the user position is usually close to the surface of the earth, it can be uniquely determined. Therefore, the general statement is that four satellites can be used to determine a user position, even though the distance measured has a bias error.

The method of solving the user position discussed in Sections 2.5 and 2.6 is through iteration. The initial position is often selected at the center of the earth. The iteration method will converge on the correct solution rather than the one in space. In the following discussion four satellites are considered the minimum number required in finding the user position.

2.4 BASIC EQUATIONS FOR FINDING USER POSITION

In this section the basic equations for determining the user position will be presented. Assume that the distance measured is accurate and under this condition three satellites are sufficient. In Figure 2.3, there are three known points at locations r_1 or (x_1, y_1, z_1), r_2 or (x_2, y_2, z_2), and r_3 or (x_3, y_3, z_3), and an unknown point at r_u or (x_u, y_u, z_u). If the distances between the three known points to the unknown point can be measured as ρ_1, ρ_2, and ρ_3, these distances can be written as

$$\rho_1 = \sqrt{(x_1 - x_u)^2 + (y_1 - y_u)^2 + (z_1 - z_u)^2}$$
$$\rho_2 = \sqrt{(x_2 - x_u)^2 + (y_2 - y_u)^2 + (z_2 - z_u)^2}$$
$$\rho_3 = \sqrt{(x_3 - x_u)^2 + (y_3 - y_u)^2 + (z_3 - z_u)^2} \tag{2.1}$$

Because there are three unknowns and three equations, the values of x_u, y_u, and z_u can be determined from these equations. Theoretically, there should be two sets of solutions as they are second-order equations. These equations can be solved relatively easily with linearization and an iterative approach. The solution of these equations will be discussed later in Section 2.6.

In GPS operation, the positions of the satellites are given. This information can be obtained from the data transmitted from the satellites and will be discussed in Chapter 5. The distances from the user (the unknown position) to the

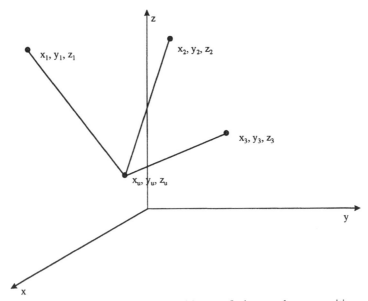

FIGURE 2.3 Use three known positions to find one unknown position.

satellites must be measured simultaneously at a certain time instance. Each satellite transmits a signal with a time reference associated with it. By measuring the time of the signal traveling from the satellite to the user the distance between the user and the satellite can be found. The distance measurement is discussed in the next section.

2.5   MEASUREMENT OF PSEUDORANGE[2]

Every satellite sends a signal at a certain time t_{si}. The receiver will receive the signal at a later time t_u. The distance between the user and the satellite i is

$$\rho_{iT} = c(t_u - t_{si}) \tag{2.2}$$

where c is the speed of light, ρ_{iT} is often referred to as the true value of pseudorange from user to satellite i, t_{si} is referred to as the true time of transmission from satellite i, t_u is the true time of reception.

From a practical point of view it is difficult, if not impossible, to obtain the correct time from the satellite or the user. The actual satellite clock time t'_{si} and actual user clock time t'_u are related to the true time as

$$t'_{si} = t_{si} + \Delta b_i$$
$$t'_u = t_u + b_{ut} \tag{2.3}$$

where Δb_i is the satellite clock error, b_{ut} is the user clock bias error. Besides the clock error, there are other factors affecting the pseudorange measurement. The measured pseudorange ρ_i can be written as[2]

$$\rho_i = \rho_{iT} + \Delta D_i - c(\Delta b_i - b_{ut}) + c(\Delta T_i + \Delta I_i + v_i + \Delta v_i) \tag{2.4}$$

where ΔD_i is the satellite position error effect on range, ΔT_i is the tropospheric delay error, ΔI_i is the ionospheric delay error, v_i is the receiver measurement noise error, Δv_i is the relativistic time correction.

Some of these errors can be corrected; for example, the tropospheric delay can be modeled and the ionospheric error can be corrected in a two-frequency receiver. The errors will cause inaccuracy of the user position. However, the user clock error cannot be corrected through received information. Thus, it will remain as an unknown. As a result, Equation (2.1) must be modified as

$$\rho_1 = \sqrt{(x_1 - x_u)^2 + (y_1 - y_u)^2 + (z_1 - z_u)^2} + b_u$$
$$\rho_2 = \sqrt{(x_2 - x_u)^2 + (y_2 - y_u)^2 + (z_2 - z_u)^2} + b_u$$
$$\rho_3 = \sqrt{(x_3 - x_u)^2 + (y_3 - y_u)^2 + (z_3 - z_u)^2} + b_u$$
$$\rho_4 = \sqrt{(x_4 - x_u)^2 + (y_4 - y_u)^2 + (z_4 - z_u)^2} + b_u \tag{2.5}$$

where b_u is the user clock bias error expressed in distance, which is related to the quantity b_{ut} by $b_u = cb_{ut}$. In Equation (2.5), four equations are needed to solve for four unknowns x_u, y_u, z_u, and b_u. Thus, in a GPS receiver, a minimum of four satellites is required to solve for the user position. The actual measurement of the pseudorange will be discussed in Chapter 9.

2.6 SOLUTION OF USER POSITION FROM PSEUDORANGES

One common way to solve Equation (2.5) is to linearize them. The above equations can be written in a simplified form as

$$\rho_i = \sqrt{(x_i - x_u)^2 + (y_i - y_u)^2 + (z_i - z_u)^2} + b_u \tag{2.6}$$

where $i = 1$, 2, 3, and 4, and x_u, y_u, z_u, and b_u are the unknowns. The pseudorange ρ_i and the positions of the satellites x_i, y_i, z_i are known.

Differentiate this equation, and the result is

$$\begin{aligned}
\delta\rho_i &= \frac{(x_i - x_u)\delta x_u + (y_i - y_u)\delta y_u + (z_i - z_u)\delta z_u}{\sqrt{(x_i - x_u)^2 + (y_i - y_u)^2 + (z_i - z_u)^2}} + \delta b_u \\
&= \frac{(x_i - x_u)\delta x_u + (y_i - y_u)\delta y_u + (z_i - z_u)\delta z_u}{\rho_i - b_u} + \delta b_u
\end{aligned} \tag{2.7}$$

In this equation, δx_u, δy_u, δz_u, and δb_u can be considered as the only unknowns. The quantities x_u, y_u, z_u, and b_u are treated as known values because one can assume some initial values for these quantities. From these initial values a new set of δx_u, δy_u, δz_u, and δb_u can be calculated. These values are used to modify the original x_u, y_u, z_u, and b_u to find another new set of solutions. This new set of x_u, y_u, z_u, and b_u can be considered again as known quantities. This process continues until the absolute values of δx_u, δy_u, δz_u, and δb_u are very small and within a certain predetermined limit. The final values of x_u, y_u, z_u, and b_u are the desired solution. This method is often referred to as the iteration method.

With δx_u, δy_u, δz_u, and δb_u as unknowns, the above equation becomes a set of linear equations. This procedure is often referred to as linearization. The above equation can be written in matrix form as

$$\begin{bmatrix} \delta\rho_1 \\ \delta\rho_2 \\ \delta\rho_3 \\ \delta\rho_4 \end{bmatrix} = \begin{bmatrix} \alpha_{11} & \alpha_{12} & \alpha_{13} & 1 \\ \alpha_{21} & \alpha_{22} & \alpha_{23} & 1 \\ \alpha_{31} & \alpha_{32} & \alpha_{33} & 1 \\ \alpha_{41} & \alpha_{42} & \alpha_{43} & 1 \end{bmatrix} \begin{bmatrix} \delta x_u \\ \delta y_u \\ \delta z_u \\ \delta b_u \end{bmatrix} \tag{2.8}$$

where

$$\alpha_{i1} = \frac{x_i - x_u}{\rho_i - b_u} \qquad \alpha_{i2} = \frac{y_i - y_u}{\rho_i - b_u} \qquad \alpha_{i3} = \frac{z_i - z_u}{\rho_i - b_u} \tag{2.9}$$

The solution of Equation (2.8) is

$$
\begin{bmatrix} \delta x_u \\ \delta y_u \\ \delta z_u \\ \delta b_u \end{bmatrix} = \begin{bmatrix} \alpha_{11} & \alpha_{12} & \alpha_{13} & 1 \\ \alpha_{21} & \alpha_{22} & \alpha_{23} & 1 \\ \alpha_{31} & \alpha_{32} & \alpha_{33} & 1 \\ \alpha_{41} & \alpha_{42} & \alpha_{43} & 1 \end{bmatrix}^{-1} \begin{bmatrix} \delta\rho_1 \\ \delta\rho_2 \\ \delta\rho_3 \\ \delta\rho_4 \end{bmatrix}
\tag{2.10}
$$

where $[\ ]^{-1}$ represents the inverse of the α matrix. This equation obviously does not provide the needed solutions directly; however, the desired solutions can be obtained from it. In order to find the desired position solution, this equation must be used repetitively in an iterative way. A quantity is often used to determine whether the desired result is reached and this quantity can be defined as

$$
\delta v = \sqrt{\delta x_u^2 + \delta y_u^2 + \delta z_u^2 + \delta b_u^2}
\tag{2.11}
$$

When this value is less than a certain predetermined threshold, the iteration will stop. Sometimes, the clock bias b_u is not included in Equation (2.11).

The detailed steps to solve the user position will be presented in the next section. In general, a GPS receiver can receive signals from more than four satellites. The solution will include such cases as when signals from more than four satellites are obtained.

2.7 POSITION SOLUTION WITH MORE THAN FOUR SATELLITES[3]

When more than four satellites are available, a more popular approach to solve the user position is to use all the satellites. The position solution can be obtained in a similar way. If there are n satellites available where $n > 4$, Equation (2.6) can be written as

$$
\rho_i = \sqrt{(x_i - x_u)^2 + (y_i - y_u)^2 + (z_i - z_u)^2} + b_u
\tag{2.12}
$$

where $i = 1, 2, 3, \ldots n$. The only difference between this equation and Equation (2.6) is that $n > 4$.

Linearize this equation, and the result is

$$
\begin{bmatrix} \delta\rho_1 \\ \delta\rho_2 \\ \delta\rho_3 \\ \delta\rho_4 \\ \vdots \\ \delta\rho_n \end{bmatrix} = \begin{bmatrix} \alpha_{11} & \alpha_{12} & \alpha_{13} & 1 \\ \alpha_{21} & \alpha_{22} & \alpha_{23} & 1 \\ \alpha_{31} & \alpha_{32} & \alpha_{33} & 1 \\ \alpha_{41} & \alpha_{42} & \alpha_{43} & 1 \\ \vdots & & & \\ \alpha_{n1} & \alpha_{n2} & \alpha_{n3} & 1 \end{bmatrix} \begin{bmatrix} \delta x_u \\ \delta y_u \\ \delta z_u \\ \delta b_u \end{bmatrix}
\tag{2.13}
$$

where

$$\alpha_{i1} = \frac{x_i - x_u}{\rho_i - b_u} \qquad \alpha_{i2} = \frac{y_i - y_u}{\rho_i - b_u} \qquad a_{i3} = \frac{z_i - z_u}{\rho_i - b_u} \qquad (2.9)$$

Equation (2.13) can be written in a simplified form as

$$\delta\rho = \alpha\delta x \qquad (2.14)$$

where $\delta\rho$ and δx are vectors, α is a matrix. They can be written as

$$\delta\rho = \begin{bmatrix} \delta\rho_1 & \delta\rho_2 & \cdots & \delta\rho_n \end{bmatrix}^T$$

$$\delta x = \begin{bmatrix} \delta x_u & \delta y_u & \delta z_u & \delta b_u \end{bmatrix}^T$$

$$\alpha = \begin{bmatrix} \alpha_{11} & \alpha_{12} & \alpha_{13} & 1 \\ \alpha_{21} & \alpha_{22} & \alpha_{23} & 1 \\ \alpha_{31} & \alpha_{32} & \alpha_{33} & 1 \\ \alpha_{41} & \alpha_{42} & \alpha_{43} & 1 \\ \vdots & & & \\ \alpha_{n1} & \alpha_{n2} & \alpha_{n3} & 1 \end{bmatrix} \qquad (2.15)$$

where $[\ ]^T$ represents the transpose of a matrix. Since α is not a square matrix, it cannot be inverted directly. Equation (2.13) is still a linear equation. If there are more equations than unknowns in a set of linear equations, the least-squares approach can be used to find the solutions. The pseudoinverse of the α can be used to obtain the solution. The solution is[3]

$$\delta x = [\alpha^T \alpha]^{-1}\alpha^T \delta\rho \qquad (2.16)$$

From this equation, the values of δx_u, δy_u, δz_u, and δb_u can be found. In general, the least-squares approach produces a better solution than the position obtained from only four satellites, because more data are used.

The following steps summarize the above approach:

A. Choose a nominal position and user clock bias x_{u0}, y_{u0}, z_{u0}, b_{u0} to represent the initial condition. For example, the position can be the center of the earth and the clock bias zero. In other words, all initial values are set to zero.

B. Use Equation (2.5) or (2.6) to calculate the pseudorange ρ_i. These ρ_i values will be different from the measured values. The difference between the measured values and the calculated values is $\delta\rho_i$.

C. Use the calculated ρ_i in Equation (2.9) to calculate α_{i1}, α_{i2}, α_{i3}.

D. Use Equation (2.16) to find δx_u, δy_u, δz_u, δb_u.

E. From the absolute values of δx_u, δy_u, δz_u, δb_u and from Equation (2.11) calculate δv.

F. Compare δv with an arbitrarily chosen threshold; if δv is greater than the threshold, the following steps will be needed.

G. Add these values δx_u, δy_u, δz_u, δb_u to the initial chosen position x_{u0}, y_{u0}, z_{u0}, and the clock bias b_{u0}; a new set of positions and clock bias can be obtained and they will be expressed as x_{u1}, y_{u1}, z_{u1}, b_{u1}. These values will be used as the initial position and clock bias in the following calculations.

H. Repeat the procedure from A to G, until δv is less than the threshold. The final solution can be considered as the desired user position and clock bias, which can be expressed as x_u, y_u, z_u, b_u.

In general, the δv calculated in the above iteration method will keep decreasing rapidly. Depending on the chosen threshold, the iteration method usually can achieve the desired goal in less than 10 iterations. A computer program (p2_1) to calculate the user position is listed at the end of this chapter. In this book, some lines in the programs are too long to be listed in one line; however, it should be easily recognized.

2.8   USER POSITION IN SPHERICAL COORDINATE SYSTEM

The user position calculated from the above discussion is in a Cartesian coordinate system. It is usually desirable to convert to a spherical system and label the position in latitude, longitude, and altitude as the every-day maps use these notations. The latitude of the earth is from -90 to 90 degrees with the equator at 0 degree. The longitude is from -180 to 180 degrees with the Greenwich meridian at 0 degree. The altitude is the height above the earth's surface. If the earth is a perfect sphere, the user position can be found easily as shown in Figure 2.4. From this figure, the distance from the center of the earth to the user is

$$r = \sqrt{x_u^2 + y_u^2 + z_u^2}$$
(2.17)

The latitude L_c is

$$L_c = \tan^{-1}\left(\frac{z_u}{\sqrt{x_u^2 + y_u^2}}\right)$$
(2.18)

The longitude l is

$$l = \tan^{-1}\left(\frac{y_u}{x_u}\right)$$
(2.19)

The altitude h is

$$h = r - r_e$$
(2.20)

where r_e is the radius of an ideal spherical earth or the average radius of the earth. Since the earth is not a perfect sphere, some of these equations need to be modified.

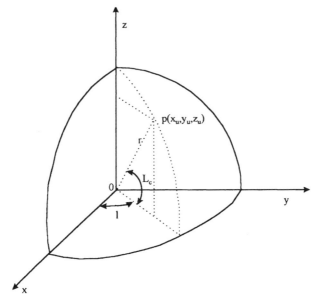

FIGURE 2.4 An octet of an ideal spherical earth.

2.9 EARTH GEOMETRY[4-6]

The earth is not a perfect sphere but is an ellipsoid; thus, the latitude and altitude calculated from Equations (2.18) and (2.20) must be modified. However, the longitude l calculated from Equation (2.19) also applies to the nonspherical earth. Therefore, this quantity does not need modification. Approximations will be used in the following discussion, which is based on references 4 through 6. For an ellipsoid, there are two latitudes. One is referred to as the geocentric latitude L_c, which is calculated from the previous section. The other one is the geodetic latitude L and is the one often used in every-day maps. Therefore, the geocentric latitude must be converted to the geodetic latitude. Figure 2.5 shows a cross section of the earth. In this figure the x-axis is along the equator, the y-axis is pointing inward to the paper, and the z-axis is along the north pole of the earth. Assume that the user position is on the x-z plane and this assumption does not lose generality. The geocentric latitude L_c is obtained by drawing a line from the user to the center of the earth, which is calculated from Equation (2.18).

The geodetic latitude is obtained by drawing a line perpendicular to the surface of the earth that does not pass the center of the earth. The angle between this line and the x is the geodetic latitude L. The height of the user is the distance h perpendicular and above the surface of the earth.

The following discussion is used to determine three unknown quantities from two known quantities. As shown in Figure 2.5, the two known quantities are the distance r and the geocentric latitude L_c and they are measured from the ideal spherical earth. The three unknown quantities are the geodetic latitude

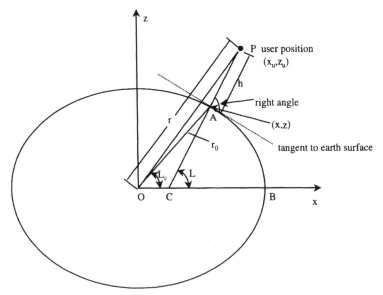

FIGURE 2.5   Geocentric and geodetic latitudes.

L, the distance r_0, and the height h. All three quantities are calculated from approximation methods. Before the actual calculations of the unknowns, let us introduce some basic relationships in an ellipse.

2.10   BASIC RELATIONSHIPS IN AN ELLIPSE[4–7]

In order to derive the relationships mentioned in the previous section, it is convenient to review the basic functions in an ellipse. Figure 2.6 shows an ellipse which can be used to represent a cross section of the earth passing through the polar axis.

Let us assume that the semi-major axis is a_e, the semi-minor axis is b_e, and the foci are separated by $2c_e$. The equation of the ellipse is

$$\frac{x^2}{a_e^2} + \frac{z^2}{b_e^2} = 1 \text{ and}$$

$$a_e^2 - b_e^2 = c_e^2 \tag{2.21}$$

The eccentricity e_e is defined as

$$e_e = \frac{c_e}{a_e} = \frac{\sqrt{a_e^2 - b_e^2}}{a_e} \quad \text{or} \quad \frac{b_e}{a_e} = \sqrt{1 - e_e^2} \tag{2.22}$$

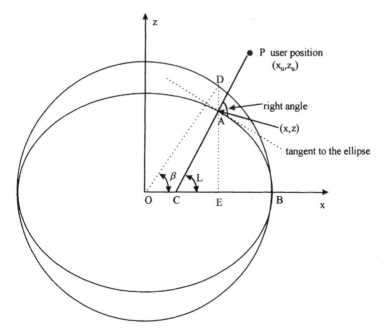

FIGURE 2.6   A basic ellipse with accessory lines.

The ellipticity e_p is defined as

$$e_p = \frac{a_e - b_e}{a_e} \tag{2.23}$$

where $a_e = 6378137 \pm 2$ m, $b_e = 6356752.3142$ m, $e_e = 0.0818191908426$, and $e_p = 0.00335281066474$.[6,7] The value of b_e is calculated from a_e; thus, the result has more decimal points.

From the user position P draw a line perpendicular to the ellipse that intercepts it at A and the x-axis at C. To help illustrate the following relation a circle with radius equal to the semi-major axis a_e is drawn as shown in Figure 2.6. A line is drawn from point A perpendicular to the x-axis and intercepts it at E and the circle at D. The position $A(x, y)$ can be found as

$$x = OE = OD \cos \beta = a_e \cos \beta$$

$$z = AE = DE \frac{b_e}{a_e} = (a_e \sin \beta) \frac{b_e}{a_e} = b_e \sin \beta \tag{2.24}$$

The second equation can be obtained easily from the equation of a circle $x^2 + z^2 = a_e^2$ and Equation (2.21). The tangent to the ellipse at A is dz/dx. Since line CP is perpendicular to the tangent,

$$\tan L = -\frac{dx}{dz} \tag{2.25}$$

From these relations let us find the relation between angle β and L. Taking the derivative of x and z of Equation (2.24), the results are

$$dx = -a_e \sin \beta d\beta$$

$$dz = b_e \cos \beta d\beta \tag{2.26}$$

Thus

$$\tan L = -\frac{dx}{dz} = \frac{a_e}{b_e} \tan \beta = \frac{\tan \beta}{\sqrt{1 - e_e^2}} \tag{2.27}$$

From these relationships let us find the three unknowns.

2.11 CALCULATION OF ALTITUDE[5]

In the following three sections the discussion is based on reference 5. From Figure 2.7 the height h can be found from the law of cosine through the triangle OPA as

$$r^2 = r_0^2 - 2r_0 h \cos(\pi - D_0) + h^2 = r_0^2 + 2r_0 h \cos D_0 + h^2 \tag{2.28}$$

where r_0 is the distance from the center of the earth to the point on the surface of the earth under the user position. The amplitude of r can be found from

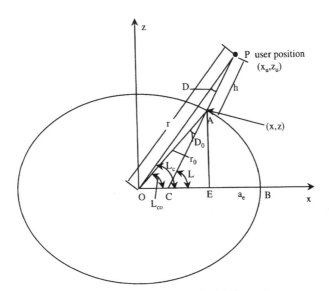

FIGURE 2.7 Altitude and latitude illustration.

completing the square for $r_0 + h$ and taking the square root as

$$r = [(r_0 + h)^2 - 2r_0h(1 - \cos D_0)]^{1/2} = (r_0 + h)\left[1 - \frac{2hr_0(1 - \cos D_0)}{(r_0 + h)^2}\right]^{1/2}$$
(2.29)

Since angle D_0 is very small, it can be approximated as

$$1 - \cos D_0 \approx \frac{D_0^2}{2}$$
(2.30)

where D_0 is the angle expressed in radians. The r value can be written as

$$r \approx (r_0 + h)\left[1 - \frac{2hr_0D_0^2/2}{(r_0 + h)^2}\right]^{1/2} = r_0 + h - \frac{hr_0D_0^2}{2(r_0 + h)}$$
(2.31)

At latitude of 45 degrees D_0 ($\approx 1/297$ radian) becomes maximum. If D_0 is neglected, the result is

$$r \approx r_0 + h - \frac{r_0hD_0^2}{2(r_0 + h)} \approx r_0 + h$$
(2.32)

Using this result, if $h = 100$ km, and $r_0 = r_e = 6368$ km (the average radius of the earth), the error term calculated is less than 0.6 m. Thus

$$h = r - r_0$$
(2.33)

is a good approximation. However, in this equation the value of r_0 must be evaluated, as discussed in Section 2.12.

2.12   CALCULATION OF GEODETIC LATITUDE[5-7]

Referring to Figure 2.7, the relation between angles L and L_c can be found from the triangle OPC. From the simple geometry it can be seen that

$$L = L_c + D$$
(2.34)

If the angle D can be found, the relation between L and L_c can be obtained. To find this angle, let us find the distance OC first. Combining Equations (2.24) and (2.27), the following result is obtained:

$$OC = OE - CE = a_e \cos \beta - \frac{AE}{\tan L} = a_e \cos \beta - \frac{b_e \sin \beta}{\tan L}$$

$$= a_e \cos \beta[1 - (1 - e_e^2)] = a_e e_e^2 \cos \beta = e_e^2 OE$$
(2.35)

where β is not shown in this figure but is shown in Figure 2.6.

From the triangle OPC and the law of sine, one can write

$$\frac{\sin D}{OC} = \frac{\sin(\pi - L)}{r} \tag{2.36}$$

From Equation (2.35),

$$OC = e_e^2 OE = e_e^2 r_0 \cos L_{co} \tag{2.37}$$

but

$$L_{co} = L - D_0 \tag{2.38}$$

Therefore,

$$OC = e_e^2 r_0 \cos(L - D_0) = e_e^2 r_0 (\cos L \cos D_0 + \sin L \sin D_0) \tag{2.39}$$

From Equation (2.23), the ellipticity e_p of the earth is

$$e_p = \frac{a_e - b_e}{a_e} \tag{2.40}$$

The eccentricity and the ellipticity can be related as

$$e_e^2 = \frac{a_e^2 - b_e^2}{a_e^2} = \frac{(a_e - b_e)}{a_e} \frac{(a_e + b_e)}{a_e} = e_p \frac{(2a_e - a_e + b_e)}{a_e} = e_p(2 - e_p) \tag{2.41}$$

Substituting Equations (2.39) and (2.41) into Equation (2.36), the result is

$$\sin D = 2e_p \left(1 - \frac{e_p}{2}\right) \frac{r_0}{r_0 + h} \left(\frac{1}{2} \sin 2L \cos D_0 + \sin^2 L \sin D_0\right) \tag{2.42}$$

In the above equation the relation $r = r_0 + h$ is used. Since D and D_0 are both very small angles, the above equation can be written as

$$D = 2e_p \left(1 - \frac{e_p}{2}\right) \frac{r_0}{r_0 + h} \left(\frac{1}{2} \sin 2L + D_0 \sin^2 L\right) \tag{2.43}$$

The relations

$$\sin D \approx D; \, s_p \sin D_0 \approx D_0; \, s_p \cos D_0 \approx 1 \tag{2.44}$$

are used in obtaining the results of Equation (2.43). If the height $h = 0$, then from Figure 2.7 $D = D_0$. Using this relation Equation (2.43) can be written as

$$D_0 \left[1 - 2e_p \left(1 - \frac{e_p}{2}\right) \sin^2 L\right] = e_p \left(1 - \frac{e_p}{2}\right) \sin 2L \text{ or}$$

$$D_0 = e_p \sin 2L + \epsilon_1 \tag{2.45}$$

where

$$\epsilon_1 = -\frac{e_p^2}{2} \sin L + 2e_p^2 \sin 2L \sin^2 L + \cdots \leq 1.6 \text{ arc} - \sec \tag{2.46}$$

Substitute the approximation of $D_0 \approx e_p \sin 2L$ into Equation (2.43) and the result is

$$D = 2e_p \left(1 - \frac{e_p}{2}\right) \left(1 - \frac{h}{r_0}\right) \left(\frac{1}{2} \sin 2L + e_p \sin 2L \sin^2 L\right) \tag{2.47}$$

or

$$D = e_p \sin 2L + \epsilon \tag{2.48}$$

where

$$\epsilon = -\frac{e_p^2}{2} \sin 2L - \frac{he_p}{r_0} \sin 2L + \cdots \tag{2.49}$$

This error is less than 4.5 arc-sec for $h = 30$ km. Using the approximate value of D, the relation between angle L and L_c can be found from Equation (2.34) as

$$L = L_c + e_p \sin 2L \tag{2.50}$$

This is a nonlinear equation that can be solved through the iteration method. This equation can be written in a slightly different form as

$$L_{i+1} = L_c + e_p \sin 2L_i \tag{2.51}$$

where $i = 0, 1, 2, \ldots$. One can start with $L_0 = L_c$. If the difference $(L_{i+1} - L_i)$ is smaller than a predetermined threshold, the last value of L_i can be considered as the desired one. It should be noted that during the iteration method L_c is a constant that is obtained from Equation (2.18).

2.13   CALCULATION OF A POINT ON THE SURFACE OF THE EARTH[5]

The final step of this calculation is to find the value r_0 in Equation (2.33). This value is also shown in Figure 2.7. The point A (x, z) is on the ellipse; therefore, it satisfies the following elliptic Equation (2.21). This equation is rewritten here for convenience,

$$\frac{x^2}{a_e^2} + \frac{z^2}{b_e^2} = 1 \tag{2.52}$$

where a_e and b_e are the semi-major and semi-minor axes of the earth. From Figure 2.7, the x and z values can be written as

$$x = r_0 \cos L_{co}$$

$$z = r_0 \sin L_{co} \tag{2.53}$$

Substituting these relations into Equation (2.52) and solving for r_0, the result is

$$r_0^2 \left(\frac{\cos^2 L_{co}}{a_e^2} + \frac{\sin^2 L_{co}}{b_e^2} \right) = r_0^2 \left(\frac{b_e^2 \cos^2 L_{co} + a_e^2 (1 - \cos^2 L_{co})}{a_e^2 b_e^2} \right) = 1 \quad \text{or}$$

$$r_0^2 = \frac{a_e^2 b_e^2}{a_e^2 \left[1 - \left(1 - \frac{b_e^2}{a_e^2} \right) \cos L_{co} \right]} = \frac{b_e^2}{1 - e_e^2 \cos L_{co}} \quad \text{or}$$

$$r_0 = b_e \left(1 + \frac{1}{2} e_e^2 \cos^2 L_{co} + \cdots \right) \tag{2.54}$$

Use Equation (2.23) to replace b_e by a_e, Equation (2.41) to replace e_e by e_p, and L to replace L_{co} because $L \approx L_{co}$, and then

$$r_0 \approx a_e (1 - e_p) \left[1 + \left(e_p - \frac{e_p^2}{2} \right) \cos^2 L + \cdots \right]$$

$$\approx a_e (1 - e_p)(1 + e_p - e_p \sin^2 L + \cdots) \tag{2.55}$$

In this equation the higher order of e_p is neglected. The value of r_0 can be found as

$$r_0 \approx a_e (1 - e_p \sin^2 L) \tag{2.56}$$

To solve for the latitude and altitude of the user, use Equation (2.51) to find the geodetic latitude L first. Then use Equation (2.56) to find r_0, and finally, use Equation (2.33) to find the altitude. The result is

$$h \approx r - r_0 \approx \sqrt{x_u^2 + y_u^2 + z_u^2} - a_e (1 - e_p \sin^2 L) \tag{2.57}$$

2.14 SATELLITE SELECTION[1,8]

A GPS receiver can simultaneously receive signals from 4 up to 11 satellites, if the receiver is on the surface of the earth. Under this condition, there are two approaches to solve the problem. The first one is to use all the satellites to calculate the user position. The other approach is to choose only four satellites from the constellation. The usual way is to utilize all the satellites to calculate the user position, because additional measurements are used. In this section and section 2.15 the selection of satellites will be presented. In order to focus on this subject only the four-satellite case will be considered.

If there are more than four satellite signals that can be received by a GPS receiver, a simple way is to choose only four satellites and utilize them to solve for the user position. Under this condition, the question is how to select the four satellites. Let us use a two-dimensional case to illustrate the situation, because it is easier to show graphically. In order to solve a position in a two-dimensional case, three satellites are required considering the user clock bias. In this discussion, it is assumed that the user position can be uniquely determined as discussed in Section 2.3. If this assumption cannot be used, four satellites are required.

Figure 2.8a shows the results measured by three satellites on a straight line, and the user is also on this line. Figure 2.8b shows that the three satellites and the user form a quadrangle. Two circles with the same center but different radii are drawn. The solid circle represents the distance measured from the user to the satellite with bias clock error. The dotted circle represents the distance after the clock error correction. From observation, the position error in Figure 2.8a is greater than that in Figure 2.8b because in Figure 2.8a all three dotted circles are tangential to each other. It is difficult to measure the tangential point accurately. In Figure 2.8b, the three circles intersect each other and the point of intersection can be measured more accurately. Another way to view this problem is to measure the area of a triangle made by the three satellites. In Figure 2.8a the total area is close to zero, while in Figure 2.8b the total area is quite large. In general, the larger the triangle area made by the three satellites, the better the user position can be solved.

The general rule can be extended to select the four satellites in a three-dimensional case. It is desirable to maximize the volume defined by the four satellites. A tetrahedron with an equilateral base contains the maximum volume and therefore can be considered as the best selection. Under this condition, one satellite is at zenith and the other three are close to the horizon and separated by 120 degrees.[8] This geometry will generate the best user position estimation. If all four satellites are close to the horizon, the volume defined by these satellites and the user is very small. Occasionally, the user position error calculated with this arrangement can be extremely large. In other words, the δv calculated from Equation (2.11) may not converge.

2.15   DILUTION OF PRECISION[1,8]

The dilution of precision (DOP) is often used to measure user position accuracy. There are several different definitions of the DOP. All the different DOPs are a function of satellite geometry only. The positions of the satellites determine the DOP value. A detailed discussion can be found in reference 8. Here only the definitions and the limits of the values will be presented.

The geometrical dilution of precision (GDOP) is defined as

$$\text{GDOP} = \frac{1}{\sigma}\sqrt{\sigma_x^2 + \sigma_y^2 + \sigma_z^2 + \sigma_b^2} \qquad (2.58)$$

where σ is the measured rms error of the pseudorange, which has a zero mean, $\sigma_x \sigma_y \sigma_z$ are the measured rms errors of the user position in the xyz directions, and σ_b is the measured rms user clock error expressed in distance.

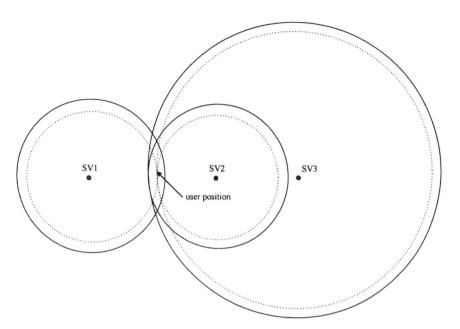

(a) Satellites and user form a straight line.

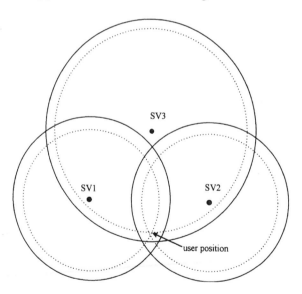

(b) Satellites and user form a quadrangle.

FIGURE 2.8   Three satellites are used to measure two-dimensional user position.

The position dilution of precision is defined as

$$PDOP = \frac{1}{\sigma}\sqrt{\sigma_x^2 + \sigma_y^2 + \sigma_z^2} \tag{2.59}$$

The horizontal dilution of precision is defined as

$$HDOP = \frac{1}{\sigma}\sqrt{\sigma_x^2 + \sigma_y^2} \tag{2.60}$$

The vertical dilution of precision is

$$VDOP = \frac{\sigma_z}{\sigma} \tag{2.61}$$

The time dilution of precision is

$$TDOP = \frac{\sigma_b}{\sigma} \tag{2.62}$$

The smallest DOP value means the best satellite geometry for calculating user position. It is proved in reference 8 that in order to minimize the GDOP, the volume contained by the four satellites must be maximized. Assume that the

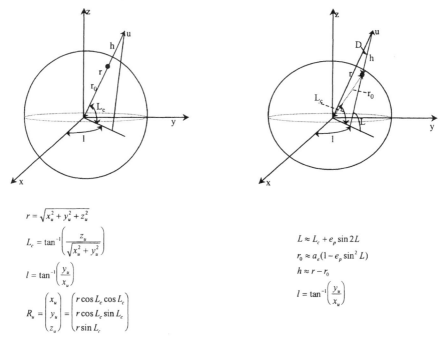

$$r = \sqrt{x_u^2 + y_u^2 + z_u^2}$$

$$L_c = \tan^{-1}\left(\frac{z_u}{\sqrt{x_u^2 + y_u^2}}\right)$$

$$l = \tan^{-1}\left(\frac{y_u}{x_u}\right)$$

$$R_u = \begin{pmatrix} x_u \\ y_u \\ z_u \end{pmatrix} = \begin{pmatrix} r\cos L_c \cos L_c \\ r\cos L_c \sin L_c \\ r\sin L_c \end{pmatrix}$$

$$L \approx L_c + e_p \sin 2L$$

$$r_0 \approx a_e(1 - e_p \sin^2 L)$$

$$h \approx r - r_0$$

$$l = \tan^{-1}\left(\frac{y_u}{x_u}\right)$$

FIGURE 2.9   Relations to change from spherical to ellipsoidal earth.

four satellites form the optimum constellation. Under this condition the elevation angle is 0 degree and three of the four satellites form an equilateral triangle. The observer is at the center of the base of the tetrahedron. Under this condition, the DOP values are: GDOP $= \sqrt{3} \approx 1.73$, PDOP $= 2\sqrt{2/3} \approx 1.63$, HDOP $=$ VDOP $= 2/\sqrt{3} \approx 1.15$, and TDOP $= 1/\sqrt{3} \approx 0.58$. These values can be considered as the minimum values (or the limits) of the DOPs. In selecting satellites, the DOP values should be as small as possible in order to generate the best user position accuracy.

2.16 SUMMARY

This chapter discusses the basic concept of solving the GPS user position. First use four or more satellites to solve the user position in terms of latitude, longitude, altitude, and the user clock bias as discussed in Section 2.5. However, the solutions obtained through this approach are for a spherical earth. Since the earth is not a perfect sphere, the latitude and altitude must be modified to reflect the ellipsoidal shape of the earth. Equations (2.51) and (2.57) are used to derive the desired values. These results are shown in Figure 2.9 as a quick reference. Finally, the selection of satellites and the DOP are discussed.

REFERENCES

1. Spilker, J. J., "GPS signal structure and performance characteristics," *Navigation*, Institute of Navigation, vol. 25, no. 2, pp. 121–146, Summer 1978.
2. Spilker, J. J. Jr., Parkinson, B. W., "Overview of GPS operation and design," Chapter 2, and Spilker, J. J. Jr., "GPS navigation data," Chapter 4 in Parkinson, B. W., Spilker, J. J. Jr., *Global Positioning System: Theory and Applications*, vols. 1 and 2, American Institute of Aeronautics and Astronautics, 370 L'Enfant Promenade, SW, Washington, DC, 1996.
3. Kay, S. M., *Fundamentals of Statistical Signal Processing Estimation Theory*, Chapter 8, Prentice Hall, Englewood Cliffs, NJ 1993.
4. Bate, R. R., Mueller, D. D., and White, J. E., *Fundamentals of Astrodynamics*, Chapter 5, Dover Publications, New York, 1971.
5. Britting, K. R., *Inertial Navigation Systems Analysis*, Chapter 4, Wiley, 1971.
6. Riggins, R. "Navigation using the global positioning system," Chapter 6, class notes, Air Force Institute of Technology, 1996.
7. "Department of Defense world geodetic system, 1984 (WGS-84), its definition and relationships with local geodetic systems," DMA-TR-8350.2, Defense Mapping Agency, September 1987.
8. Spilker, J. J. Jr., "Satellite constellation and geometric dilution of precision," Chapter 5, and Axelrad, P., Brown, R. G., "GPS navigation algorithms," Chapter 9 in Parkinson, B. W., Spilker, J. J. Jr., *Global Positioning System: Theory and Applications*, vols. 1 and 2, American Institute of Aeronautics and Astronautics, 370 L'Enfant Promenade, SW, Washington, DC, 1996.

```
% p2_1.m
% Userpos.m use pseudorange and satellite positions to
calculate user position
% JT 30 April 96

% ***** Input data *****

sp(1:3, 1:nsat); % satellite position which has the following
format
```

$$
\% \ sp = \begin{bmatrix} x_1 & y_1 & z_1 \\ x_2 & y_2 & z_2 \\ x_3 & y_3 & z_3 \\ \cdots & & \\ x_{nn} & y_{nn} & z_{nn} \end{bmatrix}
$$

```
pr(1:nsat); % is the measured pseudorange which has the
format as
% pr=[pr1 pr2 pr3 ... prnn]^T;

nn=nsat; % is the number of satellites

% ***** Select initial guessed positions and clock bias *****
x_guess = 0; y_guess = 0; z_guess = 0; bu = 0;
gu(1) = x_guess; gu(2) = y_guess; gu(3) = z_guess;

% Calculating rao the pseudorange as shown in Equation
(2.1)the
% clock bias is not included.
for j = 1:nsat
  rao(j)=((gu(1)-sp(1,j))^2+(gu(2)-sp(2,j))^2+(gu(3)-
          sp(3,j))^2)^.5;
end

% generate the fourth column of the alpha matrix in Eq. 2.15
alpha(:,4) = ones(nsat,1);

erro=1;
while erro>.01;
  for j = 1:nsat;
    for k = 1:3;
       alpha(j,k) = (gu(k)-sp(k,j))/(rao(j)); % find first
                                 %3 colums of alpha matrix
    end
  end
  drao = pr - (rao + ones(1,nsat)*bu);%** find delta rao
                                % includes clock bias

  dl = pinv(alpha)*drao'; % Equation (2.16) pinv
  (pseudoinverse)
  bu = bu + dl(4);  % new clock bias
```

```
  for k = 1:3;
     gu(k) = gu(k) + dl(k);   %** find new position
  end
  erro=dl(1)^2+dl(2)^2+dl(3)^2; % find error
  for j = 1:nsat;
    rao(j)=((gu(1)-sp(1,j))^2+(gu(2)-
           sp(2,j))^2+(gu(3)-sp(3,j))^2)^.5; % find new rao
           without clock bias
  end
end

% ***** Final result in spherical coordinate system *****
xuser = gu(1); yuser = gu(2); zuser = gu(3); bias = bu;

rsp = (xuser^2+yuser^2+zuser^2)^.5; % radius of spherical
earth
                                  % Eq 2.17
Lc = atan(zuser/(xuser^2+yuser^2)^.5); % latitude of sperical
                                  % earth Eq 2.18
lsp = atan(yuser/xuser)*180/pi; % longitude spherical and
flat
                                  % earth Eq 2.19

% ***** Converting to practical earth shape *****

e=1/298.257223563;
Ltemp=Lc;
erro1=1;

while erro1>1e-6; % calculating latitude by Eq. 2.51
    L=Lc+e*sin(2*Ltemp);
    erro1=abs(Ltemp-L);
    Ltemp=L;
end
Lflp=L*180/pi; % latitude of flat earth
re=6378137;
h=rsp-re*(1-e*(sin(L)^2)); % altitude of flat earth
lsp = lsp; % longitude of flat earth
upos = [xuser yuser zuser bias rsp Lflp lsp h]';
```

CHAPTER THREE

Satellite Constellation

3.1 INTRODUCTION

The previous chapter assumes that the positions of the satellites are known. This chapter will discuss the satellite constellation and the determination of the satellite positions. Some special terms related to the orbital mechanics, such as sidereal day, will be introduced. The satellite motion will have an impact on the processing of the signals at the receiver. For example, the input frequency shifts as a result of the Doppler effect. Such details are important for the design of acquisition and tracking loops in the receiver. However, in order to obtain some of this information a very accurate calculation of the satellite motion is not needed. For example, the actual orbit of the satellite is elliptical but it is close to a circle. The information obtained from a circular orbit will be good enough to find an estimation of the Doppler frequency. Based on this assumption the circular orbit is used to calculate the Doppler frequency, the rate of change of the Doppler frequency, and the differential power level. From the geometry of the satellite distribution, the power level at the receiver can also be estimated from the transmission power. This subject is presented in the final section in this chapter.

In order to find the location of the satellite accurately, a circular orbit is insufficient. The actual elliptical satellite orbit must be used. Therefore, the complete elliptical satellite orbit and Kepler's law will be discussed. Information obtained from the satellite through the GPS receiver via broadcast navigation data such as the mean anomaly does not provide the location of the satellite directly. However, this information can be used to calculate the precise location of the satellite. The calculation of the satellite position from these data will be discussed in detail.

Fundamentals of Global Positioning System Receivers: A Software Approach, Second Edition, by James Bao-Yen Tsui
ISBN 0-471-70647-7   Copyright © 2005 John Wiley & Sons, Inc.

3.2   CONTROL SEGMENT OF THE GPS SYSTEM[1-3]

This section will provide a very brief idea of the GPS system. The GPS system may be considered as comprising three segments: the control segment, the space segment, and the user segment. The space segment contains all the satellites, which will be discussed in Chapters 3, 4, and 5. The user segment can be considered the base of receivers and their processing, which is the focus of this text. The control segment will be discussed in this section.

The control segment consists of five control stations, including a master control station. These control stations are widely separated in longitude around the earth. The master control station is located at Falcon Air Force Base, Colorado Springs, CO. Operations are maintained at all times year round. The main purpose of the control stations is to monitor the performance of the GPS satellites. The data collected from the satellites by the control stations will be sent to the master control station for processing. The master control station is responsible for all aspects of constellation control and command. A few of the operation objectives are presented here: (1) Monitor GPS performance in support of all performance standards. (2) Generate and upload the navigation data to the satellites to sustain performance standards. (3) Promptly detect and respond to satellite failure to minimize the impact. Detailed information on the control segment can be found in reference 3.

3.3   SATELLITE CONSTELLATION[3-9]

There are a total of 24 GPS satellites divided into six orbits and each orbit has four satellites. Each orbit makes a 55-degree angle with the equator, which is referred to as the inclination angle. The orbits are separated by 60 degrees to cover the complete 360 degrees. The radius of the satellite orbit is 26,560 km and it rotates around the earth twice in a sidereal day. Table 3.1 lists all these parameters.

The central body of the Block IIR satellite is a cube of approximately 6 ft on each side.[8] The span of the solar panel is about 30 ft. The lift-off weight of the spacecraft is 4,480 pounds and the on-orbit weight is 2,370 pounds.

The four satellites in an orbit are not equally spaced. Two satellites are separated by 30.0–32.1 degrees. The other two make three angles with the first two

TABLE 3.1   Characteristics of GPS Satellites

Constellation	
Number of satellites	24
Number of orbital planes	6
Number of satellites per orbit	4
Orbital inclination	55°
Orbital radius[7]	26560 km
Period[4]	11 hrs 57 min 57.26 sec
Ground track repeat	sidereal day

satellites and these angles range 92.38–130.98 degrees.[9] The spacing has been optimized to minimize the effects of a single satellite failure on system degradation. At any time and any location on the earth, neglecting obstacles such as mountains and tall buildings, a GPS receiver should have a direct line of sight and be receiving signals from 4 to 11 satellites. A majority of the time a GPS receiver can receive signals from more than four satellites. Since four satellites are the minimum required number to find the user position, this arrangement can provide user position at any time and any location. For this 24-satellite constellation with a 5-degree elevation mask angle, more than 80% of the time seven or more satellites are in view.[9] A user at 35 degrees latitude corresponds to the approximate worst latitude where momentarily there are only four satellites in view (approximately .4% of the time).

The radius of the earth is 6,378 km around the equator and 6,357 km passing through the poles, and the average radius can be considered as 6,368 km. The radius of the satellite orbit is 26,560 km, which is about 20,192 km (26,560 − 6,368) above the earth's surface. This height agrees well with references 6 and 9. This height is approximately the shortest distance between a user on the surface of the earth and the satellite, which occurs at around zenith or an elevation angle of approximately 90 degrees. Most GPS receivers are designed to receive signals from satellites above 5 degrees. For simplicity, let us assume that the receiver can receive signals from satellites at the zero-degree point. The distance from a satellite on the horizon to the user is 25,785 km ($\sqrt{26{,}560^2 - 6{,}368^2}$). These distances are shown in Figure 3.1.

FIGURE 3.1   Earth and circular satellite orbit.

From the distances in Figure 3.1 one can see that the time delays from the satellites are in the range of 67 ms (20,192 km/c) to 86 ms (25,785 km/c), where c is the speed of light. If the user is on the surface of the earth, the maximum differential delay time from two different satellites should be within 19 $(86 - 67)$ ms. In this figure, the angle α is approximately 76.13 degrees and the angle β is approximately 13.87 degrees. Therefore, the transmitting antenna on the satellite need only have a solid angle of 13.87 degrees to cover the earth. However, the antenna for the L_1 band is 21.3 degrees and the antenna for the L_2 band is 23.4 degrees. Both are wider than the minimum required angle. The solid angle of 21.3 degrees will be used in Section 3.13 to estimate the power to the receiver. The antenna pattern will be further discussed in Section 5.2.

3.4 MAXIMUM DIFFERENTIAL POWER LEVEL FROM DIFFERENT SATELLITES

From Figure 3.1 one can calculate the relative power level of the received signals on the surface of the earth. The transmitting antenna pattern is designed to directly aim at the earth underneath it. However, the distances from the receiver to various satellites are different. The shortest distance to a satellite is at zenith and the farthest distance to a satellite is at horizon. Suppose the receiver has an omnidirectional antenna pattern. Since the power level is inversely proportional to the distance square, the difference in power level can be found as

$$\Delta p = 10 \log \left(\frac{25785^2}{20192^2} \right) \approx 2.1 \text{ dB} \tag{3.1}$$

It is desirable to receive signals from different satellites with similar strength. In order to achieve this goal, the transmitting antenna pattern must be properly designed. The beam is slightly weaker at the center to compensate for the power difference.

3.5 SIDEREAL DAY[10,11]

Table 3.1 indicates that the satellite rotates around the earth twice in a sidereal day. The sidereal day is slightly different from an apparent solar day. The apparent day has 24 hours and it is the time used daily. The apparent solar day is measured by the time between two successive transits of the sun across our local meridian, because we use the sun as our reference. A sidereal day is the time for the earth to turn one revolution. Figure 3.2 shows the difference between the apparent solar day and a sidereal day. In this figure, the effect is exaggerated and it is obvious that a sidereal day is slightly shorter than a solar day. The difference should be approximately equal to one day in one year which corresponds to about 4 min ($24 \times 60/365$) per day. The mean sidereal day is 23 hrs, 56 min, 4.09 sec.

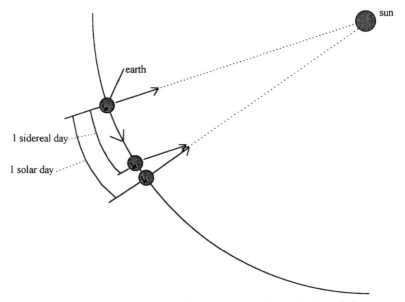

FIGURE 3.2   Configuration of apparent solar day and sidereal day.

The difference from an apparent day is 3 min, 55.91 sec. Half a sidereal day is 11 hrs, 58 min, 2.05 sec. This is the time for the satellite to rotate once around the earth. From this arrangement one can see that from one day to the next a certain satellite will be at approximately the same position at the same time. The location of the satellite will be presented in the next section.

3.6 DOPPLER FREQUENCY SHIFT

The purpose of Sections 3.6 to 3.9 is to find some coarse information on the Doppler frequency. This information will be used as guidance in the acquisition programs. More detailed information can be found in Section 12.12, where the orbits of the satellites and the Doppler frequency can be calculated for a given time and user position.

In this section, the Doppler frequency shift caused by the satellite motion both on the carrier frequency and on the coarse/acquisition (C/A) code will be discussed. This information is important for performing both the acquisition and the tracking of the GPS signal.

The angular velocity $d\theta/dt$ and the speed v_s of the satellite can be calculated from the approximate radius of the satellite orbit as

$$\frac{d\theta}{dt} = \frac{2\pi}{11 \times 3600 + 58 \times 60 + 2.05} \approx 1.458 \times 10^{-4} \text{ rad/s}$$

$$v_s = \frac{r_s d\theta}{dt} \approx 26560 \text{ km} \times 1.458 \times 10^{-4} \approx 3874 \text{ m/s} \qquad (3.2)$$

where r_s is the average radius of the satellite orbit. In 3 min, 55.91 sec, the time difference between an apparent solar day and the sidereal day, the satellite will travel approximately 914 km (3,874 m/s × 235.91 sec). Referenced to the surface of the earth with the satellite in the zenith direction, the corresponding angle is approximately .045 radian (914/20192) or 2.6 degrees. If the satellite is close to the horizon, the corresponding angle is approximately .035 radian or 2 degrees. Therefore, one can consider that the satellite position changes about 2–2.6 degrees per day at the same time with respect to a fixed point on the surface of the earth. In Figure 3.3, the satellite is at position S and the user is at position A. The Doppler frequency is caused by the satellite velocity component v_d toward the user where

$$v_d = v_s \sin \beta \qquad (3.3)$$

Now let us find this velocity in terms of angle θ. Using the law of cosine in triangle OAS, the result is

$$AS^2 = r_e^2 + r_s^2 - 2r_e r_s \cos \alpha = r_e^2 + r_s^2 - 2r_e r_s \sin \theta \qquad (3.4)$$

because of $\alpha + \theta = \pi/2$. In the same triangle, using the law of sine, the result is

$$\frac{\sin \beta}{\sin \alpha} = \frac{\sin \beta}{\cos \theta} = \frac{r_e}{AS} \qquad (3.5)$$

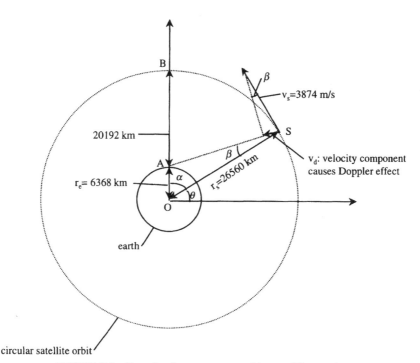

FIGURE 3.3 Doppler frequency caused by satellite motion.

Substituting these results into Equation (3.3), one obtains

$$v_d = \frac{v_s r_e \cos \theta}{AS} = \frac{v_s r_e \cos \theta}{\sqrt{r_e^2 + r_s^2 - 2r_e r_s \sin \theta}} \tag{3.6}$$

This velocity can be plotted as a function θ and is shown in Figure 3.4.

As expected, when $\theta = \pi/2$, the Doppler velocity is zero. The maximum Doppler velocity can be found by taking the derivative of v_d with respect to θ and setting the result equal to zero. The result is

$$\frac{dv_d}{d\theta} = \frac{vr_e[r_e r_s \sin^2 \theta - (r_e^2 + r_s^2) \sin \theta + r_e r_s]}{(r_e^2 + r_s^2 - 2r_e r_s \sin \theta)^{3/2}} = 0 \tag{3.7}$$

Thus $\sin \theta$ can be solved as

$$\sin \theta = \frac{r_e}{r_s} \text{ or } \theta = \sin^{-1}\left(\frac{r_e}{r_s}\right) \approx 0.242 \text{ rad} \tag{3.8}$$

At this angle θ the satellite is at the horizontal position referenced to the user. Intuitively, one expects that the maximum Doppler velocity occurs when the satellite is at the horizon position and this calculation confirms it. From the orbit

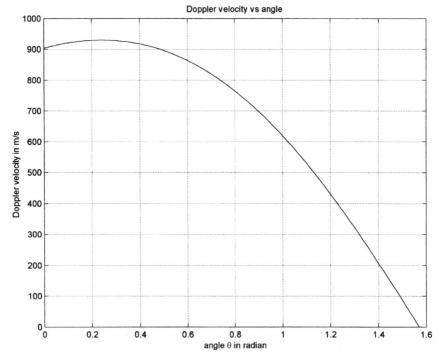

FIGURE 3.4 Velocity component toward the user versus angle θ.

speed, one can calculate the maximum Doppler velocity v_{dm}, which is along the horizontal direction as

$$v_{dm} = \frac{v_s r_e}{r_s} = \frac{3874 \times 6368}{26560} \approx 929 \text{ m/s} \approx 2078 \text{ miles/h} \tag{3.9}$$

This speed is equivalent to a high-speed military aircraft. The Doppler frequency shift caused by a land vehicle is often very small, even if the motion is directly toward the satellite to produce the highest Doppler effect. For the L_1 frequency ($f = 1575.42$ MHz), which is modulated by the C/A signal, the maximum Doppler frequency shift is

$$f_{dr} = \frac{f_r v_{dm}}{c} = \frac{1575.42 \times 929}{3 \times 10^8} \approx 4.9 \text{ kHz} \tag{3.10}$$

where c is the speed of light. Therefore, for a stationary observer, the maximum Doppler frequency shift is around ± 5 kHz.

If a vehicle carrying a GPS receiver moves at a high speed, the Doppler effect must be taken into consideration. To create a Doppler frequency shift of ± 5 kHz by the vehicle alone, the vehicle must move toward the satellite at about 2,078 miles/hr. This speed will include most high-speed aircraft. Therefore, in designing a GPS receiver, if the receiver is used for a low-speed vehicle, the Doppler shift can be considered as ± 5 kHz. If the receiver is used in a high-speed vehicle, it is reasonable to assume that the maximum Doppler shift is ± 10 kHz. These values determine the searching frequency range in the acquisition program. Both of these values are assumed an ideal oscillator and sampling frequency and further discussion is included in Section 6.15.

The Doppler frequency shift on the C/A code is quite small because of the low frequency of the C/A code. The C/A code has a frequency of 1.023 MHz, which is 1,540 (1575.42/1.023) times lower than the carrier frequency. The Doppler frequency is

$$f_{dc} = \frac{f_c v_h}{c} = \frac{1.023 \times 10^6 \times 929}{3 \times 10^8} \approx 3.2 \text{ Hz} \tag{3.11}$$

If the receiver moves at high speed, this value can be doubled to 6.4 Hz. This value is important for the tracking method (called block adjustment of synchronizing signal or BASS program), which will be discussed in Chapter 8. In the BASS program, the input data and the locally generated data must be closely aligned. The Doppler frequency on the C/A code can cause misalignment between the input and the locally generated codes.

If the data is sampled at 5 MHz (referred to as the sampling frequency), each sample is separated by 200 ns (referred to as the sampling time). In the tracking program it is desirable to align the locally generated signal and the input signal within half the sampling time or approximately 100 ns. Larger separation of these two signals will lose tracking sensitivity. The chip time of the C/A code is 977.5 ns or $1/(1.023 \times 10^6)$ sec. It takes 156.3 ms (1/6.4) to shift one cycle or 977.5 ns. Therefore, it takes approximately 16 ms ($100 \times 156.3/977.5$) to shift 100 ns. In a high-speed aircraft, a selection of a block of the input data should be checked about every 16 ms to make sure these data align well with the locally generated data. Since there is noise on the signal, using 1 ms of data may not determine the alignment accurately. One may extend the adjustment of the input data to every 20 ms. For a slow-moving vehicle, the time may extend to 40 ms.

From the above discussion, the adjustment of the input data depends on the sampling frequency. Higher sampling frequency will shorten the adjustment time because the sampling time is short and it is desirable to align the input and the locally generated code within half the sampling time. If the incoming signal is strong and tracking sensitivity is not a problem, the adjustment time can be extended. However, the input and the locally generated signals should be aligned within half a chip time or 488.75 ns (977.5/2). This time can be considered as the maximum allowable separation time. With a Doppler frequency of 6.4 Hz, the adjustment time can be extended to 78.15 ms ($1/2 \times 6.4$). Detailed discussion of the tracking program will be presented in Chapter 8.

3.7   AVERAGE RATE OF CHANGE OF THE DOPPLER FREQUENCY

In this section the rate of change of the Doppler frequency will be discussed. This information is important for the tracking program. If the rate of change of the Doppler frequency can be calculated, the frequency update rate in the tracking can be predicted. Two approaches are used to find the Doppler frequency rate. A very simple way is to estimate the average rate of change of the Doppler frequency and the other one is to find the maximum rate of change of the Doppler frequency.

In Figure 3.4, the angle for the Doppler frequency changing from maximum to zero is about 1.329 radians ($\pi/2 - \theta = \pi/2 - 0.242$). It takes 11 hrs, 58 min, 2.05 sec for the satellite to travel 2π angle; thus, the time it takes to cover 1.329 radians is

$$t = (11 \times 3600 + 58 \times 60 + 2.05)\frac{1.329}{2\pi} = 9113 \text{ sec} \qquad (3.12)$$

During this time the Doppler frequency changes from 4.9 kHz to 0, thus, the average rate of change of the Doppler frequency δf_{dr} can be simply found as

$$\delta f_{dr} = \frac{4900}{9113} \approx 0.54 \text{ Hz/s} \qquad (3.13)$$

This is a very slow rate of change in frequency. From this value a tracking program can be updated every few seconds if the frequency accuracy in the tracking loop is assumed on the order of 1 Hz. The following section shows how to find the maximum frequency rate of change.

3.8   MAXIMUM RATE OF CHANGE OF THE DOPPLER FREQUENCY

In the previous section the average rate of change of the Doppler frequency is estimated; however, the rate of change is not a constant over that time period. In this section we try to find the maximum rate of change of the frequency. The rate of change of the speed v_d can be found by taking the derivative of v_d in Equation (3.6) with respect to time. The result is

$$\frac{dv_d}{dt} = \frac{dv_d}{d\theta}\frac{d\theta}{dt} = \frac{v r_e [r_e r_s \sin^2\theta - (r_e^2 + r_s^2)\sin\theta + r_e r_s]\, d\theta}{(r_e^2 + r_s^2 - 2r_e r_s \sin\theta)^{3/2}}\frac{d\theta}{dt} \tag{3.14}$$

In deriving this equation, the result of Equation (3.7) is used. The result of this equation is shown in Figure 3.5 and the maximum rate of change of the frequency occurs at $\theta = \pi/2$.

The corresponding maximum rate of change of the speed is

$$\left.\frac{dv_d}{dt}\right|_{\max} = \left.\frac{v r_e\, d\theta/dt}{\sqrt{r_e^2 + r_s^2 - 2r_e r_s}}\right|_{\theta=\pi/2} \approx 0.178 \text{ m/s}^2 \tag{3.15}$$

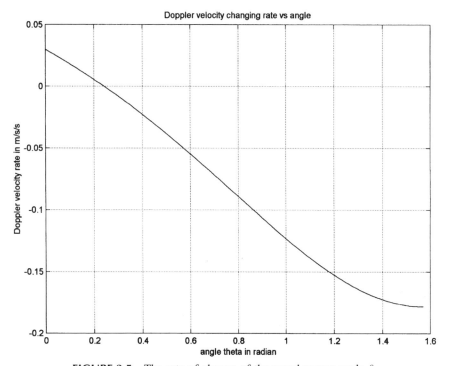

FIGURE 3.5   The rate of change of the speed versus angle θ.

In this equation, only the magnitude is of interest, thus, the sign is neglected. The corresponding rate of change of the Doppler frequency is

$$\delta f_{dr}|_{max} = \frac{dv_d}{dt}\frac{f_r}{c} = \frac{0.178 \times 1575.42 \times 10^6}{3 \times 10^8} = 0.936 \text{ Hz/s} \qquad (3.16)$$

This value is also very small. If the frequency accuracy measured through the tracking program is assumed on the order of 1 Hz, the update rate is almost one second, even at the maximum Doppler frequency changing rate.

3.9 RATE OF CHANGE OF THE DOPPLER FREQUENCY DUE TO USER ACCELERATION

From the previous two sections, it is obvious that the rate of change of the Doppler frequency caused by the satellite motion is rather low; therefore, it does not affect the update rate of the tracking program significantly.

Now let us consider the motion of the user. If the user has an acceleration of 1 g (gravitational acceleration with a value of 9.8 m/s²) toward a satellite, the corresponding rate of change of the Doppler frequency can be found from Equation (3.15) by replacing dv_d/dt by g. The corresponding result obtained from Equation (3.16) is about 51.5 Hz/s. For a high-performance aircraft, the acceleration can achieve several g values, such as 7 g. The corresponding rate of change of the Doppler frequency will be close to 360 Hz/s. Comparing with the rate of change of the Doppler frequency caused by the motions of the satellite and the receiver, the acceleration of the receiver is the dominant factor.

In tracking the GPS signal in a software GPS receiver two factors are used to update the tracking loop: the change of the carrier frequency and the alignment of the input and the locally generated C/A codes. As discussed in Section 3.5, the input data adjustment rate is about 20 ms due to the Doppler frequency on the C/A code. If the carrier frequency of the tracking loop has a bandwidth of the order of 1 Hz and the receiver accelerates at 7 g, the tracking loop must be updated approximately every 2.8 ms (1/360) due to the carrier frequency change. This might be a difficult problem because of the noise in the received signal. The operation and performance of a receiver tracking loop greatly depends on the acceleration of the receiver.

3.10 KEPLER'S LAWS[11,12]

In the previous section, the position of a satellite is briefly described. This information can be used to determine the differential power level and the Doppler frequency on the input signal. However, this information is not sufficient to calculate the position of a satellite. To find the position of a satellite, Kepler's laws are needed. The discussion in this section provides the basic equations to determine a satellite position.

Kepler's three laws are listed below (see Chapter 1 in ref. 11):

First Law: The orbit of each planet is an ellipse with the sun at a focus.
Second Law: The line joining the planet to the sun sweeps out equal areas in
    equal times.
Third Law: The square of the period of a planet is proportional to the cube
    of its mean distance from the sun.

These laws also apply to the motion of the GPS satellites. The satellite orbit
is elliptical with the earth at one of the foci. Figure 3.6 shows the orbit of a GPS
satellite. The center of the earth is at F and the position of the satellite is at S.
The angle v is called the actual anomaly. In order to illustrate the basic concept,
the shape of the ellipse is overemphasized. The actual orbit of the satellite is
very close to a circle. The point nearest to the prime focus is called the perigee
and the farthest point is called the apogee.

Kepler's second law can be expressed mathematically as (Figure 3.6)

$$\frac{t - t_p}{A_1} = \frac{T}{\pi a_s b_s} \tag{3.17}$$

where t presents the satellite position at time t, t_p is the time when the satellite
passes the perigee, A_1 is the area enclosed by the lines $t = t, t = t_p$, and the
ellipse, T is the period of the satellite, a_s and b_s are the semi-major and semi-
minor axes of the orbit, and $\pi a_s b_s$ is the total area of the ellipse. This equation
states that the time to sweep the area A_1 is proportional to the time T to sweep
the entire area of the ellipse.

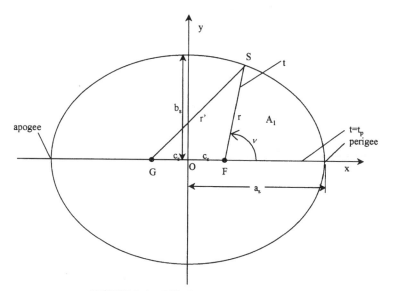

FIGURE 3.6 Elliptical orbit of a satellite.

The third law can be stated mathematically as

$$\frac{T^2}{a_s^3} = \frac{4\pi^2}{\mu} \equiv \frac{4\pi^2}{GM} \tag{3.18}$$

where $\mu = GM = 3.986005 \times 10^{14}$ meters^3/sec^2 (ref. 12) is the gravitational constant of the earth. Thus, the right-hand side of this equation is a constant. In this equation the semi-major axis a_s is used rather than the mean distance from the satellite to the center of the earth. In reference 11 it is stated that a_s can be used to replace the mean distance because the ratio of a_s to the mean distance r_s is a constant. This relationship can be shown as follows. If one considers the area of the ellipse orbit equal to the area of a circular orbit with radius r_s, then

$$\pi a_s b_s = \pi r_s^2 \quad \text{or} \quad \frac{a_s}{r_s} = \frac{r_s}{b_s} \tag{3.19}$$

Since a_s, b_s, r_s are constants, a_s and r_s is related by a constant.

3.11   KEPLER'S EQUATION[11,13]

In the following paragraphs Kepler's equation will be derived and the mean anomaly will be defined. The reason for this discussion is that the information given by the GPS system is the mean anomaly rather than the actual anomaly that is used to calculate the position of a satellite.

In order to perform this derivation, a few equations from the previous chapter will be repeated here. The eccentricity is defined as

$$e_s = \frac{\sqrt{a_s^2 - b_s^2}}{a_s} \equiv \frac{c_s}{a_s} \tag{3.20}$$

where c_s is the distance from the center of the ellipse to a focus. For an ellipse, the e_s value is $0 < e_s < 1$. When $a_s = b_s$, then $e_s = 0$, which represents a circle. The eccentricity e_s can be obtained from data transmitted by the satellite.

In Figure 3.7 an elliptical satellite orbit and a fictitious circular orbit are shown. The center of the earth is at F and the satellite is at S. The area A_1 is swept by the satellite from the perigee point to the position S. This area can be written as

$$A_1 = \text{area } PSV - A_2 \tag{3.21}$$

In the previous chapter Equation (2.24) shows that the heights of the ellipse and the circle can be related as

$$\frac{QP}{SP} = \frac{a_s}{b_s} \tag{3.22}$$

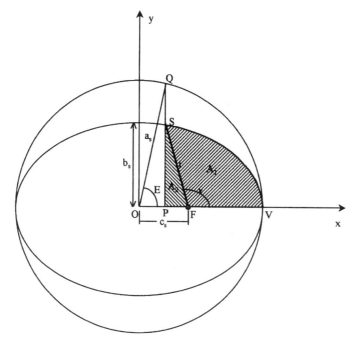

FIGURE 3.7   Fictitious and actual orbits.

Therefore, the area PSV can be obtained from area PQV as

$$\text{area } PSV = \frac{b_s}{a_s} \text{ area } PQV = \frac{b_s}{a_s}(\text{area } OQV - \text{area } OQP)$$

$$= \frac{b_s}{a_s}\left[\frac{1}{2}a_s^2 E - \frac{1}{2}a_s^2 \sin E \cos E\right] = \frac{a_s b_s}{2}(E - \sin E \cos E)$$

(3.23)

where the angle E is called eccentric anomaly. The area of triangle A_2 is

$$A_2 = \frac{1}{2}SP \times PF = \frac{1}{2}\frac{b_s}{a_s}QP \times PF = \frac{1}{2}\frac{b_s}{a_s}a_s \sin E(c_s - a_s \cos E)$$

$$= \frac{b_s}{2}\sin E(e_s a_s - a_s \cos E) = \frac{a_s b_s}{2}(e_s \sin E - \sin E \cos E)$$

(3.24)

In the above equation, the relation in Equation (3.20) is used. Substituting Equations (3.23) and (3.24) into (3.21) the area A_1 is

$$A_1 = \frac{a_s b_s}{2}(E - e_s \sin E)$$

(3.25)

Substituting this result into Equation (3.17), Kepler's second law, the result is

$$t - t_p = \frac{A_1 T}{\pi a_s b_s} = \frac{T}{2\pi}(E - e_s \sin E) = \sqrt{\frac{a_s^3}{\mu}}(E - e_s \sin E)$$

(3.26)

The next step is to define the mean anomaly M and from Equation (3.26) the result is

$$M \equiv (E - e_s \sin E) = \sqrt{\frac{\mu}{a_s^3}}(t - t_p) \qquad (3.27)$$

If one defines the mean motion n as the average angular velocity of the satellite, then from Equation (3.18) the result is

$$n = \frac{2\pi}{T} = \sqrt{\frac{\mu}{a_s^3}} \qquad (3.28)$$

Substituting this result into Equation (3.27) the result is

$$M \equiv (E - e_s \sin E) = n(t - t_p) \qquad (3.29)$$

This is referred to as Kepler's equation. From this equation one can see that M is linearly related to t; therefore, it is called the mean anomaly.

3.12   TRUE AND MEAN ANOMALY

The information obtained from a GPS satellite is the mean anomaly M. From this value, the true anomaly must be obtained because the true anomaly is used to find the position of the satellite. The first step is to obtain the eccentric anomaly E from the mean anomaly, Equation (3.29) relates M and E. Although this equation appears very simple, it is a nonlinear one; therefore, it is difficult to solve analytically. This equation can be rewritten as follows:

$$E = M + e_s \sin E \qquad (3.30)$$

In this equation, e_s is a given value representing the eccentricity of the satellite orbit. Both e_s and M can be obtained from the navigation data of the satellite. The only unknown is E. One way to solve for E is to use the iteration method. A new E value can be obtained from a previous one. The above equation can be written in an iteration format as

$$E_{i+1} = M + e_s \sin E_i \qquad (3.31)$$

where E_{i+1} is the present value and E_i is the previous value. One common choice of the initial value of E is $E_0 = M$. This equation converges rapidly because the orbit is very close to a circle. Either one can define an error signal as $E_{err} = E_{i+1} - E_i$ and end the iteration when E_{err} is less than a predetermined value, or one can just iterate Equation (3.31) a fixed number of times (i.e., from 5 to 10).

Once the E is found, the next step is to find the true anomaly v. This relation can be found by referring to Figure 3.7.

$$\cos E = \frac{OP}{a_s} = \frac{c_s - PF}{a_s} = \frac{c_s + r\cos v}{a_s} \tag{3.32}$$

Now let us find the distance r in terms of angle v. From Figure 3.6, applying the law of cosine to the triangle GSF, the following result is obtained

$$r'^2 = r^2 + 4rc_s\cos v + 4c_s^2 \tag{3.33}$$

where r and r' are the distance from the foci G and F to the point S. For an ellipse,

$$r' + r = 2a_s \tag{3.34}$$

Substituting this relation into Equation (3.33), the result is

$$r = \frac{a_s^2 - c_s^2}{a_s + c_s\cos v} = \frac{a_s(1 - e_s^2)}{1 + e_s\cos v} \tag{3.35}$$

Substituting this value of r into Equation (3.32) the result is

$$\cos E = \frac{e_s + \cos v}{1 + e_s\cos v} \tag{3.36}$$

Solve for v and the result is

$$\cos v = \frac{\cos E - e_s}{1 - e_s\cos E} \tag{3.37}$$

This solution generates multiple solutions for v because $\cos v$ is a multivalued function. One way to find the correct value of v is to keep these angles E and v in the same half plane. From Figure 3.7 one can see that the angles E and v are always in the same half plane.

Another approach to determine v is to find the $\sin v$.[13] If one takes the square on both sides of the above equation, the result is

$$\cos^2 v = 1 - \sin^2 v = \frac{(\cos E - e_s)^2}{(1 - e_s\cos E)^2} \tag{3.38}$$

Solve for $\sin v$ and the result is

$$\sin v = \frac{\sqrt{1 - e_s^2}\sin E}{1 - e_s\cos E} \tag{3.39}$$

The v can be found from Equations (3.37) and (3.39) and they are designated as v_1 and v_2 where

$$v_1 = \cos^{-1}\left(\frac{\cos E - e_s}{1 - e_s \cos E}\right)$$

$$v_2 = \sin^{-1}\left(\frac{\sqrt{1 - e_s^2}\,\sin E}{1 - e_s \cos E}\right) \tag{3.40}$$

The v_1 value calculated from Matlab is always positive for all E values and v_2 is positive for $E = 0$ to π and negative for $E = \pi$ to 2π as shown in Figure 3.8.

Thus, the true anomaly can be found as

$$v = v_1\,\text{sign}(v_2) \tag{3.41}$$

where sign (v_2) provides the sign of v_2; therefore, it is either $+1$ or -1. It is interesting to note that to find the true anomaly only M and e_s are needed. Although the semi-major axis a_s appears in the derivation, it does not appear in the final equation.

3.13   SIGNAL STRENGTH AT USER LOCATION[1,8,14-16]

In this section the signal strength at the user location will be estimated. The signal strength can be obtained from the power of the transmitting antenna, the

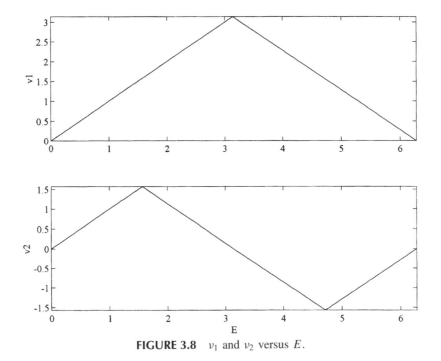

FIGURE 3.8 v_1 and v_2 versus E.

beam width of the antenna, the distance from the satellite to the user, and the effective area of the receiving antenna. The power amplifier of the transmitter is 50 w[8] (or 17 dBw). The input to the transmitting antenna is 14.3 dBw.[8] The difference might be due to impedance mismatch or circuit loss.

The gain of the transmitting antenna can be estimated from the beam width (or solid angle) of the antenna. The solid angle is denoted as θ, which is 21.3 degrees. The area on the surface of a sphere covered by the angle θ can be obtained from Figure 3.9 as

$$\text{Area} = \int_0^\theta 2\pi (r \sin \theta) r d\theta = 2\pi r^2 \int_0^\theta \sin \theta d\theta$$

$$= 2\pi r^2 (-\cos \theta)|_0^\theta = 2\pi r^2 (1 - \cos \theta) \tag{3.42}$$

The ratio of this area to the area of the sphere can be considered as the gain of the transmitting antenna, which can be written as

$$G = \frac{4\pi r^2}{2\pi r^2 (1 - \cos \theta)|_{21.3^0}} \approx \frac{2}{0.683} \approx 29.28 \approx 14.7 \text{ dB} \tag{3.43}$$

Using 14.3 dBw as the input to the antenna, the output of the antenna should be 29 dBw (14.3 + 14.7). However, the transmitting power level is listed as 478.63 w,[14,15] which corresponds to 26.8 dBw. This difference between the power levels might be due to efficiency of the antenna and the accuracy of the solid angle of the antenna because the power cannot be cut off sharply at a desired angle.

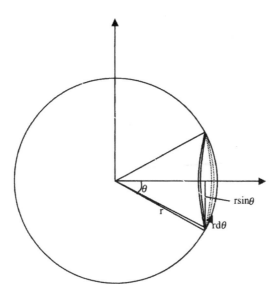

FIGURE 3.9   Area facing solid angle θ.

If the receiving antenna has a unit gain, the effective area is[16]

$$A_{\text{eff}} = \frac{\lambda^2}{4\pi} \tag{3.44}$$

where λ is the wavelength of the receiving signal.

The received power is equal to the power density multiplied by the effective area of the receiving antenna. The power density is equal to the radiating power divided by the surface of the sphere. The receiving power can be written as

$$P_r = \frac{P_t A_{\text{eff}}}{4\pi R_{su}^2} = \frac{P_t}{4\pi R_{su}^2} \frac{\lambda^2}{4\pi} = \frac{P_t \lambda^2}{(4\pi R_{su})^2} \tag{3.45}$$

where R_{su} is the distance from the satellite to the user. Assume $R_{su} = 25785 \times 10^3$ m, which is the farthest distance as shown in Figure 3.1. Using 478.63 W as the transmitting antenna and the wavelength $\lambda = 0.19$ m, the receiving power P_r calculated from the above equation is 1.65×10^{-16} w (or -157.8 dBw). If the loss through the atmosphere is taken into consideration, the received power is close to the minimum required value of -160 dBw.

The power level at the receiver is shown in Figure 3.10. It is a function of the elevation angle.[1] At zenith and horizon, the powers are at -160 dBw. The maximum power level is -158 dBw, which occurs at about 40 degrees. If the receiving antenna is taken into consideration, the received power will be modified by its antenna pattern.

3.14   SUMMARY

This chapter discusses the orbits of the GPS satellite. The orbit is elliptical but it is very close to a circle. Thus, the circular orbit is used to figure the power difference to the receiver and the Doppler frequency shift. This information is important for tracking the satellite. In order to find the position of a satellite the

FIGURE 3.10   Power level versus elevation angle.

actual elliptical satellite orbit must be used. To discuss the motion of the satellite in the elliptical-shaped orbit, Kepler's laws are introduced. Three anomalies are defined: the mean M, the eccentric E, and the true v anomalies. Mean anomaly M and eccentricity e_s are given from the navigation data of the satellite. Eccentric anomaly E can be obtained from Equation (3.30). True anomaly v can be found from Equations (3.40) and (3.41). Finally, the receiving power at the user location is estimated.

REFERENCES

1. *Global Positioning System Standard Positioning Service Signal Specification*, 2nd ed, GPS Joint Program Office, June 2, 1995.
2. Spilker, J. J., Parkinson, B. W., "Overview of GPS operation and design," Chapter 2 in Parkinson, B. W., Spilker, J. J., Jr., *Global Positioning System: Theory and Applications*, vols. 1 and 2, American Institute of Aeronautics and Astronautics, 370 L'Enfant Promenade, SW, Washington, DC, 1996.
3. Kaplan, E. D., ed., *Understanding GPS Principles and Applications*, Artech House, Norwood, MA, 1996.
4. "System specification for the NAVSTAR global positioning system," SS-GPS-300B code ident 07868, March 3, 1980.
5. Spilker, J. J., "GPS signal structure and performance characteristics," *Navigation*, Institute of Navigation, vol. 25, no. 2, pp. 121–146, Summer 1978.
6. Milliken, R. J., Zoller, C. J., "Principle of operation of NAVSTAR and system characteristics," Advisory Group for Aerospace Research and Development (AGARD), Ag-245, pp. 4-1–4-12, July 1979.
7. Misra, P. N., "Integrated use of GPS and GLONASS in civil aviation," *Lincoln Laboratory Journal*, Massachusetts Institute of Technology, vol. 6, no. 2, pp. 231–247, Summer/Fall, 1993.
8. Aparicio, M., Brodie, P., Doyle, L., Rajan, J., and Torrione, P., "GPS satellite and payload," Chapter 6 in Parkinson, B. W., Spilker, J. J. Jr., *Global Positioning System: Theory and Applications*, vols. 1 and 2, American Institute of Aeronautics and Astronautics, 370 L'Enfant Promenade, SW, Washington, DC, 1996.
9. Spilker, J. J. Jr., "Satellite constellation and geometric dilution of precision," Chapter 5 in Parkinson, B. W., Spilker, J. J. Jr., *Global Positioning System: Theory and Applications*, vols. 1 and 2, American Institute of Aeronautics and Astronautics, 370 L'Enfant Promenade, SW, Washington, DC, 1996.
10. "Reference data for radio engineers," 5th ed., Howard W. Sams & Co. (subsidiary of ITT), Indianapolis, 1972.
11. Bate, R. R., Mueller, D. D., White, J. E., *Fundamentals of Astrodynamics*, pp. 182–188, Dover Publications, New York, 1971.
12. "Department of Defense world geodetic system, 1984 (WGS-84), its definition and relationships with local geodetic systems," DMA-TR-8350.2, Defense Mapping Agency, September 1987.
13. Riggins, R., "Navigation using the global positioning system," Chapter 6, class notes, Air Force Institute of Technology, 1996.

14. Braasch, M. S., van Graas, F., "Guidance accuracy considerations for real-time GPS interferometry," *Proceedings ION-GPS*, Albuquerque, NM, September 11–13, 1991.

15. Nieuwejaar, P., "GPS signal structure," *NATO Agard lecture series No. 161*, NAV-STAR GPS system, September 1988.

16. Jordan, E. C., *Electromagnetic Waves and Radiating Systems*, Prentice Hall, Englewood Cliffs, NJ, 1950.

CHAPTER FOUR

Earth-Centered, Earth-Fixed Coordinate System

4.1 INTRODUCTION

In the previous chapter the motion of the satellite is briefly discussed. The true anomaly is obtained from the mean anomaly, which is transmitted in the navigation data of the satellite. In all discussions, the center of the earth is used as a reference. In order to find a user position on the surface of the earth, these data must be related to a certain point on or above the surface of the earth. The earth is constantly rotating. In order to reference the satellite position to a certain point on or above the surface of the earth, the rotation of the earth must be taken into consideration. This is the goal of this chapter.

The basic approach is to introduce a scheme to transform the coordinate systems. Through coordinate system transform, the reference point can be moved to the desired coordinate system. First the direction cosine matrix, which is used to transform from one coordinate system to a different one, will be introduced. Then various coordinate systems will be introduced. The final transform will put the satellite in the earth-centered, earth-fixed (ECEF) system. Finally, some perturbations will be discussed. The major portion of this discussion is based on references 1 and 2.

In order to perform the transforms, besides the eccentricity e_s and mean anomaly M, additional data are obtained from the satellite. They are the semimajor of the orbit a_s, the right ascension angle Ω, the inclination angle i, and the argument of the perigee ω. Their definitions will also be presented in this chapter.

Fundamentals of Global Positioning System Receivers: A Software Approach, Second Edition,
by James Bao-Yen Tsui
ISBN 0-471-70647-7   Copyright © 2005 John Wiley & Sons, Inc.

4.2 DIRECTION COSINE MATRIX[1-3]

In this section, the direction cosine matrix will be introduced. A simple two-dimensional example will be used to illustrate the idea, which will be extended into a three-dimensional one without further proof. Figure 4.1 shows two two-dimensional systems $(x_1,\ y_1)$ and $(x_2,\ y_2)$. The second coordinate system is obtained from rotating the first system by a positive angle α. A point p is used to find the relation between the two systems. The point p is located at $(X_1,\ Y_1)$ in the $(x_1,\ y_1)$ system and at $(X_2,\ Y_2)$ in the $(x_2,\ y_2)$ system. The relation between $(X_2,\ Y_2)$ and $(X_1,\ Y_1)$ can be found from the following equations:

$$X_2 = X_1 \cos\alpha + Y_1 \sin\alpha = X_1 \cos(X_1 \text{ on } X_2) + Y_1 \cos(Y_1 \text{ on } X_2)$$
$$Y_2 = -X_1 \sin\alpha + Y_1 \cos\alpha = X_1 \cos(X_1 \text{ on } Y_2) + Y_1 \cos(Y_1 \text{ on } Y_2) \qquad (4.1)$$

In matrix form this equation can be written as

$$\begin{bmatrix} X_2 \\ Y_2 \end{bmatrix} = \begin{bmatrix} \cos(X_1 \text{ on } X_2) & \cos(Y_1 \text{ on } X_2) \\ \cos(X_1 \text{ on } Y_2) & \cos(Y_1 \text{ on } Y_2) \end{bmatrix} \begin{bmatrix} X_1 \\ Y_1 \end{bmatrix} \qquad (4.1)$$

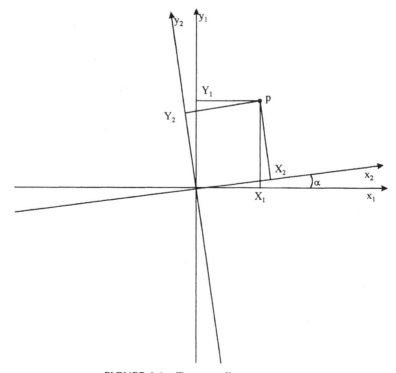

FIGURE 4.1   Two coordinate systems.

The direction cosine matrix is defined as

$$C_1^2 \equiv \begin{bmatrix} \cos(X_1 \text{ on } X_2) & \cos(Y_1 \text{ on } X_2) \\ \cos(X_1 \text{ on } Y_2) & \cos(Y_1 \text{ on } Y_2) \end{bmatrix} \tag{4.2}$$

This represents that the coordinate system is transferred from system 1 to system 2.

In a three-dimensional system, the directional cosine can be written as

$$C_1^2 \equiv \begin{bmatrix} \cos(X_1 \text{ on } X_2) & \cos(Y_1 \text{ on } X_2) & \cos(Z_1 \text{ on } X_2) \\ \cos(X_1 \text{ on } Y_2) & \cos(Y_1 \text{ on } Y_2) & \cos(Z_1 \text{ on } Y_2) \\ \cos(X_1 \text{ on } Z_2) & \cos(Y_1 \text{ on } Z_2) & \cos(Z_1 \text{ on } Z_2) \end{bmatrix} \tag{4.3}$$

Sometimes it is difficult to make one single transform from one coordinate to another one, but the transform can be achieved in a step-by-step manner. For example, if the transform is to rotate angle α around the z-axis and rotate angle β around the y-axis, it is easier to perform the transform in two steps. In other words, the directional cosine matrix can be used in a cascading manner. The first step is to rotate a positive angle α around the z-axis. The corresponding direction cosine matrix is

$$C_1^2 = \begin{bmatrix} \cos\alpha & \sin\alpha & 0 \\ -\sin\alpha & \cos\alpha & 0 \\ 0 & 0 & 1 \end{bmatrix} \tag{4.4}$$

The second step is to rotate a positive angle β around the x-axis; the corresponding direction cosine matrix is

$$C_2^3 = \begin{bmatrix} 1 & 0 & 0 \\ 0 & \cos\beta & \sin\beta \\ 0 & -\sin\beta & \cos\beta \end{bmatrix} \tag{4.5}$$

The overall transform can be written as

$$C_1^3 = C_2^3 C_1^2 = \begin{bmatrix} 1 & 0 & 0 \\ 0 & \cos\beta & \sin\beta \\ 0 & -\sin\beta & \cos\beta \end{bmatrix} \begin{bmatrix} \cos\alpha & \sin\alpha & 0 \\ -\sin\alpha & \cos\alpha & 0 \\ 0 & 0 & 1 \end{bmatrix}$$

$$= \begin{bmatrix} \cos\alpha & \sin\alpha & 0 \\ -\sin\alpha\cos\beta & \cos\alpha\cos\beta & \sin\beta \\ \sin\alpha\sin\beta & -\cos\alpha\sin\beta & \cos\beta \end{bmatrix} \tag{4.6}$$

It should be noted that the order of multiplication is very important; if the order is reversed, the wrong result will be obtained.

Suppose one wants to transform from coordinate system 1 to system n through system $2, 3, \ldots n - 1$. The following relation can be used:

$$C_1^n = C_{n-1}^n \cdots C_2^3 C_1^2 \tag{4.7}$$

In general, each C_{i-1}^i represents only one single transform. This cascade method will be used to obtain the earth-centered, earth-fixed system.

4.3 SATELLITE ORBIT FRAME TO EQUATOR FRAME TRANSFORM[1,2]

The coordinate system used to describe a satellite in the previous chapter can be considered as the satellite orbit frame because the center of the earth and the satellite are all in the same orbit plane. Figure 4.2 shows such a frame, and the x-axis is along the direction of the perigee and the z-axis is perpendicular to the orbit plane. The y-axis is perpendicular to the x and z axes to form a right-hand coordinate system. The distance r from the satellite to the center of the earth can be obtained from Equation (3.35) as

$$r = \frac{a_s(1 - e_s^2)}{1 + e_s \cos v} \tag{4.8}$$

where a_s is the semi-major of the satellite orbit, e_s is the eccentricity of the satellite orbit, v is the true anomaly, which can be obtained from previous chapter. The value of $\cos v$ can be obtained from Equation (3.37) as

$$\cos v = \frac{\cos E - e_s}{1 - e_s \cos E} \tag{4.9}$$

where E is the eccentric anomaly, which can be obtained from Equation (3.30).

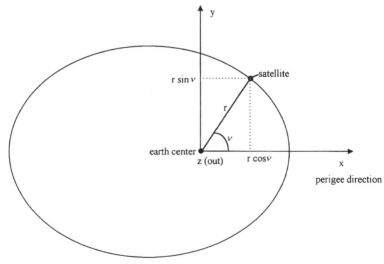

FIGURE 4.2 Orbit frame.

Substituting Equation (4.9) into Equation (4.8) the result can be simplified as

$$r = a_s(1 - e_s \cos E) \qquad (4.10)$$

The position of the satellite can be found as

$$x = r \cos v$$
$$y = r \sin v$$
$$z = 0 \qquad (4.11)$$

This equation does not reference any point on the surface of the earth but references the center of the earth. It is desirable to reference to a user position that is a point on or above the surface of the earth.

First a common point must be selected and this point must be on the surface of the earth as well as on the satellite orbit. The satellite orbit plane intercepts the earth equator plane to form a line. An ascending node is defined along this line toward the point where the satellite crosses the equator in the north (ascending) direction. The angle ω between the perigee and ascending node in the orbit plane is referred to as the argument of the perigee. This angle information can be obtained from the received satellite signal. Now let us change the x-axis from the perigee direction to the ascending node. This transform can be accomplished by keeping the z-axis unchanged and rotating the x-axis by the angle ω as shown in Figure 4.3. In Figure 4.3 the y-axis is not shown. The x_i-axis and the z_i-axis are perpendicular and the y_i-axis is perpendicular to the $x_i z_i$ plane. The corresponding direction cosine matrix is

$$C_1^2 = \begin{bmatrix} \cos\omega & -\sin\omega & 0 \\ \sin\omega & \cos\omega & 0 \\ 0 & 0 & 1 \end{bmatrix} \qquad (4.12)$$

In this equation the angle ω is in the negative direction; therefore the $\sin\omega$ has a different sign from Equation (4.4). This rotation changes the x_1-axis to x_2-axis.

The next step is to change from the orbit plane to the equator plane. This transform can be accomplished by using the x_2-axis as a pivot and rotate angle i. This angle i is the angle between the satellite orbit plane and the equator plane and is referred to as the inclination angle. This inclination angle is in the data transmitted by the satellite. The corresponding direction cosine matrix is

$$C_2^3 = \begin{bmatrix} 1 & 0 & 0 \\ 0 & \cos i & -\sin i \\ 0 & \sin i & \cos i \end{bmatrix} \qquad (4.13)$$

The angle i is also in the negative direction. After this transform, the z_3-axis is perpendicular to the equator plane rather than the orbit of the satellite and the x_3-axis is along the ascending point.

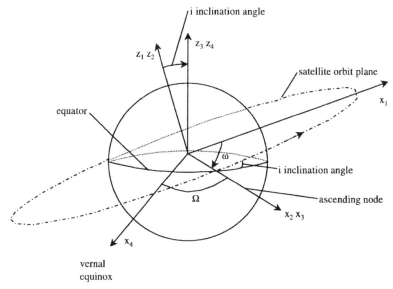

FIGURE 4.3   Earth equator and orbit plane.

There are six different orbits for the GPS satellites; therefore, there are six ascending points. It is desirable to use one x-axis to calculate all the satellite positions instead of six. Thus, it is necessary to select one x-axis; this subject will be discussed in the next section.

4.4   VERNAL EQUINOX[2]

The vernal equinox is often used as an axis in astrophysics. The direction of the vernal equinox is determined by the orbit plane of the earth around the sun (not the satellite) and the equator plane. The line of intersection of the two planes, the ecliptic plane (the plane of the earth's orbit) and the equator, is the direction of the vernal equinox as shown in Figure 4.4.

On the first day of spring a line joining from the center of the sun to the center of the earth points in the negative direction of the vernal equinox. On the first day of autumn a line joining from the center of the sun to the center of the earth points in the positive direction of the vernal equinox as shown in Figure 4.5.

The earth wobbles slightly and its axis of rotation shifts in direction slowly over the centuries. This effect is known as precession and causes the line-of-intersection of the earth's equator and the ecliptic plane to shift slowly. The period of the precession is about 26,000 years, so the equinox direction shifts westward about 50 ($360 \times 60 \times 60/26{,}000$) arc-seconds per year and this is a very small value. Therefore, the vernal equinox can be considered as a fixed axis in space.

FIGURE 4.4   Vernal equinox.

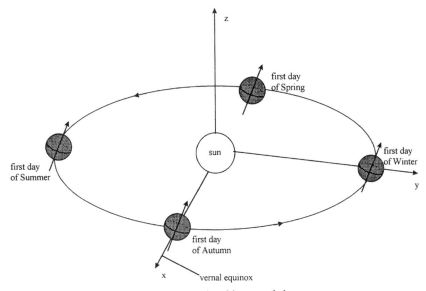

FIGURE 4.5   Earth orbit around the sun.

Again referring to Figure 4.3, the x_3-axis of the last frame discussed in the previous section will be rotated to the vernal equinox. This transform can be accomplished by rotating around the z_3-axis an angle Ω referred to as the right ascension. This angle is in plane of the equator. The direction cosine matrix is

$$C_3^4 = \begin{bmatrix} \cos\Omega & -\sin\Omega & 0 \\ \sin\Omega & \cos\Omega & 0 \\ 0 & 0 & 1 \end{bmatrix} \qquad (4.14)$$

This last frame is often referred to as the earth-centered inertia (ECI) frame. The origin of the ECI frame is at the earth's center of mass. In this frame the z_4-axis is perpendicular to the equator and the x_4-axis is the vernal equinox and in the equator plane. This frame does not rotate with the earth but is fixed with respect to stars. In order to reference a certain point on the surface of the earth, the rotation of the earth must be taken into consideration. This system is referred to as the earth-centered, earth-fixed (ECEF) frame.

4.5   EARTH ROTATION[1,2]

In this section two goals will be accomplished. The first one is to take care of the rotation of the earth. The second one is to use GPS time for the time reference.

First let us consider the earth rotation. Let the earth turning rate be $\dot{\Omega}_{ie}$ and define a time t_{er} such that at $t_{er} = 0$ the Greenwich meridian aligns with the vernal equinox. The vernal equinox is fixed by the Greenwich meridian rotates. Referring to Figure 4.6, the following equation can be obtained

$$\Omega_{er} = \Omega - \dot{\Omega}_{ie}t_{er} \tag{4.15}$$

where Ω_{er} is the angle between the ascending node and the Greenwich meridian, the earth rotation rate $\dot{\Omega}_{ie} = 7.2921151467 \times 10^{-5}$ rad/sec. When $t_{er} = 0$, $\Omega_{er} = \Omega$, this means that the Greenwich meridian and the vernal equinox are aligned.

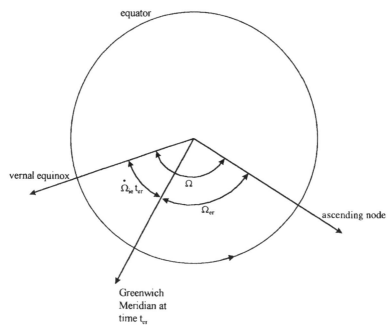

FIGURE 4.6   Rotation of the earth.

If the angle Ω_{er} is used in Equation (4.14) to replace Ω, the x-axis will be rotating in the equator plane. This x-axis is the direction of the Greenwich meridian. Using this new angle in Equation (4.14) the result is

$$
C_3^4 = \begin{bmatrix} \cos\Omega_{er} & -\sin\Omega_{er} & 0 \\ \sin\Omega_{er} & \cos\Omega_{er} & 0 \\ 0 & 0 & 1 \end{bmatrix} \tag{4.16}
$$

In this equation the rotation of the earth is included, because time is included in Equation (4.15). Using this time t_{er} in the system, every time the Greenwich meridian is aligned with the vernal equinox, $t_{er} = 0$. The maximum length of this time is a sidereal day, because the Greenwich meridian and the vernal equinox are aligned once every sidereal day.

The time t_{er} should be changed into the GPS time t. The GPS time t starts at Saturday night at midnight Greenwich time. Thus, the maximum GPS time is seven solar days. It is obvious that the time base t_{er} and the GPS time t are different. A simple way to change the time t_{er} to GPS time t is a linear shift of the time base as

$$
t_{er} = t + \Delta t \tag{4.17}
$$

where Δt can be considered as the time difference between the time based on t_{er} and the GPS time t. Substituting this equation into Equation (4.15), the result is

$$
\Omega_{er} = \Omega - \dot{\Omega}_{ie}t_{er} = \Omega - \dot{\Omega}_{ie}t - \dot{\Omega}_{ie}\Delta t \equiv \Omega - \alpha - \dot{\Omega}_{ie}t \equiv \Omega_e - \dot{\Omega}_{ie}t
$$

where $\Omega_e \equiv \Omega - \alpha$ and $\alpha \equiv \dot{\Omega}_{ie}\Delta t$ (4.18)

The reason for changing to this notation is that the angle $\Omega - \alpha$ is considered as one angle Ω_e, and this information is given in the GPS ephemeris data. However, this relation will be modified again in Section 4.7 and the final result will be used to find Ω_{er} in Equation (4.16). Before the modification of Ω_e, let us first find the overall transform.

4.6 OVERALL TRANSFORM FROM ORBIT FRAME TO EARTH-CENTERED, EARTH-FIXED FRAME

In order to transform the positions of the satellites from the satellite orbit frame to the ECEF frame, there need to be two intermediate transforms. The overall transform can be obtained from Equation (4.7). Substituting the results from Equations (4.16), (4.13), and (4.12) into (4.7), the following result is obtained:

$$
\begin{bmatrix} x_4 \\ y_4 \\ z_4 \end{bmatrix} = C_3^4 C_2^3 C_1^2 \begin{bmatrix} r\cos v \\ r\sin v \\ 0 \end{bmatrix}
$$

$$
= \begin{bmatrix} \cos\Omega_{er} & -\sin\Omega_{er} & 0 \\ \sin\Omega_{er} & \cos\Omega_{er} & 0 \\ 0 & 0 & 1 \end{bmatrix} \begin{bmatrix} 1 & 0 & 0 \\ 0 & \cos i & -\sin i \\ 0 & \sin i & \cos i \end{bmatrix} \begin{bmatrix} \cos\omega & -\sin\omega & 0 \\ \sin\omega & \cos\omega & 0 \\ 0 & 0 & 1 \end{bmatrix} \begin{bmatrix} r\cos v \\ r\sin v \\ 0 \end{bmatrix}
$$

$$
= \begin{bmatrix} \cos \Omega_{er} & -\sin \Omega_{er} \cos i & \sin \Omega_{er} \sin i \\ \sin \Omega_{er} & \cos \Omega_{er} \cos i & -\cos \Omega_{er} \sin i \\ 0 & \sin i & \cos i \end{bmatrix} \begin{bmatrix} \cos \omega & -\sin \omega & 0 \\ \sin \omega & \cos \omega & 0 \\ 0 & 0 & 1 \end{bmatrix} \begin{bmatrix} r \cos \nu \\ r \sin \nu \\ 0 \end{bmatrix}
$$

$$
= \begin{bmatrix} \cos \Omega_{er} \cos \omega - \sin \Omega_{er} \cos i \sin \omega & -\cos \Omega_{er} \sin \omega - \sin \Omega_{er} \cos i \cos \omega & \sin \Omega_{er} \sin i \\ \sin \Omega_{er} \cos \omega + \cos \Omega_{er} \cos i \sin \omega & -\sin \Omega_{er} \sin \omega + \cos \Omega_{er} \cos i \cos \omega & -\cos \Omega_{er} \sin i \\ \sin i \sin \omega & \sin i \cos \omega & \cos i \end{bmatrix}
$$

$$
\cdot \begin{bmatrix} r \cos \nu \\ r \sin \nu \\ 0 \end{bmatrix}
$$

$$
= \begin{bmatrix} r \cos \Omega_{er} \cos(\nu + \omega) - r \sin \Omega_{er} \cos i \sin(\nu + \omega) \\ r \sin \Omega_{er} \cos(\nu + \omega) + r \cos \Omega_{er} \cos i \sin(\nu + \omega) \\ r \sin i \sin(\nu + \omega) \end{bmatrix} \tag{4.19}
$$

This equation gives the satellite position in the earth-centered, earth-fixed coordinate system.

In order to calculate the results in the above equation, the following data are needed: (1) a_s: semi-major axis of the satellite orbit; (2) M: mean anomaly; (3) e_s: eccentricity of the satellite orbit; (4) i: inclination angle; (5) ω: argument of the perigee; (6) Ω-α: modified right ascension angle; (7) GPS time. The first three constants are used to calculate the distance r from the satellite to the center of the earth and the true anomaly ν as discussed in Section 3.12. The three values i, ω, and Ω-α are used to transform from the satellite orbit frame to the ECEF frame. In order to find Ω_{er} in the above equation the GPS time is needed.

4.7 PERTURBATIONS

The earth is not a perfect sphere and this phenomenon affects the satellite orbit. In addition to the shape of the earth, the sun and moon also have an effect on the satellite motion. Because of these factors the orbit of the satellite must be modified by some constants. The satellites transmit these constants and they can be obtained from the ephemeris data.

Equation (4.19) is derived based on the assumption that the orbit of the satellite is elliptical; however, the orbit is not a perfect elliptic. Thus, the parameters in the equations need to be modified. This section presents the results of the correction terms.

In Equation (4.15) the right ascension Ω will be modified as

$$
\Omega \Rightarrow \Omega + \dot{\Omega}(t - t_{oe}) \tag{4.20}
$$

where t is the GPS time, t_{oe} is the reference time for the ephemeris, and $\dot{\Omega}$ is the rate of change of the right ascension. In this equation it is implied that the right ascension is not a constant, but changes with time. The ephemeris data transmitted by the satellite contain t_{oe} and $\dot{\Omega}$. Substituting this equation into

Equation (4.18), the result is

$$\Omega_{er} = \Omega - \alpha + \dot{\Omega}(t - t_{oe}) - \dot{\Omega}_{ie}t \equiv \Omega_e + \dot{\Omega}(t - t_{oe}) - \dot{\Omega}_{ie}t \qquad (4.21)$$

where Ω_e is contained in the ephemeris data.

The mean motion in Equation (3.28) must be modified as

$$n \Rightarrow n + \Delta n = \sqrt{\frac{\mu}{a_s^3}} + \Delta n \qquad (4.22)$$

where Δn is the correction term that is contained in the ephemeris data. The mean anomaly must be modified as

$$M = M_0 + n(t - t_{oe}) \qquad (4.23)$$

where M_0 is the mean anomaly at reference time, which can be obtained from the ephemeris data. This value M will be used to find the true anomaly v.

There are six constants C_{us}, C_{uc}, C_{rs}, C_{rc}, C_{is}, and C_{ic} and they are used to modify $v + \omega$, r, and i in Equation (4.19) respectively. Let us introduce a new variable ϕ as

$$\phi \equiv v + \omega \qquad (4.24)$$

The correction term to $v + \omega$ is

$$\delta(v + \omega) \equiv \delta\phi = C_{us} \sin 2\phi + C_{uc} \cos 2\phi \qquad (4.25)$$

and the new $v + \omega$ is

$$v + \omega \Rightarrow v + \omega + \delta(v + \omega) \qquad (4.26)$$

The correction to distance r is

$$\delta r = C_{rs} \sin 2\phi + C_{rc} \cos 2\phi \qquad (4.27)$$

and the new r is

$$r \Rightarrow r + \delta r \qquad (4.28)$$

The correction to inclination i is

$$\delta i = C_{is} \sin 2\phi + C_{ic} \cos 2\phi \qquad (4.29)$$

and the new inclination i is

$$i \Rightarrow i + \delta i \qquad (4.30)$$

Substituting these new values into Equation (4.19) will produce the desired results.

4.8   CORRECTION OF GPS SYSTEM TIME AT TIME OF TRANSMISSION[5,6]

In Equations (4.21) and (4.23) the GPS time is used, and this time is often referred to at the time of transmission. (This section discusses only the correction of this time. The actual obtaining of the time of transmission will be discussed in Section 9.10.) The signals from the satellites are transmitted at the same time except for the clock error in each satellite. The time of receiving t_u is the time the signal arrives at the receiver. The relation between the t and t_u is

$$t_u = t + \rho_i/c$$
$$t = t_u - \rho_i/c \tag{4.31}$$

where ρ_i is the pseudorange from satellite i to the receiver and c is the speed of light. Since the pseudorange from each satellite to the receiver is different, the time of receiving is different. However, in calculating user position, one often uses one value for time. The time of receiving t_u is a reasonable selection. If a time of receiving t_u is used as a reference, from the above equation the time of transmission from various satellites is different. The time of transmission is the receiving time minus the transit time. This time is represented by t and is referred to as the time of transmission corrected for the transit time.

The t value must be corrected again from many other factors. However, in order to correct t, the t value must first be known. This requirement makes the correction process difficult. To simplify this process, let t_c represent the coarse GPS system time at time of transmission corrected by transit time. The value t_c can be obtained from time of the week (TOW), which will be presented in Section 5.9. For the present discussion, let us assume that the t_c value is already obtained. The time t_k shall be the actual total time difference between the time t_c and the epoch time t_{oe} and must account for the beginning or end of the week crossovers. That is, the following adjustments must be made on t_c:

If $t_k = t_c - t_{oe} > 302{,}400$ then $t_k = t_k - 604{,}800$ or $t_c \Rightarrow t_c - 604{,}800$

If $t_k = t_c - t_{oe} < -302{,}400$ then $t_k = t_k + 604{,}800$ or $t_c \Rightarrow t_c + 604{,}800$ (4.32)

where t_{oe} can be obtained from ephemeris data, 302,400 is the time of half a week in seconds. The time of a week in seconds is 604,800 ($7 \times 24 \times 3600$).

The following steps can be used to correct the GPS time t. From Equation (4.22), the mean motion is calculated as

$$n = \sqrt{\frac{\mu}{a_s^3}} + \Delta n \tag{4.33}$$

where $\mu = 3.986005 \times 10^{14}$ meters^3/sec^2 is the earth's universal gravitational parameter and is a constant, $\sqrt{a_s}$ and Δn are obtained from ephemeris data.

From this n value the mean anomaly can be found from Equation (4.23) as

$$M = M_0 + n(t_c - t_{oe}) \tag{4.34}$$

where M_0 is in the ephemeris data. In this equation t_c is used instead of t as t is not derived yet.

The eccentric anomaly E can be found from Equations (3.29) or (3.30) through iteration as

$$E = M + e_s \sin E \tag{4.35}$$

where e_s is eccentricity of the satellite orbit, which can be obtained from the ephemeris data. Let us define a constant F as

$$F = \frac{-2\sqrt{\mu}}{c^2} = -4.442807633 \times 10^{-10} \text{ sec/(meter)}^{1/2} \tag{4.36}$$

where μ is the earth's universal gravitational parameter and c is the speed of light. The relativistic correction term is

$$\Delta t_r = F e_s \sqrt{a_s} \sin E \tag{4.37}$$

The overall time correction term is

$$\Delta t = a_{f0} + a_{f1}(t_c - t_{oc}) + a_{f2}(t_c - t_{oc})^2 + \Delta t_r - T_{GD} \tag{4.38}$$

where T_{GD}, t_{oc}, a_{f0}, a_{f1}, a_{f2} are clock correction terms and T_{GD} is to account for the effect of satellite group delay differential. They can be obtained in the ephemeris data. The GPS time of transmission t corrected for transit time can be corrected as

$$t = t_c - \Delta t \tag{4.39}$$

This is the time t that will be used for the following calculations.

4.9   CALCULATION OF SATELLITE POSITION[5,6]

This section uses all the information from the ephemeris data to obtain a satellite position in the earth-centered, earth-fixed system. These calculations require the information obtained from both Chapters 3 and 4; therefore, this section can be considered as a summary of the two chapters.

Equation (4.19) is required to calculate the position of the satellite. In this equation there are five known quantities: r, $v + \omega$, i, and Ω_{er}. These quantities appear on the right side of the equation and the results represent the satellite position. Let us find these five quantities.

First let us find the value of r from Equation (4.10) as

$$r = a_s(1 - e_s \cos E) \tag{4.40}$$

In this equation, the value E must be calculated first from ephemeris data. In order to find the r value the following steps must be taken:

1. Use Equation (4.22) to calculate n where μ is a constant; a_s and Δn can be obtained from the ephemeris data.
2. Use Equation (4.34) to calculate M where M_0 and t_{oe} can be obtained from ephemeris data and t_c can be obtained from the discussion in Section 9.10.
3. The value of E can be found from Equation (4.35), where e_s can be obtained from the ephemeris data. The iteration method will be used in this operation.
4. Once E is obtained, the value of r can be found from Equation (4.40).

In the above four steps, the first three steps are to find the value of E. Once E is calculated, Equations (4.36)–(4.39) can be used to find the corrected GPS time t.

Now let us find the true anomaly v. This value can be found from Equations (3.40) and (3.41) as

$$v_1 = \cos^{-1}\left(\frac{\cos E - e_s}{1 - e_s \cos E}\right)$$

$$v_2 = \sin^{-1}\left(\frac{\sqrt{1 - e_s^2}\sin E}{1 - e_s \cos E}\right)$$

$$v = v_1 \text{sign}(v_2) \tag{4.41}$$

The argument ω can be found from the ephemeris data. Using the definition in Equation (4.24), the value of ϕ is

$$\phi \equiv v + \omega \tag{4.42}$$

The following correction terms are needed

$$\delta\phi = C_{us} \sin 2\phi + C_{uc} \cos 2\phi$$
$$\delta r = C_{rs} \sin 2\phi + C_{rc} \cos 2\phi$$
$$\delta i = C_{is} \sin 2\phi + C_{ic} \cos 2\phi \tag{4.43}$$

where the C_{us}, C_{uc}, C_{rs}, C_{rc}, C_{is}, C_{ir} are from ephemeris data:

$$\phi \Rightarrow \phi + \delta\phi$$
$$r \Rightarrow r + \delta r \tag{4.44}$$

The inclination angle i can be obtained from the ephemeris data and be corrected as

$$i \Rightarrow i + \delta i + \text{idot}(t - t_{oe}) \tag{4.45}$$

where idot can be obtained from the ephemeris data. The last term to be found is

$$\Omega_{er} = \Omega_e + \dot{\Omega}(t - t_{oe}) - \dot{\Omega}_{ie}t \tag{4.46}$$

where the earth rotation rate $\dot{\Omega}_{ie}$ is a constant, Ω_e, $\dot{\Omega}$, and t_{oe} are obtained from the ephemeris data. It should be noted that the corrected GPS time t is used in the above two equations.

Once all the necessary parameters are obtained, the position of the satellite can be found from Equation (4.19) as

$$\begin{bmatrix} x \\ y \\ z \end{bmatrix} = \begin{bmatrix} r \cos \Omega_{er} \cos \phi - r \sin \Omega_{er} \cos i \sin \phi \\ r \sin \Omega_{er} \cos \phi + r \cos \Omega_{er} \cos i \sin \phi \\ r \sin i \sin \phi \end{bmatrix} \tag{4.47}$$

The satellite position calculated in this equation is in the ECEF frame. Therefore the satellite position is a function of time. From the x, y, z, and the pseudorange of more than four satellites the user's position can be found from results in Chapter 2. The actual calculation of the pseudorange is discussed in Section 9.9.

4.10 COORDINATE ADJUSTMENT FOR SATELLITES

Using the earth-centered, earth-fixed coordinate system implies that the earth's rotation is taken into consideration. The satellite position calculated from Section 4.9 is based on the GPS time of transmission t corrected for transit time. However, the user position is calculated at the time of receiving. Since the satellite and user positions are calculated at different times, they are in different coordinate systems. This will cause an error in the user position. As discussed in Section 3.3., if the user is on the equator of the earth and an approximate signal traveling time of 76 ms is assumed, the user position is moved about 36 m $(2\pi \times 6368 \times 10^3 \times 76/(24 \times 3600 \times 10^3))$ due to the rotation of the earth.

In order to obtain the correct user position, a single coordinate system should be used. Since the user position is measured at the time of receiving, it is appropriate to use this time in the coordinate system. The satellite position calculated should be referenced to this time. This correction does not mean to change the satellite position, but only changes the coordinate system of the satellites.

In order to reference the GPS time at the time of receiving, the coordinate system of each satellite must be separately modified. Using the time of receiving as reference, the time of transmission of satellite i in the new coordinate system is the time of receiving minus the transition time as shown in Equation (4.31). The transit time cannot be determined before the user position is calculated because of the unknown user clock bias. Only when the user position is obtained can the pseudorange be found. Once the user position is found, the pseudorange can be found from the user position to the satellite position. The earth rotation appears only in Equation (4.46). Equations (4.46) and (4.47) are used in the operation.

The following steps can be taken to improve user position accuracy.

1. From the satellite and user position the transit time t_t can be found as

$$t_t = \sqrt{(x - x_u)^2 + (y - y_u)^2 + (z - z_u)^2}/c \tag{4.48}$$

where x, y, z, and x_u, y_u, z_u are the coordinates of the satellite and the user respectively, c is the speed of light.

2. Use the transit time to modify the angle Ω_{er} in Equation (4.46) as

$$\Omega_{er} \Rightarrow \Omega_{er} - \dot{\Omega}_{ie}\, t_t \tag{4.49}$$

3. Use the new value of Ω_{er} in Equation (4.47) to calculate the position of the satellite x, y, and z.
4. The above operations should be performed on every satellite. From these values a new user position x_u, y_u, z_u will be calculated.
5. Repeat steps 1, 2, 3, and 4 again to obtain a new set of x, y, and z. When the old and new sets are within a predetermined value, the new set can be considered as the position of the satellite in the new coordinate system. It usually requires calculating the x, y, and z values only twice.
6. These new x, y, and z values will be used to find the user position.

4.11   EPHEMERIS DATA[4-6]

In the previous sections several constants and many ephemeris data are used in the calculations. This section lists all these constants and the ephemeris data used in the calculations. The details of the ephemeris data transmitted by the satellites will be presented in the next chapter.

The constants are listed as follows:[4]

$\mu = GM = 3.986005 \times 10^{14}$ meters^3/sec^2, which is the WGS-84 value of the earth's universal gravitational parameter.

$\dot{\Omega}_{ie} = 7.2921151467 \times 10^{-5}$ rad/sec, which is the WGS-84 value of the earth's rotational rate.

$\pi = 3.1415926535898$.

$c = 2.99792458 \times 10^8$ meter/sec, which is the speed of light.

The ephemeris data are:

M_0: mean anomaly at reference time.

Δn: mean motion difference from computed value.

$\sqrt{a_s}$: square root of the semi-major axis of the satellite orbit.

e_s: eccentricity of the satellite orbit.

$T_{GD}, t_{oc}, a_{f0}, a_{f1}, a_{f2}$: clock correction parameters.

t_{oe}: reference time ephemeris.

C_{us}, C_{uc}: amplitude of the sine and cosine harmonic correction term to the argument of latitude, respectively.

C_{rs}, C_{rc}: amplitude of the sine and cosine harmonic correction term to the orbit radius, respectively.

C_{is}, C_{ic}: amplitude of the sine and cosine harmonic correction term to the angle of inclination, respectively.

Ω_e: longitude of ascending node of orbit plane at weekly epoch.

$\dot{\Omega}$: rate of the right ascension.

i: inclination angle at reference time.

ω: argument of perigee.

idot: rate of inclination angle.

4.12 SUMMARY

This chapter takes the satellite position calculated in Chapter 3 and transforms it into an earth-centered, earth-fixed coordinate system because this coordinate references a fixed position on or above the earth. Since the satellite orbit cannot be described perfectly by an elliptic, corrections must be made to the position of the satellite. The information for correction is contained in the ephemeris data transmitted by the satellite. This information can be obtained if the GPS signal is decoded. The GPS time t at time of transmission needs to be corrected for the transit time as well as from the ephemeris data. Obtaining the coarse GPS time t_c at time of transmission corrected for transit time will be discussed in Section 9.7. Finally, the coordinate system of the satellite must be adjusted to accommodate the transit time.

REFERENCES

1. Riggins, R. "Navigation using the global positioning system," Chapter 6, class notes, Air Force Institute of Technology, 1996.

2. Bate, R. R., Mueller, D. D., White, J. E., *Fundamentals of Astrodynamics*, pp. 182–188, Dover Publications, New York, 1971.

3. Britting, K. R., *Inertial Navigation Systems Analysis*, Chapter 4, Wiley, New York, 1971.

4. "Department of Defense world geodetic system, 1984 (WGS-84), its definition and relationships with local geodetic systems," DMA-TR-8350.2, Defense Mapping Agency, September 1987.

5. *Global Positioning System Standard Positioning Service Signal Specification*, 2nd ed., GPS Joint Program Office, June 1995.

6. Spilker, J. J. Jr., "GPS signal structure and theoretical performance," Chapter 3 in Parkinson, B. W., Spilker, J. J. Jr., *Global Positioning System: Theory and Applications*, vols. 1 and 2, *American Institute of Aeronautics and Astronautics*, 370 L'Enfant Promenade, SW, Washington, DC, 1996.

CHAPTER FIVE

GPS C/A Code Signal Structure

5.1 INTRODUCTION[1,2]

In the previous chapters user positions are calculated. In order to perform the user position calculation, the positions of the satellites and pseudoranges to the satellites must be measured. Many parameters are required to calculate the positions of the satellites and they are transmitted in the satellite signals.

This chapter provides the details associated with the GPS signals. Spilker[1,2] not only gives a very good discussion on the signal, it also gives the reasons these signals are selected. The discussion in this chapter is limited to the fundamentals of the signals, such that a receiver design can be based on the signals.

There are basically two types of signals: the coarse (or clear)/acquisition (C/A) and the precision (P) codes. The actual P code is not directly transmitted by the satellite, but it is modified by a Y code, which is often referred to as the P(Y) code. The P(Y) code is not available to civilian users and is primarily used by the military. In other words, the P(Y) code is classified. The P(Y) code has similar properties of the P code. In order to receive the P(Y) code, one must have the classified code. Therefore, only the fundamentals of the P code will be mentioned in this book. The discussion will be focused on the C/A code. In general, in order to acquire the P(Y) code, the C/A code is usually acquired first. However, in some applications it is desirable to acquire the P(Y) code directly, which is known as direct Y acquisition.

The radio frequency (RF) of the C/A code will be presented first, then the C/A code. The generation of the C/A code and its properties will be presented because they are related closely to acquiring and tracking the GPS signals. Finally, the

Fundamentals of Global Positioning System Receivers: A Software Approach, Second Edition, by James Bao-Yen Tsui
ISBN 0-471-70647-7 Copyright © 2005 John Wiley & Sons, Inc.

data carried by the signals will be presented. The applications of the data will be briefly discussed.

5.2 TRANSMITTING FREQUENCY[1-4]

The GPS signal contains two frequency components: link 1 (L1) and link 2 (L2). The center frequency of L1 is at 1575.42 MHz and L2 is at 1227.6 MHz. These frequencies are coherent with a 10.23 MHz clock. These two frequencies can be related to the clock frequency as

$$L1 = 1575.42 \ \text{MHz} = 154 \times 10.23 \ \text{MHz}$$

$$L2 = 1227.6 \ \text{MHz} = 120 \times 10.23 \ \text{MHz}$$

These frequencies are very accurate as their reference is an atomic frequency standard. When the clock frequency is generated, it is slightly lower than 10.23 MHz to take the relativistic effect into consideration. The reference frequency is off by[3] -4.567×10^{-3} Hz, which corresponds to a fraction of -4.4647×10^{-10} $(-4.567 \times 10^{-3}/10.23 \times 10^6)$. Therefore, the reference frequency used by the satellite is 10.229999995433 MHz $(10.23 \times 10^6 - 4.567 \times 10^{-3})$ rather than 10.23 MHz. When a GPS receiver receives the signals, they are at the desired frequencies. However, the satellite and receiver motions can produce a Doppler effect as discussed in Section 3.5. The Doppler frequency shift produced by the satellite motion at L1 frequency is approximately ± 5 kHz.

The signal structure of the satellite may be modified in the future. However, at the present time, the L1 frequency contains the C/A and P(Y) signals, while the L2 frequency contains only the P(Y) signal. The C/A and P(Y) signals in the L1 frequency are in quadrant phase of each other and they can be written as:

$$S_{L1} = A_p P(t) D(t) \cos(2\pi f_1 t + \phi) + A_c C(t) D(t) \sin(2\pi f_1 t + \phi) \quad (5.1)$$

where S_{L1} is the signal at L1 frequency, A_p is the amplitude of the P code, $P(t) = \pm 1$ represents the phase of the P code, $D(t) = \pm 1$ represents the data code, f_1 is the L1 frequency, ϕ is the initial phase, A_c is the amplitude of the C/A code, $C(t) = \pm 1$ represents the phase of the C/A code. These terms will be further discussed in the following sections. In this equation the P code is used instead of the P(Y) code. The P(Y), C/A, and the carrier frequencies are all phase locked together.

The minimum power levels of the signals must fulfill the values listed in Table 5.1 at the receiver. These power levels are very weak and the spectrum is spread, therefore they cannot be directly observed from a spectrum analyzer. Even when the signal is amplified to a reasonable power level, the spectrum of the C/A code cannot be observed because the noise is stronger than the signal. When a high gain antenna is used to track a certain satellite, its signal structure can be displayed clearly on a scope or a spectrum analyzer.

As discussed in Section 3.3, the received power levels at various points on the earth are different. The maximum difference is about 2.1 dB between a point just under the satellite and a point tangential to the surface of the earth. In order

TABLE 5.1 Power Level of GPS Signals

	P	C/A
L1	−133 dBm	−130 dBm
L2	−136 dBm	−136 dBm*

*Presently not in L2 frequency.

to generate a uniform power over the surface of the earth, the main beam pattern of the transmitting antenna is slightly weaker at the center to compensate for the user at the edge of the beam. The resulting power level versus elevation angle is shown in Figure 3.10. The maximum power is −128 dBm, which occurs at about 40 degrees. Of course, the receiving antenna pattern also contributes to the power level of the receiver. Usually the receiving antenna has a higher gain in the zenith direction. This incorporates the ability of attenuating multipath but loses gain to signals from lower elevation angles. As discussed in Sections 3.3 and 3.10, the minimum required beam width of the transmitting antenna to cover the earth is 13.87 degrees. The beam width of the antenna[2] is 21.3 degrees, which is wider than needed to cover the earth as shown in Figure 5.1.

If the user is in an aircraft, as long as it is in the main beam of the GPS signal and not in the shadow of the earth it can receive the signal. The signals generated by the satellite transmitting antenna are right-hand polarized. Therefore, the receiver antenna should be right-hand polarized to achieve maximum efficiency.

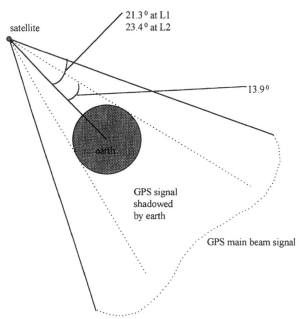

FIGURE 5.1 GPS signal main beam.

5.3   CODE DIVISION-MULTIPLE ACCESS (CDMA) SIGNALS

A signal S can be written in the following form:

$$S = A \sin (2\pi f t + \phi) \qquad (5.2)$$

where A is the amplitude, f is the frequency, ϕ is the initial phase. These three parameters can be modulated to carry information. If A is modulated, it is referred to as amplitude modulation. If f is modulated, it is frequency modulation. If ϕ is modulated, it is phase modulation.

The GPS signal is a phase-modulated signal with $\phi = 0, \pi$; this type of phase modulation is referred to as bi-phase shift keying (BPSK). The phase change rate is often referred to as the chip rate. The spectrum shape can be described by the sinc function $(\sin x/x)$ with the spectrum width proportional to the chip rate. For example, if the chip rate is 1 MHz, the main lobe of the spectrum has a null-to-null width of 2 MHz. Therefore, this type of signal is also referred to as a spread-spectrum signal. If the modulation code is a digital sequence with a frequency higher than the data rate, the system can be called a direct-sequence modulated system.

A code division multiple access (CDMA) signal in general is a spread-spectrum system. All the signals in the system use the same center frequency. The signals are modulated by a set of orthogonal (or near-orthogonal) codes. In order to acquire an individual signal, the code of that signal must be used to correlate with the received signal. The GPS signal is CDMA using direct sequence to bi-phase modulate the carrier frequency. Since the CDMA signals all use the same carrier frequency, there is a possibility that the signals will interfere with one another. This effect will be more prominent when strong and weak signals are mixed together. In order to avoid the interference, all the signals should have approximately the same power levels at the receiver. Sometimes in the acquisition one finds that a cross-correlation peak of a strong signal is stronger than the desired peak of a weak signal. Under this condition, the receiver may obtain wrong information.

5.4   P CODE[1,2]

The P code is bi-phase modulated at 10.23 MHz; therefore, the main lobe of the spectrum is 20.46 MHz wide from null to null. The chip length is about 97.8 ns (1/10.23 MHz). The code is generated from two pseudorandom noise (PRN) codes with the same chip rate. One PRN sequence has 15,345,000 chips, which has a period of 1.5 seconds, the other one has 15,345,037 chips, and the difference is 37 chips. The two numbers, 15,345,000 and 15,345,037, are relative prime, which means there are no common factors between them. Therefore, the code length generated by these two codes is 23,017,555.5 ($1.5 \times 15,345,037$) seconds, which is slightly longer than 38 weeks. However, the actual length of the P code is 1 week as the code is reset every week. This 38-week-long code can be divided into 37 different P codes and each satellite can use a different portion of the code. There are a total of 32 satellite identification numbers although only 24 of them are in the orbit. Five of the P code signals (33–37) are reserved for other uses such as ground transmission. In order to perform acquisition on the signal, the time of the week must be known very accurately. Usually this time is found from the C/A code signal that will be discussed in the next section. The navigation data rate carried by the P code through phase modulation is at a 50 Hz rate.

5.5 C/A CODE AND DATA FORMAT[1,2,5]

The C/A code is a bi-phase modulated signal with a chip rate of 1.023 MHz. Therefore, the null-to-null bandwidth of the main lobe of the spectrum is 2.046 MHz. Each chip is about 977.5 ns (1/1.023 MHz) long. The transmitting bandwidth of the GPS satellite in the L1 frequency is approximately 20 MHz to accommodate the P code signal; therefore, the C/A code transmitted contains the main lobe and several sidelobes. The total code period contains 1,023 chips. With a chip rate of 1.023 MHz, 1,023 chips last 1 ms; therefore, the C/A code is 1 ms long. This code repeats itself every millisecond. The spectrum of a C/A code is shown in Figure 5.2.

In order to find the beginning of a C/A code in the received signal only a very limited data record is needed such as 1 ms. If there is no Doppler effect

FIGURE 5.2 Spectrum of a C/A code.

on the received signal, then one millimeter of data contains all the 1,023 chips. Different C/A codes are used for different satellites. The C/A code belongs to the family of Gold codes,[5] which will be discussed in the next section.

Figure 5.3 shows the GPS data format. The first row shows a C/A code with 1,023 chips; the total length is 1 ms. The second row shows a navigation data bit that has a data rate of 50 Hz; thus, a data bit is 20 ms long and contains 20 C/A codes. Thirty data bits make a word that is 600 ms long as shown in the third row. Ten words make a subframe that is 6 seconds long as shown in row four. The fifth row shows a page that is 30 seconds long and contains 5 subframes. Twenty-five pages make a complete data set that is 12.5 minutes long as shown in the sixth row. The 25 pages of data can be referred to as a superframe.

The parameters mentioned in Section 4.10 are contained in the first three subframes of a page. If one can receive the information of these three subframes from four or more satellites, the user location can be found. Theoretically, one can take a minimum of about 18 seconds of data from four satellites and be able to calculate the user position. However, the subframes from each satellite will not reach the receiver at the same time. Besides, one does not know when the beginning of subframe 1 will be received. A guaranteed way to receive the first three subframes is to take 30 seconds (or one page) of data. Thus, one can take a minimum of 30 seconds of data and calculate the user position.

5.6 GENERATION OF C/A CODE[1,2,6]

The GPS C/A signals belong to the family of Pseudorandom noise (PRN) codes known as the Gold codes. The signals are generated from the product of two 1,023-bit PRN sequence G1 and G2. Both G1 and G2 are generated by a maximum-length linear shift register of 10 stages and are driven by a 1.023 MHz clock. Figure 5.4 shows the G1 and G2 generators. Figure 5.4a shows the G1 generator and Figures 5.4b and 5.4c show the G2 generator. Figure 5.4c is a simplified notation of Figure 5.4b.

The basic operating principles of these two generators are similar; therefore, only G2 will be discussed in detail. A maximum-length sequence (MLS) generator can be made from a shift register with proper feedback. If the shift register has n bits, the length of the sequence generated is $2^n - 1$. Both shift generators in G1 and G2 have 10 bits, thus, the sequence length is 1,023 ($2^{10} - 1$). The feedback circuit is accomplished through modulo-2 adders.

The operating rule of the modulo-2 adder is listed in Table 5.2. When the two inputs are the same the output is 0, otherwise it is 1. The positions of the feedback circuit determine the output pattern of the sequence. The feedback of G1 is from bits 3 and 10 as shown in Figure 5.4a and the corresponding polynomial can be written as G1: $1 + x^3 + x^{10}$. The feedback of G2 is from bits 2, 3, 6, 8, 9, 10 as shown in Figure 5.4b and the corresponding polynomial is G2: $1 + x^2 + x^3 + x^6 + x^8 + x^9 + x^{10}$.

In general, the output from the last bit of the shift register is the output of the sequence as shown in Figure 5.4a. Let us refer to this output as the MLS

FIGURE 5.3 GPS data format.

(a) G1 generator.

(b) G2 generator.

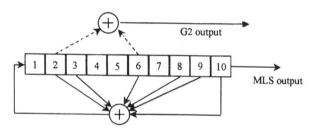

(c) Simplified notation of G2.

FIGURE 5.4   G1, G2 maximum-length sequence generators.

TABLE 5.2   Modulo-2 Addition

Input 1	Input 2	Output
0	0	0
0	1	1
1	0	1
1	1	0

output. However, the G2 generator does not use the MLS output as the output. The output is generated from two bits which are referred to as the code phase selections through another modulo-2 adder as shown in Figures 5.4b and c. This G2 output is a delayed version of the MLS output. The delay time is determined by the positions of the two output points selected.

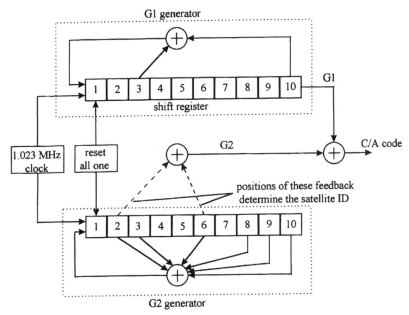

FIGURE 5.5 C/A code generator.

Figure 5.5 shows the C/A code generator. Another modulo-2 adder is used to generate the C/A code, which uses the outputs from G1 and G2 as inputs. The initial values of the two shift registers G1 and G2 are all 1's and they must be loaded in the registers first. The satellite identification is determined by the two output positions of the G2 generator. There are 37 unique output positions. Among these 37 outputs, 32 are utilized for the C/A codes of 32 satellites, but only 24 satellites are in orbit. The other five outputs are reserved for other applications such as ground transmission.

Table 5.3 lists the code phase assignments. In this table there are five columns and the first column gives the satellite ID number, which is from 1 to 32.

The second column gives the PRN signal number; and it is from 1 to 37. It should be noted that the C/A codes of PRN signal numbers 34 and 37 are the same. The third column provides the code phase selections that are used to form the output of the G2 generator. The fourth column provides the code delay measured in chips. This delay is the difference between the MLS output and the G2 output. This is redundant information of column 3, because once the code phase selections are chosen this delay is determined. The last column provides the first 10 bits of the C/A code generated for each satellite. These values can be used to check whether the generated code is wrong. This number is in an octal format.

The following example will illustrate the use of the information listed in Table 5.3. For example, in order to generate the C/A code of satellite 19, the 3 and 6 tabs must be selected for the G2 generator. With this selection, the G2

TABLE 5.3   Code Phase Assignments

Satellite ID Number	GPS PRN Signal Number	Code Phase Selection	Code Delay Chips	First 10 Chips C/A Octal
1	1	$2 \oplus 6$	5	1440
2	2	$3 \oplus 7$	6	1620
3	3	$4 \oplus 8$	7	1710
4	4	$5 \oplus 9$	8	1744
5	5	$1 \oplus 9$	17	1133
6	6	$2 \oplus 10$	18	1455
7	7	$1 \oplus 8$	139	1131
8	8	$2 \oplus 9$	140	1454
9	9	$3 \oplus 10$	141	1626
10	10	$2 \oplus 3$	251	1504
11	11	$3 \oplus 4$	252	1642
12	12	$5 \oplus 6$	254	1750
13	13	$6 \oplus 7$	255	1764
14	14	$7 \oplus 8$	256	1772
15	15	$8 \oplus 9$	257	1775
16	16	$9 \oplus 10$	258	1776
17	17	$1 \oplus 4$	469	1156
18	18	$2 \oplus 5$	470	1467
19	19	$3 \oplus 6$	471	1633
20	20	$4 \oplus 7$	472	1715
21	21	$5 \oplus 8$	473	1746
22	22	$6 \oplus 9$	474	1763
23	23	$1 \oplus 3$	509	1063
24	24	$4 \oplus 6$	512	1706
25	25	$5 \oplus 7$	513	1743
26	26	$6 \oplus 8$	514	1761
27	27	$7 \oplus 9$	515	1770
28	28	$8 \oplus 10$	516	1774
29	29	$1 \oplus 6$	859	1127
30	30	$2 \oplus 7$	860	1453
31	31	$3 \oplus 8$	861	1625
32	32	$4 \oplus 9$	862	1712
**	33	$5 \oplus 10$	863	1745
**	34*	$4 \oplus 10$	950	1713
**	35	$1 \oplus 7$	947	1134
**	36	$2 \oplus 8$	948	1456
**	37*	$4 \oplus 10$	950	1713

*34 and 37 have the same C/A code.
**GPS satellites do not transmit these codes; they are reserved for other uses.

output sequence is delayed 471 chips from the MLS output. The last column is 1633, which means 1 110 011 011 in binary form. If the first 10 bits generated for satellite 19 do not match this number, the code is incorrect.

A computer program (p5_1) is listed at the end of this chapter to generate both the MLS and G2 output sequences. The program takes columns 3 and 4 of Table 5.3 as inputs and checks the time delay. If the correct data are used as inputs, the output will show "OK," otherwise, it will show "not match."

A program (p5_2) can be used to generate the C/A code. The program is an extension of the program (p5_1) to include the two maximum-length sequence generators. In the program, the delay time listed in Table 5.3 is used as input to generate the G2 signal rather than using the code phase selections in column 3. The first 10 bits of the generated C/A code should be compared with the result listed in the last column of Table 5.3.

5.7 CORRELATION PROPERTIES OF C/A CODE[1,2]

One of the most important properties of the C/A codes is their correlation result. High autocorrelation peak and low cross-correlation peaks can provide a wide dynamic range for signal acquisition. In order to detect a weak signal in the presence of strong signals, the autocorrelation peak of the weak signal must be stronger than the cross-correlation peaks from the strong signals. If the codes are orthogonal, the cross correlations will be zero. However, the Gold codes are not orthogonal but near orthogonal, implying that the cross correlations are not zero but have small values.

The cross correlation of the Gold code is listed in Table 5.4.[1]

For the C/A code $n =$ even $= 10$, thus, $P = 1023$. Using the relations in this table, the cross-correlation values are: $-65/1023$ (occurrence 12.5%), $-1/1023$ (75%), and $63/1023$ (12.5%). The autocorrelation of the C/A codes of satellite 19 and the cross correlation of satellites 19 and 31 are shown in Figures 5.6a and 5.6b respectively. These satellites are arbitrarily chosen.

TABLE 5.4   Cross Correlation of Gold Code

Code Period	Number of Shift Register Stages	Normalized Cross Correlation Level	Probability of Level
$P = 2^n - 1$	$n =$ odd	$-\dfrac{2^{(n+1)/2} + 1}{P}$	0.25
		$-\dfrac{1}{P}$	0.5
		$\dfrac{2^{(n+2)/2} - 1}{P}$	0.24
$P = 2^n - 1$	$n =$ even	$-\dfrac{2^{(n+2)/2} + 1}{P}$	0.125
		$-\dfrac{1}{P}$	0.75
		$\dfrac{2^{(n+2)/2} - 1}{P}$	0.125

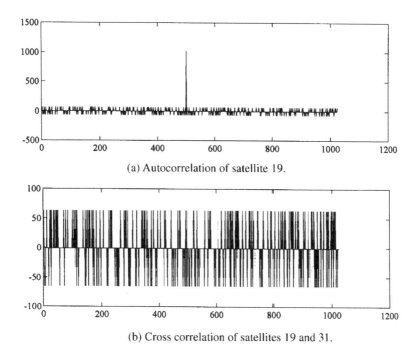

(a) Autocorrelation of satellite 19.

(b) Cross correlation of satellites 19 and 31.

FIGURE 5.6   Auto and cross correlation of C/A code.

In Figure 5.6a, the maximum of the autocorrelation peak is 1023, which equals the C/A code length. The position of the maximum peak is deliberately shifted to the center of the figure for a clear view. The rest of the correlation has three values 63, −1, and −65. The cross-correlation shown in Figure 5.6b also has three values 63, −1, −65.

These are the values calculated by using equations in Table 5.4. The difference between the maximum of the autocorrelation to the cross correlation determines the processing gain of the signal. In order to generate these figures, the outputs from the C/A code generator must be 1 and −1, rather than 1 and 0. The mathematical operation to generate these figures will be discussed in the Section 7.7.

5.8   NAVIGATION DATA BITS[2,3,7]

The C/A code is a bi-phase coded signal which changes the carrier phase between 0 and π at a rate of 1.023 MHz. The navigation data bit is also bi-phase code, but its rate is only 50 Hz, or each data bit is 20 ms long. Since the C/A code is 1 ms, there are 20 C/A codes in one data bit. Thus, in one data bit all 20 C/A codes have the same phase. If there is a phase transition due to the data bit, the phases of the two adjacent C/A codes are different by $\pm\pi$. This information is important in signal acquisition. One can perform signal acquisition on two consecutive 10 ms of data. Between two consecutive sets of 10 ms of data there is at most one navigation data bit phase transition. Therefore, one set of these data will have no data bit phase transition and coherent acquisition should produce the desired result. Thirty data bits make a navigation word and 10 words make a subframe. Figure 5.3 shows these relations.

The GPS time is given by the number of seconds in one week and this value is reset every week at the end/start of a week. At end/start of a week the cyclic paging to subframes 1 through 5 will restart with subframe 1 regardless of which subframe was last transmitted prior to end/start of week. The cycling of the 25 pages will restart with page 1 of each of the subframes, regardless of which page was the last to be transmitted prior to the end/start of week. All upload and page cutovers will occur on frame boundaries (i.e., modulo 30 seconds relative to end/start of week). Accordingly, new data in subframes 4 and 5 may start to be transmitted with any of the 25 pages of these subframes.

In the following sections the navigation data will be discussed. Only the limited information used to determine the user position will be included. Detailed information can be found in references 3 and 7.

5.9   TELEMETRY (TLM) AND HAND OVER WORD (HOW)[2,3,7]

As previously mentioned, five subframes make a page. The first two words of all the subframes are the telemetry (TLM) and hand over word (HOW). Each word contains 30 bits and the message is transmitted from bit 1 to bit 30. These two words are shown in Figure 5.7. The TLM word begins with an 8-bit preamble, followed by 16 reserved bits and 6 parity bits. The bit pattern of the preamble is shown in this figure. The bit pattern of the preamble will be used to match the navigation data to detect the beginning of a subframe.

The HOW word can be divided into four parts.

1. The first 17 bits (1–17) are the truncated time of week (TOW) count that provides the time of the week in units of 6 seconds. The TOW is the truncated LSB of the Z count, which will be discussed in the next section.

2. The next two bits (18, 19) are flag bits. For satellite configuration 001 (block II satellite) bit 18 is an alert bit and bit 19 is antispoof. Satellites are procured in blocks. Most block I satellites are experimental ones and all the satellites in orbit are from block II. When bit 18 = 1, it indicates that the satellite user range accuracy may be worse than indicated in subframe 1 and the user uses the satellite at the user's own risk. Bit 19 = 1 indicates the antispoof mode is on.

3. The following three bits (20–22) are the subframe ID and their values are 1, 2, 3, 4, and 5 or (001, 010, 011, 100, and 101) to identify one of the five subframes. These data will be used for subframe matching.

4. The last 8 bits (23–30) are used for parity bits.

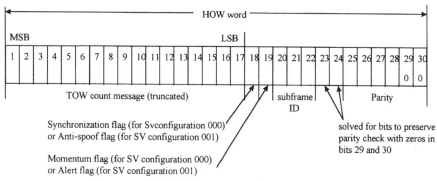

FIGURE 5.7   TLM and HOW words.

5.10   GPS TIME AND THE SATELLITE Z COUNT[3,16]

GPS time is used as the primary time reference for all GPS operation. GPS time is referenced to a universal coordinated time (UTC). The acronym UTC is an English-French mixture for Coordinated Universal Time (CUT) in English or Temps Universal Coordonne (TUC) in French[16]. It was internationally agreed to write Universal Coordinated Time as UTC, rather than CUT or TUC, making it language-independent. The GPS zero time is defined as midnight on the night of January 5/morning of January 6, 1980. The largest unit used in stating GPS time is one week, defined as 604,800 seconds ($7 \times 24 \times 3600$). The GPS time may differ from UTC because GPS time is a continuous time scale, while UTC is corrected periodically with an integer number of leap seconds. The GPS time scale is maintained to be within one μs of UTC (modulo of one second). This means the two times can be different by an integer number of seconds. A history of the difference of UTC and GPS time will be shown in Section 5.14.

In each satellite, an internally derived 1.5-second epoch, the Z count, provides a convenient unit for precise counting and communication time. The Z count has 29 bits consisting of two parts: the 19 least-significant bits (LSBs) referred to as the time of the week (TOW) and the 10 most-significant bits (MSBs) as the week number. In the actual data transmitted by the satellite, there are only 27 Z count bits. The 10-bit week number is in the third word of subframe 1. The 17-bit

TOW is in the HOW in every subframe as discussed in the previous section. The two LSBs are implied through multiplication of the truncated Z count.

The TOW count has a time unit of 1.5 sec and covers one week of time. Since one week has 604,800 seconds, the TOW count is from 0 to 403,199 because 604,800/1.5 = 403,200. The epoch occurs at approximately midnight Saturday night/Sunday morning, where midnight is defined as 0000 hours on the UTC scale, which is nominally referenced to the Greenwich Meridian. Over the years, the occurrence of the zero-state epoch differs by a few seconds from 0000 hours on the UTC scale. The 17-bit truncated version of the TOW count covers a whole week and the time unit is 6 sec (1.5 sec × 4), which equals one subframe time. This truncated TOW is from 0 to 100,799, because 604,800/6 = 100,800.

The timeline is shown in Figure 5.8. In Figure 5.8 the Z count is at the end and start of a week as shown in the upper part of the figure. The TOW count consists of the 17 MSBs of the actual 19-bit TOW count at the start of the next subframe as shown in the lower part of the figure. It is important to note that since the TOW count shows the start of the next subframe its value is 1 rather than 0 at the end and start of the week. Multiplying the truncated 17-bit TOW count by 4 converts to the actual 19-bit TOW count as shown in Figure 5.8. This operation changes the truncated TOW from 0 to 100,799 to from 0 to 403,199, the full range of the Z count.

The 10 MSBs of the Z count is the week number (WN). It represents the number of weeks from midnight on the night of January 5, 1980/morning of January 6, 1980. The total range of WN is from 0 to 1023. At the expiration of GPS week, the GPS week number will roll over to zero. Users must add the previous 1,024 weeks into account when converting from GPS time to a calendar date.

5.11   PARITY CHECK ALGORITHM[3,7]

In this section the operation of parity bits will be discussed. From Figure 5.9 (in the following section) one can see that each word has 30 bits and 6 of these are parity bits. These parity bits are used for parity check and to correct the polarity of the navigation bits. If the parity check fails, the data should not be used. In order to check parity, 8 parity bits are used. The additional two bits are the last two bits (also the last two parity bits) from the previous word.

Let D_i represent the data bits in a word received by a receiver where $i = 1, 2, 3, \ldots, 24$ represent the source data and $i = 25, 26, \ldots, 30$ represent the parity bits. The parity encoding equations are listed in Table 5.5, where D_{29}^* and D_{30}^* are the twentyninth and thirtieth data of the previous word, $\oplus$ is the modulo-2 addition and its operation rule is listed in Table 5.2, D_{25} through D_{30} are the parity data.

In using Table 5.5, the first 24 calculations must be carried out first. The purpose is to generate a new set of data d_i for $i = 1$ to 24. If $D_{30}^* = 0$, then from the relation in Table 5.2 $d_i = D_i$ (for $i = 1$ to 24), which means there is no

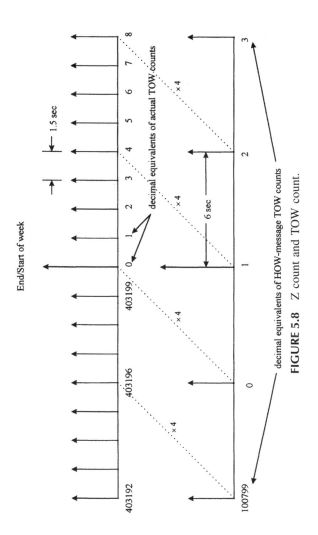

FIGURE 5.8   Z count and TOW count.

Subframe 1

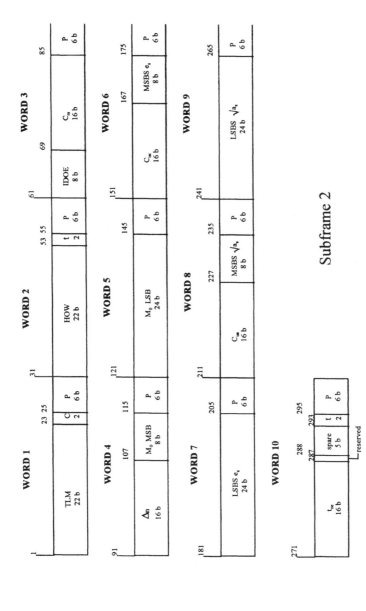

FIGURE 5.9 Data in subframes 1, 2, and 3. ***Reserved. p: Parity bits. t: Two noninformation-bearing bits used for parity computation.

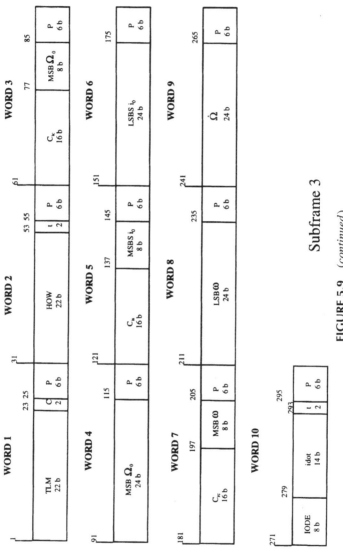

FIGURE 5.9 (*continued*)

Subframe 3

TABLE 5.5   Parity Encoding Equations

$$d_1 = D_1 \oplus D_{30}^*$$
$$d_2 = D_2 \oplus D_{30}^*$$
$$d_3 = D_3 \oplus D_{30}^*$$
$$\vdots$$
$$d_{24} = D_{24} \oplus D_{30}^*$$
$$D_{25} = D_{29}^* \oplus d_1 \oplus d_2 \oplus d_3 \oplus d_5 \oplus d_6 \oplus d_{10} \oplus d_{11} \oplus d_{12} \oplus d_{13} \oplus d_{14} \oplus d_{17} \oplus d_{18} \oplus d_{20} \oplus d_{23}$$
$$D_{26} = D_{30}^* \oplus d_2 \oplus d_3 \oplus d_4 \oplus d_6 \oplus d_7 \oplus d_{11} \oplus d_{12} \oplus d_{13} \oplus d_{14} \oplus d_{15} \oplus d_{18} \oplus d_{19} \oplus d_{21} \oplus d_{24}$$
$$D_{27} = D_{29}^* \oplus d_1 \oplus d_3 \oplus d_4 \oplus d_5 \oplus d_7 \oplus d_8 \oplus d_{12} \oplus d_{13} \oplus d_{14} \oplus d_{15} \oplus d_{16} \oplus d_{19} \oplus d_{20} \oplus d_{22}$$
$$D_{28} = D_{30}^* \oplus d_2 \oplus d_4 \oplus d_5 \oplus d_6 \oplus d_8 \oplus d_9 \oplus d_{13} \oplus d_{14} \oplus d_{15} \oplus d_{16} \oplus d_{17} \oplus d_{20} \oplus d_{21} \oplus d_{23}$$
$$D_{29} = D_{30}^* \oplus d_1 \oplus d_3 \oplus d_5 \oplus d_6 \oplus d_7 \oplus d_9 \oplus d_{10} \oplus d_{14} \oplus d_{15} \oplus d_{16} \oplus d_{17} \oplus d_{18} \oplus d_{21} \oplus d_{22} \oplus d_{24}$$
$$D_{30} = D_{29}^* \oplus d_3 \oplus d_5 \oplus d_6 \oplus d_8 \oplus d_9 \oplus d_{10} \oplus d_{11} \oplus d_{13} \oplus d_{15} \oplus d_{19} \oplus d_{22} \oplus d_{23} \oplus d_{24}$$

sign change. If $D_{30}^* = 1$, then $D_i = 0$ changes to $d_i = 1$ and $D_i = 1$ changes to $d_i = 0$ (for $i = 1$ to 24). This operation changes the signs of the source bits. These values of d_i are used to check the parity relation given through D_{25} to D_{30}.

In a receiver the polarity of the navigation data bits is usually arbitrarily assigned. The operations listed in Table 5.5 can automatically correct the polarity. If $D_{30}^* = 0$, the polarity of the next 24 data bits does not change. If the $D_{30}^* = 1$, the polarity of the next 24 data will change. This operation takes care of the polarity of the bit pattern.

The equations listed in Table 5.5 can be calculated from a matrix operation. This matrix is often referred as the parity matrix and defined as[7]

$$H = \begin{array}{c} \begin{array}{cccccccccccccccccccccccc} 1 & 2 & 3 & 4 & 5 & 6 & 7 & 8 & 9 & 10 & 11 & 12 & 13 & 14 & 15 & 16 & 17 & 18 & 19 & 20 & 21 & 22 & 23 & 24 \end{array} \\ \begin{bmatrix} 1 & 1 & 1 & 0 & 1 & 1 & 0 & 0 & 0 & 1 & 1 & 1 & 1 & 1 & 0 & 0 & 1 & 1 & 0 & 1 & 0 & 0 & 1 & 0 \\ 0 & 1 & 1 & 1 & 0 & 1 & 1 & 0 & 0 & 0 & 1 & 1 & 1 & 1 & 1 & 0 & 0 & 1 & 1 & 0 & 1 & 0 & 0 & 1 \\ 1 & 0 & 1 & 1 & 1 & 0 & 1 & 1 & 0 & 0 & 0 & 1 & 1 & 1 & 1 & 1 & 0 & 0 & 1 & 1 & 0 & 1 & 0 & 0 \\ 0 & 1 & 0 & 1 & 1 & 1 & 0 & 1 & 1 & 0 & 0 & 0 & 1 & 1 & 1 & 1 & 1 & 0 & 0 & 1 & 1 & 0 & 1 & 0 \\ 1 & 0 & 1 & 0 & 1 & 1 & 1 & 0 & 1 & 1 & 0 & 0 & 0 & 1 & 1 & 1 & 1 & 1 & 0 & 0 & 1 & 1 & 0 & 1 \\ 0 & 0 & 1 & 0 & 1 & 1 & 0 & 1 & 1 & 1 & 1 & 0 & 1 & 0 & 1 & 0 & 0 & 0 & 1 & 0 & 0 & 1 & 1 & 1 \end{bmatrix} \end{array}$$

(5.3)

This matrix matches the last six equations in Table 5.5. If a certain d_i is present a 1 will be placed in the matrix. If a certain d_i does not exist, a zero will be placed in the matrix. Note that each row in H is simply a cyclic shift of the previous row except for the last row. In order to use the parity matrix, the following property must be noted. The similarity between modulo-2 and multiplication of $+1$ and -1 must be found first. The results in Table 5.2 are listed in Table 5.6 again for comparison.

TABLE 5.6 Comparison of Modulo-2 Addition and +1, −1 Multiplication

Modulo-2 Addition			Multiplication		
Input 1	Input 2	Output	Input 1	Input 2	Output
0	0	0	+1	+1	+1
0	1	1	+1	−1	−1
1	0	1	−1	+1	−1
1	1	0	−1	−1	+1

It appears that in order to use $+1$, -1 multiplication to replace the modulo-2 addition the input should be converted as $0 \Rightarrow +1$ and $1 \Rightarrow -1$. This operation can extend to more than two inputs. This designation contradicts the conventional approach from $0 \Rightarrow -1$ and $1 \Rightarrow +1$. The following steps can be taken to check parity:

1. Arbitrarily represent the data D_i by 1 and 0 and change $1 \Rightarrow -1$ and $0 \Rightarrow +1$.
2. Change the signs of D_i ($i = 1$ to 24) by multiplying them with D_{30}^*. These new data are as d_i for $i = 1$ to 24.
3. These values of d_i for $i = 1$ to 24 are used to multiply each row of the H matrix element by element. The results are 6 rows and each row has 24 elements. Each element can be one of the three values $+1$, 0, and -1. The nonzero terms, which are $+1$ and -1, are multiplied together and the new results should be either $+1$ or -1. These new results are multiplied either by D_{29}^* or D_{30}^* according to last six equations in Table 5.5.
4. The final results should equal to $[D_{25}\ D_{26}\ D_{27}\ D_{28}\ D_{29}\ D_{30}]$.
5. The last step is to convert $+1$, -1 back to 0 and 1 for further processing.

Subframe matching will be discussed in Section 9.4. A program with subframe matching and parity check will be listed at the end of Chapter 9.

5.12 NAVIGATION DATA FROM SUBFRAME 1[3,7]

The data contained in the first three subframes are shown in Figure 5.9. The minimal parameters required to calculate the user position are contained in these three subframes.

The data used for calculations of locations of the satellites and the user are discussed below.

1. *Week number (61–70)*: These ten bits are discussed in Section 5.9. Thus represents the MSB of the Z counts and indicates the number of weeks from

midnight on the night of January 5, 1980/morning of January 6, 1980. Users must count the rollover if it is over 1,023 weeks.

2. *User range accuracy (73–76)*: These four bits give the predicted user range accuracy and its value N ranges from 0–15. The accuracy value X is:

   - If N is 6 or less, $X = 2^{(1+N/2)}$ (rounded-of-values $N = 1$, $X = 2.8$; $N = 3$, $X = 5.7$; $N = 5$, $X = 11.3$).

   - If N is 6 or more, but less than 15, $X = 2^{(N-2)}$.

   - $N = 15$ will indicate the absence of an accuracy prediction and will advise the user to use that satellite at the user's risk.

3. *Satellite health (77–82)*: These six bits represent the health indication of the transmitting satellite. The MSB (bit 77) indicates a summary of the health of the navigation data, where bit 77 equals:

   $0 =$ All navigation data are OK.

   $1 =$ Some or all navigation data are bad.

   The five LSBs indicate the health of the signal components in Table 5.7. Additional satellite health data are given in subframes 4 and 5. The data given in subframe 1 may differ from that shown in subframes 4 and/or 5 of other satellites, since the latter may be updated at a different time.

4. *Issue of data, clock (IODC) (83–84 MSB, 211–218 LSB)*: These 10-bit IODC data indicate the issue number of the data set and thereby provide the user with a convenient means of detecting any change in the correction parameters. The transmitted IODC will be different from any value transmitted by the satellite during the preceding seven days. The relationship between IODC and IODE (in both subframes 2 and 3) will be discussed in the next section.

5. *Estimated group delay differential T_{GD} (197–204)*: This eight-bit information is a clock correction term to account for the effect of satellite group delay differential. It is used in Equation (4.38).

TABLE 5.7 Codes for Health of Satellite Signal Components

MSB						LSB
0	0	0	0	0	$\Rightarrow$	All signals OK.
1	1	1	0	0	$\Rightarrow$	Satellite *is* temporarily out – do not use this satellite during current pass.
1	1	1	0	1	$\Rightarrow$	Satellite *will be* temporarily out–use with caution.
1	1	1	1	0	$\Rightarrow$	Spare
1	1	1	1	1	$\Rightarrow$	More than one combination would be required to describe anomalies.
All other combinations					$\Rightarrow$	Satellite experiencing code modulation and/or signal level transmission problem – modulation navigation data valid; however, user may experience intermittent tracking problems if satellite is acquired.

TABLE 5.8   Ephemeris Parameters in Subframe 1

Parameter	Location	Number of Bits	Scale Factor (LSB)	Effective Range**	Units
WN: Week number	61–70	10	1		week
Satellite accuracy	73–76	4			
Satellite health	77–82	6	1		
IODC: Issue of data, clock	83–84	10			
	211–218				
T_{GD}: Satellite group delay differential	197–204	8*	2^{-31}		seconds
t_{oe}: Satellite clock correction	219–234	16	2^4	604,784	seconds
a_{f2}: Satellite clock correction	241–248	8*	2^{-55}		sec/sec^2
a_{f1}: Satellite clock correction	249–264	16*	2^{-43}		sec/sec
a_{f0}: Satellite clock correction	271–292	22*	2^{-31}		seconds

*Parameters so indicated are two's complement, with the sign bit (+ or −) occupying the MSB.
**Unless otherwise indicated in this column, effective range is the maximum range attainable with indicated bit allocation and scale factor.

6. *Satellite clock correction parameters*: This subframe also contains the four additional satellite clock correction parameters: t_{oc} (219–234), a_{f0} (271–292), a_{f1} (249–264), and a_{f2} (241–248). They are used in Equation (4.37).

7. In subframe 1 there are some reserved data fields and their locations are 71–72; 91–114; 121–144; 151–174; 181–196. All reserved data fields support valid parity within their respective words.

The ephemeris parameters in subframe 1 are listed in Table 5.8.

5.13   NAVIGATION DATA FROM SUBFRAMES 2 AND 3[3,7]

Figures 5.9b and c show the following ephemeris data contained in subframes 2 and 3:

1. *The issue of data, ephemeris (IODE)*: This parameter has 8 bits and is in both subframes 2 (61–68) and 3 (271–278). The IODE equals the 8 LSB of the IODC, which has 10 bits. The IODE provides the user with a convenient means for detecting any change in the ephemeris representation parameters. The transmitted IODE will be different from any value transmitted by the satellite during the preceding six hours. Whenever these three terms, two IODEs from subframes 2, 3 and the 8 LSBs of the IODC, do not match, a data set cutover has occurred and new data must be collected.

   Any change in the subframe 2 and 3 data will be accomplished in concert with a change in both IODE words. Cutovers to new data will occur only on hour boundaries except for the first data set of a new upload. The first data set may be cut in at any time during the hour and therefore may be transmitted by the satellite for less than one hour. Additionally, the t_{oe} value for at least the first data set transmitted by a satellite after an upload will be different from that transmitted prior to the cutover.

2. *The rest of the ephemeris data*: These are listed in Tables 5.9 and 5.10.

3. *Spare and reserved data fields*: In subframe 2 bit 287 is reserved and bits 288–292 are spared. All spare and reserved data fields support valid parity within their respective words. Contents of spare data fields are alternating ones and zeros until they are allocated for a new function. Users are cautioned that the contents of spare data fields can change without warning.

TABLE 5.9 Ephemeris Parameters in Subframe 2

Parameter	Location	Number of Bits	Scale Factor (LSB)	Effective Range	Units
IODE	61–68	8			(see text)
C_{rs}: Amplitude of the sine harmonic correction terms to the orbit radius	69–84	16	2^{-5}		meters
Δn: Mean motion difference from computed value	91–106	16*	2^{-43}		semicircles/sec
M_0: Mean anomaly at reference time	107–114; 121–144	32*	2^{-31}		semicircle
C_{uc}: Amplitude of the cosine harmonic correction term to the argument of latitude of argument of latitude	151–166	16*	2^{-29}		radians
e_s: Eccentricity	167–174; 181–204	32	2^{-33}	0.03	dimensionless
C_{us}: Amplitude of the sine harmonic correction term to the argument of latitude	211–226	16*	2^{-29}		radians
$\sqrt{a_s}$: Square root of the semimajor axis	227–234; 241–264	32	2^{-19}		meters$^{1/2}$
t_{oe}: Reference time ephemeris	277–286	16	2^4	604,784	seconds

TABLE 5.10 Ephemeris Parameters in Subframe 3

Parameter	Location	Number of Bits	Scale Factor (LSB)	Effective Range	Units
C_{ic}: Amplitude of the cosine harmonic correction term to angle of inclination	61–76	16*	2^{-29}		radians
Ω_e: Longitude of ascending node of orbit plane at weekly epoch	77–84; 91–114	32*	2^{-31}		semicircles
C_{is}: Amplitude of the sine harmonic correction term to angle of inclination	121–136	16*	2^{-29}		radians
i_0: Inclination angle at reference time	137–144; 151–174	32*	2^{-31}		semicircles
C_{rc}: Amplitude of the sine harmonic correction term to the orbit radius	181–196	16*	2^{-5}		meters
ω: Argument of perigee	197–204; 211–234	32*	2^{31}		semicircles
$\dot{\Omega}$: Rate of right ascension	241–264	24*	2^{-43}		semicircles/sec
IDOE	271–278				(see text)
idot: Rate of inclination angle	279–292	14*	2^{-43}		semicircles/sec

5.14 NAVIGATION DATA FROM SUBFRAMES 4 AND 5–SUPPORT DATA[3,7,8]

Both subframes 4 and 5 are subcommutated 25 times each. The 25 versions of these subframes are referred to as pages 1 to 25 of each superframe. With the possible exception of "spare" pages and explicit repeats, each page contains different data in words 3 through 10, which are from bits 91–300. Subframe 4 has six different formats but only five of them are shown in Figure 5.10a. Five pages, 1, 6, 11, 16, 21, are in one format. Six pages, 12, 19, 20, 22, 23, 24, are in one format. Page 18 is in one format. Page 25 is in one format, and pages 13, 14, 15, and 17 are in one format. There are a total of 17 pages. Pages 2, 3, 4, 5, 7, 8, 9, and 10 are not shown because they have the same format as page 1 through 24 of subframe 5. Subframe 5 has two different formats as shown in Figure 5.10b.

FIGURE 5.10   Data format for subframes 4 and 5. **The indicated portions of words 3 through 10 of pages 13, 14, and 15 are spares, while those of page 17 are reserved for special messages. ***Reserved. P: Six parity bits. t: Two non-information-bearing bits used for parity computation. C: TLM bits 23 and 24 are reserved. Note: Pages 2, 3, 4, 5, 7, 8, 9, and 10 of subframe 4 have the same format as pages 1 through 24 of subframe 5.

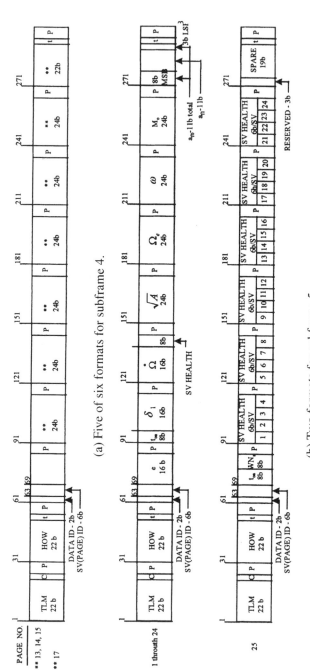

(a) Five of six formats for subframe 4.

(b) Two formats for subframe 5.

FIGURE 5.10 (*continued*)

The information in subframes 4 and 5 and its applications are listed below:

1. *Subframe 4*:
   - Pages 2, 3, 4, 5, 7, 8, 9, and 10 contain the almanac data for satellite 25 through 32. These pages may be designated for other functions. The satellite ID of that page defines the format and content.
   - Page 17 contains special messages.
   - Page 18 contains ionospheric and universal coordinated time (UTC).
   - Page 25 contains antispoof flag, satellite configuration for 32 satellites, and satellite health for satellites 25–32.
   - Pages 1, 6, 11, 12, 16, 19, 20, 21, 22, 23, and 24 are reserved.
   - Pages 13, 14, and 15 are spares.

2. *Subframe 5*:
   - Pages 1–24 contain almanac data for satellites 1 through 24.
   - Page 25 contains satellite health for satellites 1 through 24, the almanac reference time, and the almanac reference week number.

3. *Almanac data*: The almanac parameters provided in subframes 4 and 5 are: e_s, t_{oa}, $\dot{\Omega}$, $\sqrt{a_s}$, Ω_e, ω, M_0, a_{f0}, and a_{f1}. The almanac data are much less accurate than the detailed ephemeris data of subframes 2 and 3. However, the almanac data are valid for longer periods of time and do not require frequency updates. The almanac data and their application will be further discussed in Section 12.12.

4. *Translation of GPS time to UTC time*: In page 18 of subframe 4 the parameters in Table 5.11 are included.

The GPS/UTC time relationship is given by:[3,7]

$$t_{UTC} = (t_E - \Delta t_{UTC})\{\text{modulo } 86{,}400 \text{ seconds}\} \qquad (5.4)$$

TABLE 5.11 UTC Parameters

Parameter**	Number of Bits	Scale Factor	Effective Range***	Units
A_0	32*	2^{-30}		seconds
A_1	24*	2^{-50}		sec/sec
Δt_{LS}	8	1		seconds
t_{ot}	8	2^{12}	602,112	seconds
WN_r	8	1		weeks
WN_{LSF}	8	1		weeks
DN	8****	1	7	days
Δt_{LSF}	8*	1		seconds

*Parameters so indicated are two's complement, with the sign bit (+ or −) occupying the MSB.
**See Figure 5.9a for bit allocation in subframe 4.
***Unless otherwise indicated in this column, effective range is the maximum range attainable with indicated bit allocation and scale factor.
****Right justified.

where t_{UTC} is in seconds and

$$\Delta t_{UTC} = \Delta t_{LS} + A_0 + A_1[t_E - t_{ot} + 604{,}800(WN - WN_t)] \text{ seconds}$$
(5.5)

t_E: GPS time as estimated by the user on the basis of correcting t_{SV} for factors given in the subframe 1 clock correction discussion as well as for ionospheric and satellite (dither) effects.

t_{SV}: effective satellite pseudorange code phase time at message of transmission time.

Δt_{LS}: delta time due to leap seconds.

A_0, A_1: constant and first-order terms of polynomial.

t_{ot}: reference time for UTC data.

WN: current week number (derived from subframe 1).

WN_t: UTC reference week number.

The estimated GPS time (t_E) is in seconds relative to end/start of week. The reference time for UTC data (t_{ot}) is referenced to the start of that week whose number (WN_t) is given in bits (227–234) of page 18 subframe 4 representing the 8 LSB of the week. The user must account for the truncated nature of the week number.

Whenever the user's current time falls within the time span of $DN + 3/4$ to $DN + 5/4$, proper accommodation of the leap second event with a possible week number transition is provided by the following expression for UTC:

$$t_{UTC} = W[\text{modulo}(86{,}400 + \Delta t_{LSF} - \Delta t_{LS})] \text{ seconds} \qquad (5.6)$$

where

$$W = (t_E - \Delta t_{UTC} - 43{,}200)[\text{modulo } 86{,}400] + 43{,}200 \text{ seconds} \quad (5.7)$$

The definition of Δt_{UTC} given in Equation (5.4) applies throughout the transition period. Note that when a leap second is added, unconventional time values of the form 23:59:60.xxx are encountered. Some user equipment may be designed to approximate UTC by decrementing the running count of time within several seconds after the event, thereby promptly returning to a proper time indication. Whenever a leap second event is encountered, the user equipment must consistently implement carries or borrows into any year/week/day counts. Table 5.12 gives the past history of the difference between the GPS and the UTC times.[8] In 19 years the difference is 13 seconds.

The tendency is that most of the modern navigation equipment uses GPS time as the time base. Therefore, the translation from GPS time to UTC time may no longer be needed in modern equipment.

5. *Ionospheric data*: In page 18 subframe 4, there are eight ionospheric data: α_0 (69–76), α_1 (77–84), α_2 (91–98), α_3 (99–106), β_0 (107–114), β_1

TABLE 5.12   Difference Between GPS and UTC Times

Date	GPS-UTC Time (sec)
6 Jan 1980	0 (Start of GPS system time)
1 Jul 1981	1
1 Jul 1982	2
1 Jul 1983	3
1 Jul 1985	4
1 Jan 1988	5
1 Jan 1990	6
1 Jan 1991	7
1 Jul 1992	8
1 Jul 1993	9
1 Jul 1994	10
1 Jan 1996	11
1 Jul 1997	12
1 Jan 1999	13

$(121-128)$, β_2 $(129-136)$, β_3 $(137-144)$. These data can be used to correct the time received from the satellite for ionospheric effect. The applications of these data are discussed in the next section.

5.15   IONOSPHERIC MODEL[3,7-10]

The atmosphere around the earth will affect the traveling speed of the GPS signal and cause measurement errors. These errors should be corrected. For GPS application, the atmosphere is usually divided into two portions: the ionosphere and the troposphere. Troposphere is the closer of the two to the surface of the earth while ionosphere is above the troposphere. The troposphere contains neutral particles and ionosphere contains free ions. The ionosphere will cause a code delay but a carrier phase advance.[10] This section presents a correction model for the ionospheric error.

Besides the selectivity availability (SA), which will be discussed in the next section, the ionospheric effect can cause one of the most significant position errors in a GPS receiver. If a receiver operates on both the L1 and L2 frequencies, such as in a military receiver, the time delay Δt_1 at frequency L1 caused by the ionospheric effect can be calculated as[9]

$$\Delta t_1 = \frac{f_2^2}{f_1^2 - f_2^2} \delta(\Delta t) \tag{5.8}$$

where f_1 and f_2 are the frequencies at L1 and L2 respectively, $\delta(\Delta t)$ is the measured time difference between frequencies f_1 and f_2 from the same satellite. This Δt_1 can be considered as the measured value.

In most commercial GPS receivers only the L1 frequency is available. The ionospheric data collected from subframe 4 can be used to reduce the ionospheric effect; this is often referred to as the single-frequency ionospheric model. Using this model one can reduce the user root mean square (rms) position error caused by ionospheric effect at least by 50 percent.[7]

The ionospheric model is[3,7]

$$
T_{iono} = \begin{cases} T * \left[5.0 * 10^{-9} + (AMP)\left(1 - \dfrac{x^2}{2} + \dfrac{x^4}{24}\right)\right] & \text{if } |x| < 1.57 \\ T * (5.0 * 10^{-9}) & \text{if } |x| \geq 1.57 \end{cases} \text{(sec)}
$$

$$(5.9)$$

where T_{iono} is the addition delay time and

$$
AMP = \begin{cases} \displaystyle\sum_{n=0}^{3} \alpha_n \phi_m^n & \text{if } AMP \geq 0 \\ AMP = 0 & \text{if } AMP < 0 \end{cases} \text{(sec)} \tag{5.10}
$$

$$
x = \frac{2\pi(t - 50,400)}{PER} \text{(radians)} \tag{5.11}
$$

$$
PER = \begin{cases} \displaystyle\sum_{n=0}^{3} \beta_n \phi_m^n & \text{if } PER \geq 72,000 \\ PER = 72,000 & \text{if } PER < 72,000 \end{cases} \text{(sec)} \tag{5.12}
$$

$$
T = 1.0 + 16.0[0.53 - \xi]^3 \tag{5.13}
$$

where α_n and β_n with ($n = 0, 1, 2, 3$) are the ionospheric data obtained from the satellite and ξ is the elevation angle between the user and satellite.

Other equations that must be solved are

$$
\phi_m = \phi_i + 0.064 \cos(\lambda_i - 1.617) \text{(semicircles)} \tag{5.14}
$$

$$
\lambda_i = l_u + \frac{\psi \cos A}{\cos \phi_i} \text{(semicircle)} \tag{5.15}
$$

$$
\phi_i = \begin{cases} L_u + \psi \cos A & \text{if } |\phi_i| \leq 0.416 \\ +0.416 & \text{if } \phi_i > +0.416 \\ -0.416 & \text{if } \phi_i < -0.401 \end{cases} \text{(semicircle)} \tag{5.16}
$$

$$
\psi = \frac{0.00137}{\xi + 0.11} + 0.022 \text{(semicircle)} \tag{5.17}
$$

$$
t = 4.32 * 10^4 \lambda_i + GPS \text{ time(sec)} \tag{5.18}
$$

where $0 \leq t < 86,400$; therefore, if $t \geq 86,400$ seconds, subtract 86,400 seconds; if $t < 0$ seconds, add 86,400 seconds.

The terms used in computation of ionospheric delay are as follows:

Satellite-Transmitted Terms

α_n: The coefficients of a cubic equation representing the amplitude of the vertical delay (4 coefficients – 8 bits each).
β_n: the coefficients of a cubic equation representing the period of the model (4 coefficients – 8 bits each).

Receiver-Generated Terms

ξ: elevation angle between the user and satellite (semicircles).
A: azimuth angle between the user and satellite, measured clockwise positive from true North (semicircles).
L_u: user geodetic latitude (semicircle).
l_u: user geodetic longitude (semicircle).
GPS time: receiver-computed system time.

Computed Terms

x: phase (radians).
T: obliquity factor (dimensionless).
t: local time (sec).
ϕ_m: geomagnetic latitude of the earth projection of the ionospheric intersection point (mean ionospheric height assumed 350 km) (semicircle).
λ_i: geomagnetic latitude of the earth projection of the ionospheric intersection point (semicircle).
ϕ_i: geomagnetic latitude of the earth projection of the ionospheric intersection point (semicircle).
ψ: earth's central angle between user position and earth projection of ionospheric intersection point (semicircles).

5.16 TROPOSPHERIC MODEL[11]

Compared with the ionospheric effect, the tropospheric effect is about an order of magnitude less. The satellites do not transmit any data to correct for the tropospheric effect. There are many models to correct the error. Here only a simple model will be presented. The delay in meters is given by:[11]

$$\Delta = \frac{2.47}{\sin \xi + 0.0121} \text{meters} \tag{5.19}$$

where ξ is the elevation angle between the user and satellite.

TABLE 5.13   Observed GPS Positioning Errors with Typical Standard Positioning Service (SPS) Receiver[15]

Error Source	Typical Range Error Magnitude (meters 1σ)
Selective availability	24.0
Ionospheric*	7.0
Tropospheric**	0.7
Satellite clock & ephemeris	3.6
Receiver noise	1.5

*After applying Ionospheric model. Actual values can range between approximately 1–30 m.
**After applying tropospheric model.

5.17   SELECTIVITY AVAILABILITY (SA) AND TYPICAL POSITION ERRORS[8,12–15]

The selectivity availability is aimed to degrade the performance of the GPS. It was put in effect on March 25, 1990. In accordance with the current policy of the U.S. Department of Defense, the signal available from the GPS is actually a purposefully degraded version of the C/A code. The signal degradation is achieved by dithering the satellite clock frequency and providing only a coarse description of the satellite ephemeris. This policy, known as selective availability, effectively raises the value of the user range error by a factor of four or more. The selectivity availability affects only the performance of a GPS receiver, it does not impact the design of the receiver. A Presidential Decision Directive (PDD) released in March 1996 states that the selectivity availability will be turned off within 10 years. The SA signal was terminated on May 1, 2000 and this information was obtained from the internet.

Some typical position errors caused by different effects are listed in Table 5.13. The distance error given in meters for one standard deviation (1σ) is listed in this table.

5.18   SUMMARY

In this chapter the C/A code signal of the L1 frequency is discussed. The radio frequency and the C/A code length are important information for performing acquisition and tracking by a receiver. The navigation data are in five subframes. In order to obtain the data, the beginning of the subframe must be found. There are parity data that must be checked before the data can be used. The information in the first three subframes is enough to find the user position. The information in the fourth and fifth subframes is support data. Ionospheric and tropospheric models are introduced to improve receiver accuracy. Selectivity availability is introduced in the L1 C/A code signal to deliberately degrade the user position accuracy.

REFERENCES

1. Spilker, J. J., "GPS signal structure and performance characteristics," *Navigation, Institute of Navigation*, vol. 25, no. 2, pp. 121–146, Summer 1978.

2. Spilker, J. J. Jr., "GPS signal structure and theoretical performance," Chapter 3 in Parkinson, B. W., Spilker, J. J. Jr., *Global Positioning System: Theory and Applications*, vols. 1 and 2, American Institute of Aeronautics and Astronautics, 370 L'Enfant Promenade, SW, Washington, DC, 1996.

3. *Global Positioning System Standard Positioning Service Signal Specification*, 2nd ed., GPS Joint Program Office, June 2, 1995.

4. Aparicio, M., Brodie, P., Doyle, L., Rajan, J., Torrione, P., "GPS satellite and payload," Chapter 6 in Parkinson, B. W., Spilker, J. J. Jr., *Global Positioning System: Theory and Applications*, vols. 1 and 2, American Institute of Aeronautics and Astronautics, 370 L'Enfant Promenade, SW, Washington, DC, 1996.

5. Dixon, R. C., *Spread Spectrum Systems*, Wiley, New York, 1976.

6. Gold, R., "Optimal binary sequences for spread spectrum multiplexing," *IEEE Trans. on Information Theory*, vol. 13, pp. 619–621, October 1967.

7. Spilker, J. J. Jr., "GPS navigation data," Chapter 4 in Parkinson, B. W., Spilker, J. J. Jr., *Global Positioning System: Theory and Applications*, vols. 1 and 2, American Institute of Aeronautics and Astronautics, 370 L'Enfant Promenade, SW, Washington, DC, 1996.

8. Raquet, J., "Navigation using GPS," class notes, Air Force Institute of Technology, Dayton OH, 1999.

9. Klobuchar, J. A., "Ionospheric effects on GPS," Chapter 12 in Parkinson, B. W., Spilker, J. J. Jr., *Global Positioning System: Theory and Applications*, vols. 1 and 2, American Institute of Aeronautics and Astronautics, 370 L'Enfant Promenade, SW, Washington, DC, 1996.

10. Kaplan, E. D., ed., *Understanding GPS Principles and Applications*, Artech House, Norwood, MA, 1996.

11. Spilker, J. J. Jr., "Tropospheric effects on GPS," Chapter 13 in Parkinson, B. W., Spilker, J. J. Jr., *Global Positioning System: Theory and Applications*, vols. 1 and 2, American Institute of Aeronautics and Astronautics, 370 L'Enfant Promenade, SW, Washington, DC, 1996.

12. van Graas, F., Braasch, M. S., "Selective availability," Chapter 17 in Parkinson, B. W., Spilker, J. J. Jr., *Global Positioning System: Theory and Applications*, vols. 1 and 2, American Institute of Aeronautics and Astronautics, 370 L'Enfant Promenade, SW, Washington, DC, 1996.

13. Misra, P. N., "Integrated use of GPS and GLONASS in civil aviation," *Lincoln Laboratory Journal*, Massachusetts Institute of Technology, vol. 6, no. 2, pp. 231–247, Summer/Fall, 1993.

14. Kovach, K. L., Van Dyke, K. L., "GPS in 10 years," *Microwave*, p. 22, February 1998.

15. Raquet, J., "Navigation using GPS," Air Force Institute of Technology course EENG 533, Spring 1999.

16. Agilent AN 1289 "The Science of timekeeping" Application Note, Agilent Technologies. 1997, 2000.

```
% p5_1.m generates MLS and G2 outputs and check their delay
time

% input
% k1, k2: positions of two taps
% k3: delay time
%
k=input('enter [k1 k2 k3] = ');
inp1=-ones(1,10); % initial condition of register
for j=1:1023;
  mlsout(j)=inp1(10); % MLS output
  modulo=inp1(2)*inp1(3)*inp1(6)*inp1(8)*inp1(9)*inp1(10);
  inp1(2:10)=inp1(1:9);
  inp1(1)=modulo;
  g2(j)=inp1(k(1))*inp1(k(2)); % G2 output
end
if mlsout==g2([k(3):1023 1:k(3)-1])
  disp('OK')
  else
  disp('not match')
end
```

```
% p5_2.m generate one of the 32 C/A codes written by D.Akos
modified by J.Tsui

svnum=input('enter the satellite number = ');   % the
Satellite's ID number

% ca : a vector containing the desired output sequence

% the g2s vector holds the appropriate shift of the g2 code
to generate
% the C/A code (ex. for SV#19 - use a G2 shift of
g2s(19)=471)

g2s = [5;6;7;8;17;18;139;140;141;251;252;254;255;256;257;258;
        469;470;471; ...
        472;473;474;509;512;513;514;515;516;859;860;861;862];

g2shift=g2s(svnum,1);

% Generate G1 code
      %    load shift register
           reg = -1*ones(1,10);
      %
      for i = 1:1023,
          g1(i) = reg(10);
          slave1 = reg(3)*reg(10);
          reg(1,2:10) = reg(1:1:9);
          reg(1) = save1;
      end,
%
% Generate G2 code
%
      %    load shift register
           reg = -1*ones(1,10);
      %
      for i = 1:1023,
          g2(i) = reg(10);
          save2 = reg(2)*reg(3)*reg(6)*reg(8)*reg(9)*reg(10);
          reg(1,2:10) = reg(1:1:9);
          reg(1) = save2;
      end,
      %
      %    Shift G2 code
      %
      g2tmp(1,1:g2shift)=g2(1,1023-g2shift+1:1023);
      g2tmp(1,g2shift+1:1023)=g2(1,1:1023-g2shift);
      %
      g2 = g2tmp;
%
```

```
%  Form single sample C/A code by multiplying G1 and G2
%
ss_ca = g1.*g2;
ca = ss_ca;
%  Change to 1 0 outputs

ind1=find(ca==-1);
ind2=find(ca==1);
ca(ind1)=ones(1,length(ind1));
ca(ind2)=zeros(1,length(ind2));
ca(1:10) %print first 10 bits
```

CHAPTER SIX

Receiver Hardware Considerations

6.1 INTRODUCTION[1]

This chapter discusses the hardware of a GPS receiver. Since the basic design of GPS receiver in this book is software oriented, the hardware presented here is rather simple. The only information needed for a software receiver is the sampled data. These sampled or digitized data will be stored in memory to be processed. For postprocessing the memory size dictates the length of data record. A minimum of 30 seconds of data is needed to find the user position as mentioned in Section 5.5. In real-time processing the memory serves as a buffer between the hardware and the software signal processing. The hardware includes the radio frequency (RF) chain and analog-to-digital converter (ADC). Thus, the signal processing software must be capable of processing the digitized data in the memory at a real-time rate. Under this condition, the size of the memory determines the latency allowable for the signal processing software.

This chapter will include the discussion of the antenna, the RF chain, and the digitizers. Two types of designs will be discussed. One is a single channel to collect real data and the other is an in-phase and quadrature phase (I-Q) channel to collect complex data. In both approaches, the input signals can be either down-converted to a lower intermediate frequency (IF) before digitization or directly digitized at the transmitted frequency. The relation between the sampling frequency and the input frequency will be presented. Some suggestions on the sampling frequency selection will be included. Two hardware setups to collect

Fundamentals of Global Positioning System Receivers: A Software Approach, Second Edition,
by James Bao-Yen Tsui
ISBN 0-471-70647-7   Copyright © 2005 John Wiley & Sons, Inc.

real data will be discussed in detail as examples. The impact of the number of digitized bits will also be discussed.

A digital band folding technique will be discussed that can alias two or more narrow frequency bands into the baseband. This technique can be used to alias the L1 and L2 bands of the GPS into the baseband, or to alias the GPS L1 frequency and the Russian Global Navigation Satellite System (GLONASS) signals into the baseband. If one desires, all three bands, L1, L2, and the GLONASS, can be aliased into the baseband. With this arrangement the digitized signal will contain the information from all three input bands.

One of the advantages of a software receiver is that the receiver can process data collected with various hardware. For example, the data can be real or complex with various sampling frequencies. A simple program modification in the receiver should be able to use the data. Or the data can be changed from real to complex and complex to real such that the receiver can process them.

6.2 ANTENNA[2–4]

A GPS antenna should cover a wide spatial angle to receive the maximum number of signals. The common requirement is to receive signals from all satellites about 5 degrees above the horizon. Combining satellites at low elevation angles and high elevation angles can produce a low value of geometric dilution of precision (GDOP) as discussed in Section 2.15. A jamming or interfering signal usually comes from a low elevation angle. In order to minimize the interference, sometimes an antenna will have a relatively narrow spatial angle to avoid signals from a low elevation angle. Therefore, in selecting a GPS antenna a trade-off between the maximum number of receiving satellites and interference must be carefully evaluated.

If an antenna has small gain variation from zenith to azimuth, the strength of the received signals will not separate far apart. In a code division multiple access (CDMA) system it is desirable to have comparable signal strength from all the received signals. Otherwise, the strong signals may interfere with the weak ones and make them difficult to detect. Therefore, the antenna should have uniform gain over a very wide spatial angle.

If an antenna is used to receive both the L1 (1575.42 MHz) and the L2 (1227.6 MHz), the antenna can either have a wide bandwidth to cover the entire frequency range or have two narrow bands covering the desired frequency ranges. An antenna with two narrow bands can avoid interference from the signals in between the two bands.

The antenna should also reject or minimize multipath effect. Multipath effect is the GPS signal reflections from some objects that reach the antenna indirectly. Multipath can cause error in the user position calculation. The reflection of a right-handed circular polarized signal is a left-handed polarized signal. A right-handed polarized receiving antenna has higher gain for the signals from the satellites. It has a lower gain for the reflected signals because the polarization is in the opposite direction. In general it is difficult to suppress the multipath because it can come from any direction. If the direction of the reflected signal is known, the antenna can be designed to suppress it. One common multipath is the reflection from the ground below the antenna. This multipath can be reduced because the direction of the incoming signal is known. Therefore, a GPS antenna should have a low back lobe. Some techniques such as a specially designed ground plane can be used to minimize the multipath from the ground below. The multipath requirement usually complicates the antenna design and increases its size.

Since the GPS receivers are getting smaller as a result of the advance of integrated circuit technology, it is desirable to have a small antenna. If an antenna is used for airborne applications, its profile is very important because it will be installed on the surface of an aircraft. One common antenna design to receive a circular polarized signal is a spiral antenna, which inherently has a wide bandwidth. Another type of popular design is a microstrip antenna, sometimes also referred to as the patch antenna. If the shape is properly designed and the feed point properly selected, a patch antenna can produce a circular polarized wave. The advantage of the patch antenna is its simplicity and small size.

In some commercial GPS receivers the antenna is an integral part of the receiver unit. Other antennas are integrated with an amplifier. These antennas can be connected to the receiver through a long cable because the amplifier gain can compensate the cable loss. A patch antenna (M/A COM ANP-C-114-5) with an integrated amplifier is used in the data collection system discussed in this chapter. The internal amplifier has a gain of 26 dB with a noise figure of 2.5 dB. The overall size of the antenna including the amplifier is diameter of 3″ and thickness about 0.75″. The antenna pattern is measured in an anechoic chamber and the result is shown in Figure 6.1a. Figure 6.1b shows the frequency response of the antenna. The beam of this antenna is rather broad. The gain in the zenith direction is about +3.5 dBic where ic stands for isotropic circular polarization. The gain at 10 degrees is about −3 dBic.

6.3 AMPLIFICATION CONSIDERATION[5-7,10]

In this section the signal level and the required amplification will be discussed. The C/A code signal level at the receiver set should be at least −130 dBm[5] as discussed in Section 5.2. The available thermal noise power N_i at the input of a receiver is:[6]

$$N_i = kTB \text{ watts} \tag{6.1}$$

where k is the Boltzmann's constant ($= 1.38 \times 10^{-23}$ J/°K) T is the temperature of resistor R (R is not included in the above equation) in Kelvin, B is the bandwidth of the receiver in hertz, N_i is the noise power in watts. The thermal noise at room temperature where $T = 290$°K expressed in dBm is

$$N_i(\text{dBm}) = -174 \text{ dBm/Hz or } N_i(\text{dBm}) = -114 \text{ dBm/MHz} \tag{6.2}$$

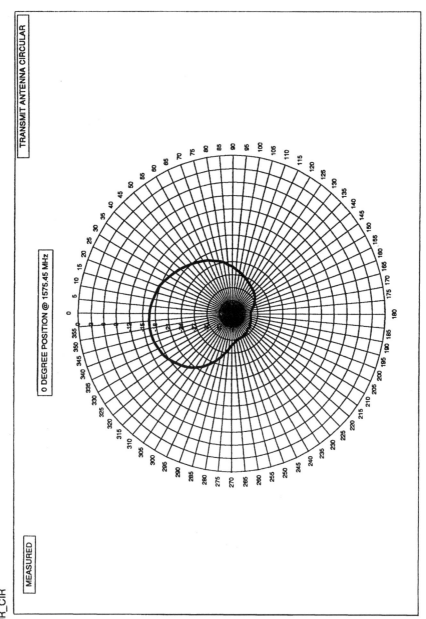

(a) Spatial pattern.

FIGURE 6.1 Antenna measurements of an M/A COM ANP-C-114-5 antenna.

(b) Frequency response.

FIGURE 6.1 (*continued*)

If the input to the receiver is an antenna pointing at the sky, the thermal noise is lower than room temperature, such as $50°K$.

For the C/A code signal, the null-to-null bandwidth is about 2 (or 2.046) MHz, thus, the noise floor is at -111 dBm ($-114 + 10$ log2). Supposing that the GPS signal is at -130 dBm, the signal is 19 dB ($-130 + 111$) below the noise floor. One cannot expect to see the signal in the collected data. The amplification needed depends on the analog-to-digital converter (ADC) used to generate the data. A simple rule is to amplify the signal to the maximum range of the ADC. However, this approach should not be applied to the GPS signal, because the signal is below the noise floor. If the signal level is brought to the maximum range of the ADC, the noise will saturate the ADC. Therefore, in this design the noise floor rather than the signal level should be raised close to the maximum range of the ADC.

A personal computer (PC)-based card[7] with two ADCs is used to collect data. This card can operate at a maximum speed of 60 MHz with two 12-bit ADCs. If both ADCs operate simultaneously, the maximum operating speed is 50 MHz. The maximum voltage to exercise all the levels of the ADC is about 100 mv and the corresponding power is:

$$P = \frac{(0.1)^2}{2 \times 50} = 0.0001 \text{ watt} = 0.1 \text{ mw} = -10 \text{ dBm} \qquad (6.3)$$

It is assumed that the system has a characteristic impedance of 50 Ω. A simple way to estimate the gain of the amplifier chain is to amplify the noise floor to this level, thus, a net gain of about 101 dB ($-10 + 111$) is needed. Since in the RF chain there are filters, mixer, and cable loss, the insertion loss of these components must be compensated with additional gain. The net gain must be very close to the desired value[10] of 101 dB. Too low a gain value will not activate all the possible levels of the ADC. Too high a gain will saturate some components or the ADC and create an adverse effect.

6.4   TWO POSSIBLE ARRANGEMENTS OF DIGITIZATION BY FREQUENCY PLANS[8,9]

Although many possible arrangements can be used to collect digitized GPS signal data, there are two basic approaches according to the frequency plan. One approach is to digitize the input signal at the L1 frequency directly, which can be referred to as direct digitization. The other one is to down-convert the input signal to a lower frequency, called the intermediate frequency (IF), and digitize it. This approach can be referred to as the down-converted approach.

The direct digitization approach has a major advantage; that is, in this design the mixer and local oscillator are not needed. A mixer is a nonlinear device, although in receiver designs it is often treated as a linear device. A mixer usually generates spurious (unwanted) frequencies, which can contaminate the output. A local oscillator can be expensive and any frequency error or impurity produced by the local oscillator will appear in the digitized signal. However, this arrangement does not eliminate the oscillator (or clock) used for the ADC.

The major disadvantage of direct digitization is that the amplifiers used in this approach must operate at high frequency and they can be expensive. The ADC must have an input bandwidth to accommodate the high input frequency. In general, ADC operating at high frequency is difficult to build and has fewer effective bits. The number of effective bits can be considered as the useful bits, which are fewer than the designed number of bits. Usually, the number of effective bits decreases at higher input frequency. In this approach the sampling frequency must be very accurate, which will be discussed in Section 6.15. Another problem is that it is difficult to build a narrow-band filter at a higher frequency, and usually this kind of filter has relatively high insertion loss.

In the down-converted approach the input frequency is converted to an IF, which is usually much lower than the input frequency. It is easy to build a narrow-band filter with low insertion loss and amplifiers at a lower frequency are less expensive. The mixer and the local oscillator must be used and they can be expensive and cause frequency errors.

Both approaches will be discussed in the following sections. Some considerations are common to both designs and these will be discussed first.

6.5   FIRST COMPONENT AFTER THE ANTENNA[6]

The first component following the antenna can be either a filter or an amplifier. If the antenna is integrated with an amplifier, the first component after the antenna is the amplifier. Both arrangements have advantages and disadvantages, which will be discussed in this section.

The noise figure of a receiver can be expressed as:[6]

$$F = F_1 + \frac{F_2 - 1}{G_1} + \frac{F_3 - 1}{G_1 G_2} + \cdots + \frac{F_N - 1}{G_1 G_2 \cdots G_{N1}} \tag{6.4}$$

where F_i and G_i ($i = 1, 2, \ldots N$) are the noise figure and gain of each individual component in the RF chain.

If the amplifier is the first component, the noise figure of the receiver is low and is approximately equal to the noise figure of the first amplifier, which can be less than 2 dB. The overall noise figure of the receiver caused by the second component, such as the filter, is reduced by the gain of the amplifier. The potential problem with this approach is that strong signals in the bandwidth of the amplifier may drive it into saturation and generate spurious frequencies.

If the first component is a filter, it can stop out-of-band signals from entering the input of the amplifier. If the filter only passes the C/A band, the bandwidth is around 2 MHz. A filter with 2 MHz bandwidth with a center frequency at 1575.42 MHz is considered high Q. Usually, the insertion loss of such a filter is relatively high, about 2–3 dB, and the filter is bulky. The receiver noise figure with the filter as the first component is about 2–3 dB higher than the previous arrangement. Usually, a GPS receiver without special interfering signals in the neighborhood uses an amplifier as the first component after the antenna to obtain a low noise figure.

6.6   SELECTING SAMPLING FREQUENCY AS A FUNCTION OF THE C/A CODE CHIP RATE

An important factor in selecting the sampling frequency is related to the C/A code chip rate. The C/A code chip rate is 1.023 MHz and the sampling frequency should not be a multiple number of the chip rate. In other words, the sampling should not be synchronized with the C/A code rate. For example, using a sampling frequency of 5.115 MHz (1.023 × 5) is not a good choice. With this sampling rate the time between two adjacent samples is 195.5 ns (1/5.115 MHz). This time resolution is used to measure the beginning of the C/A code. The corresponding distance resolution is 58.65 m ($195.5 \times 10^{-9} \times 3 \times 10^8$ m). This distance resolution is too coarse to obtain the desired accuracy of the user position. Finer distance resolution should be obtained from signal processing. With

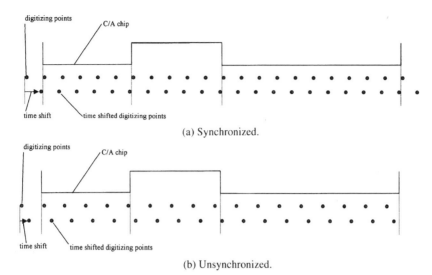

(a) Synchronized.

(b) Unsynchronized.

FIGURE 6.2 Relation between sampling rate and C/A code.

synchronized sampling frequency, it is difficult to obtain fine distance resolution. This phenomenon is illustrated as follows.

Figure 6.2 shows the C/A code chip rate and the sampled data points. Figures 6.2a and 6.2b show the synchronized and the unsynchronized sampling, respectively. In each figure there are two sets of digitizing points. The lower row is a time-shifted version of the top row.

In Figure 6.2a, the time shift is slightly less than 195.5 ns. These two sets of digitizing data are exactly the same as shown in this figure. This illustrates that shifting time by less than 195.5 ns produces the same output data, if the sampling frequency is synchronized with the C/A code. Since the two digitized data are the same, one cannot detect the time shift. As a result, one cannot derive finer time resolution (or distance) better than 195.5 ns through signal processing.

In Figure 6.2b the sampling frequency is lower than 5.115 MHz; therefore, it is not synchronized with the C/A code. The output data from the time-shifted case are different from the original data as shown in the figure. Under this condition, a finer time resolution can be obtained through signal processing to measure the beginning of the C/A code. This fine time resolution can be converted into finer distance resolution.

As discussed in Chapter 3, the Doppler frequency on the C/A code is about ± 6 Hz, which includes the speed of a high-speed aircraft. Therefore, the code frequency should be considered as in the range of $1.023 \times 10^6 \pm 6$ Hz. The sampling frequency should not be a multiple of this range of frequencies. In general, even in the sampling frequency is close to the multiple of this range of frequencies, the time-shifted data can be the same as the original data for a period of time. Under this condition, in order to generate a fine time resolution, a relatively long record of data must be used, which is not desirable.

6.7 SAMPLING FREQUENCY AND BAND ALIASING FOR REAL DATA COLLECTION[(10)]

If only one ADC is used to collect digitized data from one RF channel, the output data are often referred to as real data (in contrast to complex data). The input signal bandwidth is limited by the sampling frequency. If the sampling frequency is f_s, the unambiguous bandwidth is $f_s/2$. As long as the input signal bandwidth is less than $f_s/2$, the information will be maintained and the Nyquist sampling rate will be fulfilled. Although for many low-frequency applications the input signal can be limited to 0 to $f_s/2$, in general, the sampling frequency need not be twice the highest input frequency.

If the input frequency is f_i, and the sampling frequency is f_s, the input frequency is aliased into the baseband and the output frequency f_o is

$$f_o = f_i - nf_s/2 \quad \text{and} \quad f_o < f_s/2 \tag{6.5}$$

where n is an integer. The relationship between the input and the output frequency is shown in Figure 6.3.

When the input is from nf_s to $(2n + 1)f_s/2$, the frequency is aliased into the baseband in a direct transition mode, which means a lower input frequency translates into a lower output frequency. When the input is from $(2n + 1)f_s/2$ to $(n + 1)f_s$, it is aliased into the baseband in an inverse transition mode, which means a lower input frequency translates into a higher output frequency. Either case can be implemented if the frequency translation is properly monitored.

If the input signal bandwidth is Δf, it is desirable to have the minimum sampling frequency f_s higher than the Nyquist requirement of $2\Delta f$. Usually, $2.5\Delta f$ is used because it is impractical to build a filter with very sharp skirt (or a brick wall filter) to limit the out-of-band signals. Thus, for the C/A code the required minimum sampling rate is about 5 MHz. This sampling frequency is adequately separated from the undesirable frequency of 5.115 MHz. The sampling frequency must be properly selected. Figure 6.4a shows the desired frequency aliasing. The input band is placed approximately at the center of the output band and the input and output bandwidths are equal.

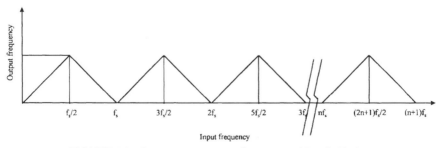

FIGURE 6.3 Input versus output frequency of band aliasing.

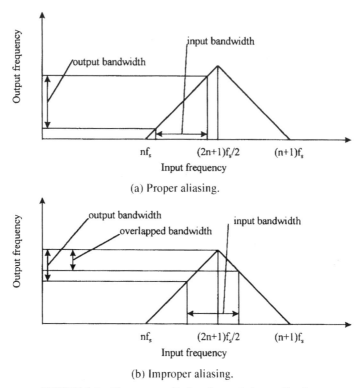

(a) Proper aliasing.

(b) Improper aliasing.

FIGURE 6.4   Frequency aliasing for real data collection.

Figure 6.4b shows improper frequency aliasing. In Figure 6.4b, the center frequency of the input signal does not alias to the center of the baseband. The frequency higher than $(2n + 1)f_s/2$ and the portion of the frequency lower than $(2n + 1)f_s/2$ are aliased on top of each other. Therefore, portion of the output band contains an overlapping spectrum, which is undesirable. When there is a spectrum overlapping in the output, the output bandwidth is narrower than the input bandwidth.

In order to alias the input frequency near the center of the baseband, the following relation must hold,

$$f_o = f_i - n(f_s/2) \approx f_s/4 \quad \text{and} \quad f_s > 2\Delta f \tag{6.6}$$

where Δf is the bandwidth of input signal. The first part of this equation is to put the aliasing signal approximately at the center of the output band. The second part states that the Nyquist sampling requirement must hold. If the frequency of the input signal f_i is known, this equation can be used to find the sampling frequency. Examples will be presented in Sections 6.8 and 6.9.

6.8 DOWN-CONVERTED RF FRONT END FOR REAL DATA COLLECTION[8-10]

In this section a down-converted approach to digitize the signal will be discussed. The IF and sampling frequency will be determined, followed by some general discussion. A set of hardware to collect data for user location calculation will be presented.

In this approach the input signal is down converted to an IF, then digitized by an ADC. In Equation (6.6) there are three unknowns: n, f_i, and f_s; therefore, the solutions are not unique. Many possible solutions can be selected to build a receiver. In the hardware design, the sampling frequency of $f_s = 5$ MHz is selected. From Equation (6.6) $f_i = \text{IF} = 5n + 1.25$ MHz, where n is an integer. The value of $n = 4$ is arbitrarily selected and the corresponding IF $= 21.25$ MHz, which can be digitized by an ADC.

Of course, one can choose $n = 0$ and down convert the input frequency to 1.25 MHz directly. In this approach the mixer generates more spurious frequencies. The input signal is down-converted from 0.25 to 2.25 MHz, which covers more than an octave bandwidth. An octave bandwidth means that the highest frequency in the band is equal to twice the lowest frequency in the band. A common practice in receiver design is to keep the IF bandwidth under an octave to avoid generation of in-band second harmonics.

There are many different ways to build an RF front end. The two important factors are the total gain and filter installations. Filters can be used to reject out-of-band signals and limit the noise bandwidth, but they add insertion loss. If multiple channels are used, such as in the I-Q channels, filters may increase the difficulty of amplitude and phase balancing. The locations of filters in a receiver affect the performance of the RF front end.

The personal computer–based ADC card discussed in Section 6.3 is used as the ADC. It requires about 100 mv input voltage or -10 dBm to activate all the bits. A net gain of 101 dB is required to achieve this level. If a digital scope is used as the ADC[8,9] because of the built-in amplifiers in the scope, it can digitize a rather weak signal. In this kind of arrangement, only about 90 dB gain is used.

Two RF front-end arrangements are shown in Figure 6.5. The major difference between Figures 6.5a and b is in the amplifiers. In Figure 6.5a amplifiers 2, 3, and 4 operate at IF, which costs less than amplifiers operating at RF. Filter 1 is used to limit the input bandwidth. Filter 2 is used to limit the spurious frequencies generated by the mixer, and filter 3 is used to limit noise generated by the three amplifiers. Although Figure 6.5a is the preferred approach, in actual laboratory experiments Figure 6.5b is used because of the availability of amplifiers.

In Figure 6.5b, the M/A COM ANP-C-114-5 antenna with amplifier is used. Amplifier 1 is an integrated part of the antenna with a 26 dB gain and a 2.5 dB noise figure. The bias T is used to supply 5-volt dc to the amplifier at the antenna. Filter 1 is centered at 1575.42 MHz with a 3 dB bandwidth of 3.4 MHz, which is wider than the desired value of 2 MHz. Amplifiers 2 and 3 provide a total of 60 dB gain. The frequency of the local oscillator is at 1554.17 MHz. The mixer-down converts the input frequency from 1575.42 to 21.25 MHz. In this frequency conversion, high input frequency transforms to high output frequency. The attenuator placed

(a) Most gain in IF stage.

(b) Most gain in RF stage.

FIGURE 6.5 Two arrangements of data collection.

between the mixer and the oscillator is used to improve impedance matching and it reduces the power to the mixer. After the mixer an IF amplifier with 24 dB of gain is used to further amplify the signal. Finally, filter 2 is used to reject spurious frequencies generated by the mixer and limit the noise bandwidth. Filter 2 has a center frequency of 21.25 MHz and bandwidth of 2 MHz. If filter 2 is not used all the noise will alias into the output band and be digitized by the ADC as shown in Figure 6.3. The overall gain from the four amplifiers is 110 dB $(26 + 30 + 30 + 24)$. Subtracting the insertion losses from the filters, bias T, and mixer, the gain is slightly over 100 dB. There is no filter after the mixer because it is not available in our laboratory. In general, the local oscillator and the digitizing clock are phase locked. In our special design, the local oscillator is a crystal based commercial oscillator and the digitizing clock is a signal generator and they are not phase locked. In a later design, the local oscillator and the digitizing clock are phase locked.

6.9 DIRECT DIGITIZATION FOR REAL DATA COLLECTION[8,9]

Direct digitization at RF is a straightforward approach. The only components required are amplifiers and two filters. The amplifiers must provide the desired

FIGURE 6.6 Arrangement for Direct Digitization.

RF gain. One filter is used after the first amplifier to limit out-of-band signal and the second filter is placed in front of the ADC to limit the noise bandwidth. A direct digitization arrangement is shown in Figure 6.6. In this arrangement the second filter is very important. Without this filter the noise in the collected data can be very high and it will affect signal detection.

In the direct sampling case, the frequency of the input signal is fixed; one must find the correct sampling frequency f_s to avoid band overlapping in the output. In this approach there are two unknowns: f_s and n, in Equation (6.6). An exact solution is somewhat difficult to obtain. However, the problem can be easily solved if the approximate sampling frequency is known.

Let us use an example to illustrate the operation. In this example, the input GPS L1 signal is at 1575.42 MHz, and the sampling frequency is about 5 MHz. First use Equation (6.5) with $f_o = 1.25$ MHz, $f_i = 1575.42$ MHz, and $f_s = 5$ MHz to find $n = 629.66$. Round off $n = 630$ and use $f_o = f_s/4$ in Equation (6.6). The result is $f_s = 5.009$ MHz and the center is aliased to 1.252 MHz.

In one arrangement a scope is used to collect digitized data because the personal computer–based ADC card cannot accommodate the frequency of the input signal. The scope has a specified bandwidth of dc-1000 MHz, but it can digitize a signal at 1600 MHz with less sensitivity. The scope can operate at 5 MHz, but the sampling frequency cannot be fine-tuned. If 5 MHz is used to sample the input frequency, the center frequency will be aliased to 420 kHz. Since the bandwidth of the C/A code is 2 MHz, there is band overlapping with the center frequency at 420 kHz as shown in Figure 6.7.

Actual GPS data were collected through this arrangement. Although there is band overlapping, the data could still be processed, because the overlapping range is close to the edge of the signal where the spectrum density is low. In this arrangement, the overall amplification is reduced because the scope can digitize weak signals.

In another arrangement, an experimental ADC built by TRW is used. The ADC can sample only between approximately 80 to 120 MHz, limited by the circuit around the ADC. In order to obtain digitized data at 5 MHz, the output from the ADC is decimated. For example, if the ADC operates at 100 MHz and one data point is kept out of every 20 data points, the equivalent sampling rate is 5 MHz. The actual sampling frequency is selected to be 5.161 MHz; the input signal is aliased to 1.315 MHz, which is close to the center of the output band at about 1.29 (5.161/4) MHz.

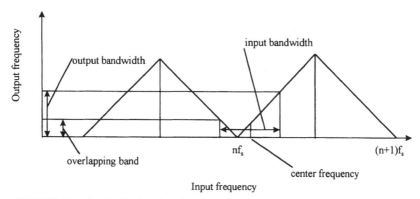

n=315
f_s=5 MHz
f_c=1575.42 MHz

FIGURE 6.7 Band aliasing for direct sampling of the L1 frequency at 5 MHz.

With today's technology, it is easier to build a down-converted approach, but the direct digitization is attractive for its simplicity. There is another advantage for direct digitization, which is to alias more than one desired signal into the baseband. This approach will be discussed in Section 6.11.

6.10 IN-PHASE (I) AND QUADRANT-PHASE (Q) DOWN CONVERSION[10]

In many commercial GPS receivers, the input signal is down converted into I-Q channels. The data collected through this approach are complex and the two sets of data are often referred to as real and imaginary. Since there are two channels, the Nyquist sampling is $f_s = \Delta f$. A common practice is to choose $f_s > 1.25\Delta f$ to accommodate the skirt of the filter. The relation between the input and the output frequencies is

$$f_o = f_i - nf_s \quad \text{and} \quad f_o < f_s \tag{6.7}$$

where n is a positive integer. The relation between the input and output band is shown in Figure 6.8. In the I-Q channel digitization, as long as $\Delta f < f_s$ there is no spectrum overlapping in the output baseband.

A common frequency selection is shown in Figure 6.8. The center of the output frequency is placed at zero. This approach usually can be achieved only through a down-converted design, and the input frequency f_i is usually set to zero or to a multiple of the sampling frequency f_s. In this arrangement, the input frequency is divided into two equal bands. The lower input frequency is aliased to a higher frequency in the baseband and the higher input frequency is

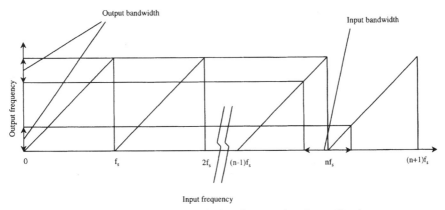

FIGURE 6.8   Frequency aliasing for complex data collection.

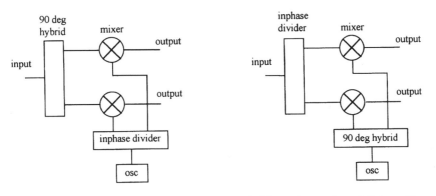

(a) Ninety-degree phase shift in input.   (b) Ninety-degree phase shift in oscillator.

FIGURE 6.9   I and Q down converter.

aliased to a lower frequency as shown in Figure 6.8. This phenomenon affects
the data conversion procedure discussed in Section 6.14.

There are two ways to build an I-Q down converter as shown in Figure 6.9.
In Figure 6.9a, a 90-degree phase shift is introduced in the input circuit. In
Figure 6.9b, the 90-degree phase shift is introduced in the oscillator circuit. Both
approaches are popularly used. If one wants the output frequency to be zero,
the local oscillator is often selected as the input signal or at 1575.42 MHz.
For a wideband receiver the I-Q approach can double the input bandwidth with
the same sampling frequency. Since the GPS receiver bandwidth is relatively
narrow, this approach is not needed to improve the bandwidth. This approach
uses more hardware because one additional channel is required. The amplitude
and phase of the two outputs are difficult to balance accurately. From the software
receiver point of view, there is no obvious advantage of using an I-Q channel
down converter.

Actual complex data with zero center frequency have been collected. Since the acquisition and tracking programs used in this book can process only real data, the complex data are converted into real data through software. The details will be presented in Section 6.14.

6.11 ALIASING TWO OR MORE INPUT BANDS INTO A BASEBAND[(8,9)]

If one desires to receive signals from two separate bands, the straightforward way is to use two mixers and two local oscillators to convert the two input bands into desired IF ranges such as adjacent bands, combine, and digitize them. If direct digitization is used and the correct sampling frequency is selected, two input bands can be aliased into a desired output band. Figure 6.10 shows the arrangement of aliasing two input bands into the baseband for a real data collection system.

The aliased signals in the baseband can be either overlapped or separated. In Figure 6.10 the two signals in the baseband are separated. Separated bands have better signal-to-noise ratio because the noise in the two bands is separated. Separated spectra occupy a wider frequency range and require a higher sampling rate. The overlapped bands have lower signal-to-noise ratio because the noise of two bands is added together. Overlapped spectra occupy a narrower bandwidth and require a lower sampling rate. The aliasing scheme can be used to fold more than two input bands together and it also applies to complex data collection. Before the input bands can be folded together, analog filters must be used to properly filter the desired input bands.

Let us use three examples to illustrate the band aliasing idea. In the first example the two P code channels from both L1 and L2 frequencies are aliased into two separated bands in the baseband. Since the P code has a bandwidth of approximately 20 MHz, two P code bands will occupy 40 MHz. The minimum sampling frequency to cover these bands is about 100 MHz (40×2.5), if 2.5 rather than the Nyquist sampling rate of 2 is used as the minimum required sampling rate. The two input frequency ranges are 1565.42–1585.42 MHz and

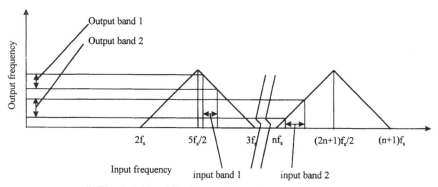

FIGURE 6.10 Aliasing two input bands to baseband.

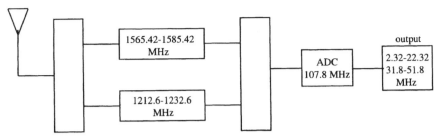

FIGURE 6.11   Aliasing of L1 and L2 bands to baseband.

1217.6–1237.6 MHz for L1 and L2 bands, respectively. It is tedious to solve the desired sampling frequency. It is easier to solve the sampling frequency f_s through trial and error. The output frequency can be obtained from Equation (6.5) by increasing the sampling frequency in 100 kHz steps starting from 100 MHz. When the two output frequency ranges are properly aliased into the baseband, the sampling frequency is the desired one. There are many sampling frequencies that can fulfill this requirement. One of the lower sampling frequencies is arbitrarily selected, such as $f_s = 107.8$ MHz. With this sampling frequency, the L1 band is aliased to 2.32–22.32 MHz, and the L2 band is aliased to 31.8–51.8 MHz. These two bands are not overlapped and they are within the baseband of 0–53.9 ($f_s/2$) MHz. Figure 6.11 shows such an arrangement.

In the second example, the same two bands are allowed to partially overlap after they are aliased into the baseband. The sampling frequency can be found through the same approach. The output bandwidth can be as narrow as 20 MHz, when the input bands are totally overlapped. Therefore, the minimum sampling frequency is about 50 MHz. if the sampling frequency $f_s = 57.8$ MHz, the L1 frequency aliases to 3.8–23.8 MHz and L2 aliases to 4.82–24.82 MHz and they are partially overlapped. The output bandwidth is from 0 to 28.9 ($f_s/2$) MHz.

The third example is to alias the C/A band of the GPS signal and Russia's GLONASS signal into separate bands in the baseband. The GLONASS is Russia's standard position system, which is equivalent to the GPS system of the United States. The GLONASS uses bi-phase coded signals with 24 channels at frequencies $1602 + 0.5625\,n$ where n is an integer representing the channel number. A future plan to revise the frequency channels used will eliminate a number of the upper channels. Therefore, only the 1–12 channels will be considered. The center frequency of these 12 channels is at 1605.656 (1602 + 0.5625 × 6.5) MHz. The total bandwidth is 7.3125 (0.5625 × 13) MHz. For simplicity let us use a 7.5 MHz bandwidth, the signal frequency is approximately 1601.9–1609.4 MHz. Including the C/A code of the GPS signal, the overall bandwidth is about 9.5 MHz. The minimum sampling frequency can be 23.75 (9.5 × 2.5) MHz. If $f_s = 35.1$ MHz, the GPS channel is aliased to 12.47–14.47 MHz, the GLONASS to 4.8–12.35 MHz, and both signals are within the 0 to 17.55 ($f_s/2$) MHz baseband. Hardware has been built to test this idea. The collected data contain both the GPS and the GLONASS signals.

6.12   QUANTIZATION LEVELS[11-13]

As discussed in Section 5.3, GPS is a CDMA signal. In order to receive the maximum of signals, it is desirable to have comparable signal strength from all visible satellites at the receiver. Under this condition, the dynamic range of a GPS receiver need not be very high. An ADC with a few bits is relatively easy to fabricate and may operate at high frequency. Another advantage of using fewer bits is that it is easier to process the digitized data, especially when they are processed through hardware. The disadvantage of using fewer bits is the degradation of the signal-to-noise ratio. Spilker[11] indicated that a 1-bit ADC degrades the signal-to-noise ratio by 1.96 dB and a 2-bit ADC degrades the signal-to-noise by 0.55 dB. Many commercial GPS receivers use only 1- or 2-bit ADCs.

Chang[12] claims that the degradation due to the number of bits of the ADC is a function of input signal-to-noise ratio and sampling frequency. Low signal-to-noise ratio signal sampled at a higher frequency causes less degradation in a receiver. The GPS signal should belong to the low signal-to-noise ratio because the signal is below the noise. At a Nyquist sampling rate, the minimum degradation is about 3.01 and 0.72 dB for 1- and 2-bit quantizers, respectively. At five times the Nyquist sampling rate, the minimum degradation is 2.18 and .60 dB for 1- and 2-bit quantizers, respectively. These values are slightly higher than the results in reference 11.

The only time that a high number of bits in ADC is required in a GPS receiver is to build a receiver with antijamming capability. Usually, the jamming signal is much stronger than the desired GPS signals. An ADC with a small number of bits will be easily saturated by the jamming signal. Under this condition, the GPS signals might be masked by the jamming signal and the receiver cannot detect the desired signals. If an ADC with a large number of bits is used, the dynamic range of the receiver is high. Under this condition, the jamming signal can still disturb the operation; however, the weak GPS signals are preserved in the digitized data. If proper digital signal processing is applied, the GPS signals should be recovered. This problem can be considered in the frequency domain. Assume that there are two signals, a strong one and a weak one, and they are close in frequency. In order to receive both signals, the receiver must have enough instantaneous dynamic range, which is defined as the capability to receive strong and weak signals simultaneously. If the ADC does not have enough dynamic range, the weak signal may not be received. Reference 12 provides more information on this subject. In Section 12.9, the number of ADC bit required under jamming is discussed.

6.13   HILBERT TRANSFORM[10]

In this book a single channel is used to collect data and the software is written to process real data. If a software receiver is designed to process complex data and only real data are available, the real data can be changed to complex data through the Hilbert transform.[10] A detailed discussion on the Hilbert transform will not be included. Only the procedure will be presented here.

First the Hilbert transform from Matlab will be presented. The approach is through discrete Fourier transform (DFT) or fast Fourier transform (FFT). The following steps are taken:

1. The DFT result can be written as:

$$X(k) = \sum_{n=0}^{N-1} x(n)e^{\frac{-j2\pi nk}{N}} \tag{6.8}$$

where $x(n)$ is the input data, $X(k)$ is the output frequency components, $k = 0, 1, 2, \ldots, N - 1$, and $n = 0, 1, 2, \ldots, N - 1$. Since the input data are real, the frequency components have the following properties:

$$X(k) = X(N - k)^* \quad \text{for} \quad k = 1 \sim \frac{N}{2} - 1 \tag{6.9}$$

where * represents complex conjugate. If the input data are complex the relationship in this equation does not exist.

2. Find a new set of frequency components $X_1(k)$. They have the following values:

$$X_1(0) = \frac{1}{2}X(0)$$

$$X_1(k) = X(k) \quad \text{for} \quad k = 1 \sim \frac{N}{2} - 1$$

$$X_1\left(\frac{N}{2}\right) = \frac{1}{2}X\left(\frac{N}{2}\right)$$

$$X_1(k) = 0 \quad \text{for} \quad k = \frac{N}{2} + 1 \sim N - 1 \tag{6.10}$$

These new frequency components also have N values from $k = 0$ to $N - 1$.

3. The new data $x_1(n)$ in time domain can be obtained from the inverse DFT of the $X_1(k)$ as:

$$x_1(n) = \frac{1}{N} \sum_{k=0}^{N-1} X_1(k)e^{\frac{j2\pi nk}{N}} \tag{6.11}$$

From this approach, if there are N points of real input data, the result will be N points of complex data. Obviously, additional information is generated through this operation. This is caused by padding the $X_1(k)$ values with zeros as shown in Equation (6.10). Padding with zeros in the frequency domain is equivalent to interpolating in the time domain.[10]

The above method generates N points of complex data from N points of real data. The new data may increase the processing load without gaining significant receiver performance improvement. Therefore, another approach is presented, which is similar to the above Matlab approach, but generates only $N/2$ points of complex data. In taking real digitized data the sampling frequency $f_s \approx 2.5\Delta f$ is used and the input signal is aliased close to the center of the baseband. Under this condition, the frequency component $X(N/2)$ should be very small. The following steps can be taken to obtain complex data:

1. The first step is the same as step 1 (Equation 6.8) in the Matlab approach to take the FFT of the input signal.
2. The new $X_1(k)$ can be obtained as

$$X_1(k) = X(k) \quad \text{for} \quad k = 0, 1, 2, \ldots, \frac{N}{2} - 1 \qquad (6.12)$$

Therefore, only half of the frequency components are kept.
3. The new data in time domain can be obtained as

$$x_1(n) = \frac{2}{N} \sum_{k=0}^{\frac{N}{2}-1} X_1(k) e^{\frac{j4\pi nk}{N}} \qquad (6.13)$$

The final results are $N/2$ points of complex data in the time domain and they contain the same information as the N points of real data. These data cover the same length of time; therefore, the equivalent sampling rate of the complex data is $f_{s1} = f_s/2$. The argument is reasonable because for complex data the Nyquist sampling rate is $f_{s1} = \Delta f$.

6.14 CHANGE FROM COMPLEX TO REAL DATA

In this section changing complex data to real data will be discussed. The approach basically reverses the operation in Section 6.13. However, the IF of the down conversion is very important in this operation. The detail operation depends on this frequency. One of the common I-Q converter designs is to make the IF at zero frequency as shown in Figure 6.8. Under this condition, the center frequency of the input signal is determined by the Doppler shift. For this arrangement the following steps can be taken:

1. Take the DFT of $x(n)$ to generate $X(k)$ as shown in Equation (6.8),

$$X(k) = \sum_{n=0}^{N-1} x(n) e^{\frac{-j2\pi nk}{N}} \qquad (6.14)$$

where $x(n)$ is complex, $k = 0, 1, 2, \ldots, N - 1$, and $n = 0, 1, 2, \ldots, N - 1$.

2. Generate a new set of frequency components $X_1(k)$ from $X(k)$ as

$$X_1(k) = X\left(\frac{N}{2} + k\right) \quad \text{for} \quad k = 0 \sim \frac{N}{2} - 1$$

$$X_1(k) = X\left(-\frac{N}{2} + k\right) \quad \text{for} \quad k = \frac{N}{2} \sim N - 1 \qquad (6.15)$$

In Figure 6.8, the lower input frequency is converted into a higher output frequency and the higher frequency is converted into a lower frequency. This operation puts the two separate bands in Figure 6.8 into the correct frequency range as the input signals. Or one can consider that it shifts the center of the input signal from zero to $f_s/2$. If the IF is not at zero frequency a different shift is required. If the IF of the I-Q channels is at $f_s/2$, no shift is required and this step can be omitted, because the input signal will not split into two separate bands.

3. Generate additional frequency components for $X_1(k)$ as

$$X_1(N) = 0$$

$$X_1(N + k) = X_1(N - k)^* \quad \text{for} \quad k = 1 \sim N - 1 \qquad (6.16)$$

Including the results from Equation (6.15) there are total $2N$ frequency components from $k = 0 \sim 2N - 1$.

4. The final step is to find the new data in the time domain through inverse FFT as

$$x_1(n) = \frac{1}{2N} \sum_{k=0}^{2N-1} X_1(k) e^{\frac{j\pi nk}{N}} \qquad (6.17)$$

The N points of complex data generate $2N$ points of real data and they contain the same amount of information. These $2N$ points cover the same time period; therefore, the equivalent sampling frequency is doubled or $f_{s1} = 2f_s$.

Actual complex data are collected from satellites with I-Q channels of zero IF from Xetron Corporation. The sampling frequency is 3.2 MHz and the Nyquist bandwidth is also 3.2 MHz. The operations from Equations (6.14) to (6.17) are used to change these data to real data with IF = 1.6 MHz. The number of real data is double the number of complex data and the equivalent sampling frequency is 6.4 MHz. These data are processed and the user position has been found.

6.15   EFFECT OF SAMPLING FREQUENCY ACCURACY

Although the sampling frequency discussed in this chapter is given a certain mathematical value, the actual frequency used in the laboratory usually has limited accuracy. The effect of this inaccuracy will be discussed as follows.

The first impact to be discussed is on the center frequency of the digitized signal. For the down-converted approach, the sampling frequency inaccuracy causes a small error in the output frequency. For example, if the IF is at 21.25 MHz and the sampling frequency is at 5 MHz, the digitized output should be at 1.25 MHz. This value can be found from Equation (6.5) and the corresponding $n = 4$. If the true sampling frequency $f_s = 5,000,100$ Hz, there is an error frequency of 100 Hz. The center frequency of the digitized signal is at 1,249,600 Hz. The error frequency is 400 Hz, which is four times the error in the sampling frequency because in this case $n = 4$. This frequency error will affect the search range of the acquisition procedure.

For a direct digitization system, the error in the sampling frequency will create a larger error in the output frequency. As discussed in Section 6.9, if the sampling frequency $f_s = 5,161,000$ Hz, with the input signal at 1575.42 MHz, the output will be 1.315 MHz with $n = 305$ from Equation (6.5). If $f_s = 5,161,100$ Hz, which is off by 100 Hz from the desired value, the input will be aliased to 1,284,500 Hz, which is off by 30,500 Hz because $n = 305$. The frequency error will have a severe impact on the acquisition procedure. Therefore, for the direct digitization approach the accuracy of the sampling frequency is rather important.

The second impact of inaccurate sampling frequency is on the processing of the signal. In a software receiver, both the acquisition and tracking programs take the sampling frequency as input. If the actual sampling frequency is off by too much the acquisition program might not cover the anticipated frequency range and would not find the signal. For a small error in sampling frequency, it will not have a significant effect on the acquisition and the tracking programs. For example, the sampling frequency of 3.2 MHz used to collect the complex data must be off slightly because all the Doppler frequencies calculated are of one sign. If the correct sampling frequency is used in the program, the Doppler frequency should have both positive and negative values because the receiving antenna is stationary. From these experimental results, no obvious adverse effect on the acquisition and tracking is discovered due to the slight inaccuracy of the sampling frequency.

The most important effect of sampling frequency inaccuracy may be the pseudorange measured. The differential pseudorange is measured by sampling time, which will be discussed in Section 9.9. If the sampling frequency is not accurate, the sampling time will be off. The inaccuracy in the pseudorange will affect the accuracy of the user position measured.

6.16 SUMMARY

This chapter discusses the front end of a GPS receiver. The antenna should have a broad beam to receive signals from the zenith to the horizon. It should be right handed circularly polarized to reduce reflected signals. The overall gain of the amplifier chain depends on the input voltage of the ADC. Usually the overall gain is about 100 dB. The input signal can either be down converted and then digitized or directly digitized without frequency translation. Although direct digitization seems to have some advantages, with today's technology a down conversion approach is simpler to build. It appears that I-Q channel down conversion does not have much advantage over a single channel conversion for a software GPS receiver. Direct digitization can be used to alias several narrow input signals into the same baseband. Several experimental setups to collect data are presented. The number of quantization bits is discussed. One or two bits may be enough for GPS application with degradation of receiver sensitivity. If antijamming is of concern, a large number of bits needed. The conversion of data from real to complex and from complex to real are discussed. Finally, the impact of sampling frequency accuracy is discussed.

REFERENCES

1. Van Dierendonck, A. J., "GPS receivers," Chapter 8 in Parkinson, B. W., Spilker, J. J. Jr., *Global Positioning System: Theory and Applications*, vols. 1 and 2, American Institute of Aeronautics and Astronautics, 370 L'Enfant Promenade, SW, Washington, DC, 1996.

2. Bahl, I. J., Bhartia, P., *Microstrip Antennas*, Artech House, Dedham, MA, 1980.

3. Johnson, R. C., Jasik, H., eds., *Antenna Engineering handbook*, McGraw-Hill, New York, 1984.

4. Braasch, M. S., "Multipath effects," Chapter 14 in Parkinson, B. W., Spilker, J. J. Jr., *Global Positioning System: Theory and Applications*, vols. 1 and 2, American Institute of Aeronautics and Astronautics, 370 L'Enfant Promenade, SW, Washington, DC, 1996.

5. Spilker, J. J. Jr., "GPS signal structure and theoretical performance," Chapter 3 in Parkinson, B. W., Spilker, J. J. Jr., *Global Positioning System: Theory and Applications*, vols. 1 and 2, American Institute of Aeronautics and Astronautics, 370 L'Enfant Promenade, SW, Washington, DC, 1996.

6. Tsu, J. B. Y., *Microwave Receivers with Electronic Warfare Applications*, Wiley, New York, 1986.

7. *GaGe Scope Technical Reference and User's Guide*, GaGe Applied Sciences, 5610 Bois Franc, Montreal, Quebec, Canada.

8. Tsui, J. B. Y., Akos, D. M., "Comparison of direct and downconverted digitization in GPS receiver front end designs," *IEEE MTT-S International Microwave Symposium*, pp. 1343–1346, San Francisco, CA, June 17–21, 1996.

9. Akos, D. M., Tsui, J. B. Y., "Design and implementation of a direct digitization GPS receiver front end," *IEEE Trans. Microwave Theory and Techniques*, vol. 44, no. 12, pp. 2334–2339, December 1996.

10. Tsui, J. B. Y., *Digital Techniques for Wideband Receivers*, Artech House, Boston, 1995.

11. Spilker, J. J. Jr., *Digital Communication by Satellite*, pp. 550–555, Prentice Hall, Englewood Cliffs, NJ, 1995.

12. Chang, H., "Presampling filtering, sampling and quantization effects on the digital matched filter performance," *Proceedings of International Telemetering Conference*, pp. 889–915, San Diego, CA, September 28–30, 1982.

13. Moulin, D., Solomon, M. N., Hopkinson, T. M., Capozza, P. T., Psilos, J., "High performance RF-to-digital translators for GPS anti-jam applications," *ION GPS-98*, pp. 233–239, Nashville, TN, September 15–18, 1998.

CHAPTER SEVEN

Acquisition of GPS C/A Code Signals

7.1 INTRODUCTION

In order to track and decode the information in the GPS signal, an acquisition method must be used to detect the presence of the signal. Once the signal is detected, the necessary parameters must be obtained and passed to a tracking program. From the tracking program information such as the navigation data can be obtained. As mentioned in Section 3.5, the acquisition method must search over a frequency range of ±10 kHz to cover all of the expected Doppler frequency range for high-speed aircraft. In order to accomplish the search in a short time, the bandwidth of the searching program cannot be very narrow. Using a narrow bandwidth for searching means taking many steps to cover the desired frequency range and it is time consuming. Searching through with a wide bandwidth filter will provide relatively poor sensitivity. On the other hand, the tracking method has a very narrow bandwidth; thus high sensitivity can be achieved.

In this chapter three acquisition methods will be discussed: conventional, fast Fourier transform (FFT), and delay and multiplication. The concept of acquiring a weak signal using a relatively long record will also be discussed. In this edition, two chapters (10 and 11) are devoted to weak signals. The FFT method and the conventional method generate the same results. The FFT method can be considered as a reduced computational version of the conventional method. The delay and multiplication method can operate faster than the FFT method with inferior performance, that is, lower signal-to-noise ratio. In other words, there is a trade-off between these two methods that is speed versus sensitivity. If the signal is

Fundamentals of Global Positioning System Receivers: A Software Approach, Second Edition, by James Bao-Yen Tsui
ISBN 0-471-70647-7 Copyright © 2005 John Wiley & Sons, Inc.

strong, the fast, low-sensitivity acquisition method can find it. If the signal is weak, the low-sensitivity acquisition will miss it but the conventional method will find it. If the signal is very weak, the long data length acquisition should be used. A proper combination of these approaches should achieve fast acquisition. However, a discussion on combining these methods is not included in this book.

Once the signals are found, two important parameters must be measured. One is the beginning of the C/A code period and the other one is the carrier frequency of the input signal. A set of collected data usually contains signals of several satellites. Each signal has a different C/A code with a different starting time and different Doppler frequency. The acquisition method is to find the beginning of the C/A code and use this information to despread the spectrum. Once the spectrum is despread, the output becomes a continuous wave (cw) signal and its carrier frequency can be found. The beginning of the C/A code and the carrier frequency are the parameters passed to the tracking program.

In this and the following chapters, the data used are collected from the down-converted system. The intermediate frequency (IF) is at 21.25 MHz and sampling frequency is 5 MHz. Therefore, the center of the signal is at 1.25 MHz. The data are collected through a single-channel system by one analog-to-digital converter (ADC). Thus, the data are considered real in contrast to complex data. The hardware arrangement is discussed in the previous chapter.

7.2   ACQUISITION METHODOLOGY

One common way to start an acquisition program is to search for satellites that are visible to the receiver. If the rough location (say Dayton, Ohio, U.S.A.) and the approximate time of day are known, information is available on which satellites are available, such as on some Internet locations, or can be computed from a recently recorded almanac broadcast. If one uses this method for acquisition, only a few satellites (a maximum of 11 satellites if the user is on the earth's surface) need to be searched. In Section 12.2 it is illustrated that coarse information for acquisition can be obtained from almanac data. However, in case the wrong location or time is provided, the time to locate the satellites increases as the acquisition process may initially search for the wrong satellites.

The other method to search for the satellites is to perform acquisition on all the satellites in space; there are 24 of them. This method assumes that one knows which satellites are in space. If one does not even know which satellites are in space and there could be 32 possible satellites, the acquisition must be performed on all the satellites. This approach could be time consuming; a fast acquisition process is always preferred.

The conventional approach to perform signal acquisition is through hardware in the time domain. The acquisition is performed on the input data in a continuous manner. Once the signal is found, the information will immediately pass to the tracking hardware. In some receivers the acquisition can be performed on many satellites in parallel.

When a software receiver is used, the acquisition is usually performed on a block of data. When the desired signal is found, the information is passed on to the tracking program. If the receiver is working in real time, the tracking program will work on data currently collected by the receiver. Therefore, there is a time elapse between the data used for acquisition and the data being tracked. If the acquisition is slow, the time elapse is long and the information passed to the tracking program obtained from old data might be out-of-date. In other words, the receiver may not be able to track the signal. If the software receiver does not operate in real time, the acquisition time is not critical because the tracking program can process stored data. It is desirable to build a real-time receiver; thus, the speed of the acquisition is very important.

7.3   MAXIMUM DATA LENGTH FOR ACQUISITION

Before the discussion of the actual acquisition methods, let us find out the length of the data used to perform the acquisition. The longer the data record used the higher the signal-to-noise ratio that can be achieved. Using a long data record requires increased time of calculation or more complicated chip design if the acquisition is accomplished in hardware. There are two factors that can limit the length of the data record. The first one is whether there is a navigation data transition in the data. The second one is the Doppler effect on the C/A code.

Theoretically, if there is a navigation data transition, the transition will spread the spectrum and the output will no longer be a cw signal. The spectrum spread will degrade the acquisition result. Since navigation data is 20 ms or 20 C/A code long, the maximum data record for acquisition should be 10 ms. However, data longer than 10 ms are also used for acquisition, which will be discussed in Sections 12.5 and 12.8. The reasoning is as follows. In 20 ms of data at most there can be only one data transition. If one takes the first 10 ms of data and there is a data transition, the next 10 ms will not have one.

In actual acquisition, even if there is a phase transition caused by a navigation data in the input data, the spectrum spreading is not very wide. For example, if 10 ms of data are used for acquisition and there is a phase transition at 5 ms, the width of the peak spectrum is about 400 Hz ($2/(5 \times 10^{-3})$). This peak usually can be detected, therefore, the beginning of the C/A code can be found. However, under this condition the carrier frequency is suppressed. Carrier frequency suppression is well known in bi-phase shift keying (BPSK) signal. In order to simplify the discussion let us assume that there is no navigation data phase transition in the input data. The following discussion will be based on this assumption.

Since the C/A code is 1 ms long, it is reasonable to perform the acquisition on at least 1 ms of data. Even if only one millisecond of data is used for acquisition, there is a possibility that a navigation data phase transition may occur in the data set. If there is a data transition in this set of data, the next 1 ms of data will not have a data transition. Therefore, in order to guarantee there is no data transition in the data, one should take two consecutive data sets to perform acquisition. This data length is up to a maximum of 10 ms. If one takes two consecutive 10 ms of data to perform acquisition, it is guaranteed that in one data set there is no transition. In reality, there is a good probability that a data record more than 10 ms long does not contain a data transition.

The second limit of data length is from the Doppler effect on the C/A code. If a perfect correlation peak is 1, the correlation peak decreases to 0.5 when a C/A code is off by half a chip. This corresponds to 6 dB decrease in amplitude. Assume that the maximum allowed C/A code misalignment is half a chip (0.489 us) for effective correlation. The chip frequency is 1.023 MHz and the maximum Doppler shift expected on the C/A code is 6.4 Hz as discussed in Section 3.6. It takes about 78 ms $(1/(2 \times 6.4))$ for two frequencies different by 6.4 Hz to change by half a chip. This data length limit is much longer than the 10 ms; therefore, 10 ms of data should be considered as the longest data used for acquisition, in this book, although other researchers used longer data.

7.4 FREQUENCY STEPS IN ACQUISITION

Another factor to be considered is the carrier frequency separation needed in the acquisition. As discussed in Section 3.6, the Doppler frequency range that needs to be searched is ± 10 kHz. It is important to determine the frequency steps needed to cover this 20 kHz range. The frequency step is closely related to the length of the data used in the acquisition. When the input signal and the locally generated complex signal are off by 1 cycle there is no correlation. When the two signals are off less than 1 cycle there is partial correlation. It is arbitrarily chosen that the maximum frequency separation allowed between the two signals is 0.5 cycle. If the data record is 1 ms, a 1 kHz signal will change 1 cycle in the 1 ms. In order to keep the maximum frequency separation at 0.5 cycle in 1 ms, the frequency step should be 1 kHz. Under this condition, the furthest frequency separation between the input signal and the correlating signal is 500 Hz or 0.5 Hz/ms and the input signal is just between two frequency bins. If the data record is 10 ms, a searching frequency step of 100 Hz will fulfill this requirement. A simpler way to look at this problem is that the frequency separation is the inverse of the data length, which is the same as a conventional FFT result.

The above discussion can be concluded as follows. When the input data used for acquisition is 1 ms long, the frequency step is 1 kHz. If the data is 10 ms long, the frequency is 100 Hz. From this simple discussion, it is obvious that the number of operations in the acquisition is not linearly proportional to the total number of data points. When the data length is increased from 1 ms to 10 ms, the number of operations required in the acquisition is increased more than 10 times. The length of data is increased 10 times and the number of frequency bins is also increased 10 times. Therefore, if the speed of acquisition is important, the length of data should be kept at a minimum. The increase in operation depends on the actual acquisition methods, which are discussed in the following sections.

7.5   C/A CODE MULTIPLICATION AND FAST FOURIER TRANSFORM (FFT)

The basic idea of acquisition is to despread the input signal and find the carrier frequency. If the C/A code with the correct phase is multiplied on the input signal, the input signal will become a cw signal as shown in Figure 7.1. The top plot is the input signal, which is a radio frequency (RF) signal phase coded by a C/A code. It should be noted that the RF and the C/A code are arbitrarily chosen for illustration and they do not represent a signal transmitted by a satellite. The second plot is the C/A code, which has values of ±1. The bottom plot is a cw signal representing the multiplication result of the input signal and the C/A code, and the corresponding spectrum is no longer spread, but becomes a cw signal. This process is sometimes referred to as stripping the C/A code from the input.

Once the signal becomes a cw signal, the frequency can be found from the FFT operation. If the input data length is 1 ms long, the FFT will have a frequency resolution of 1 kHz. A certain threshold can be set to determine whether a frequency component is strong enough. The highest-frequency component crossing the threshold is the desired frequency. If the signal is digitized at 5 MHz, 1 ms of data contain 5,000 data points. A 5,000-point FFT generates 5,000 frequency components. However, only the first 2,500 of the 5,000 frequency components contain useful information. The last 2,500 frequency components are the complex conjugate of the first 2,500 points. The frequency resolution is 1 kHz; thus, the total frequency range covered by the FFT is 2.5 MHz, which is half of the sampling frequency. However, the frequency range of interest is only 20 kHz,

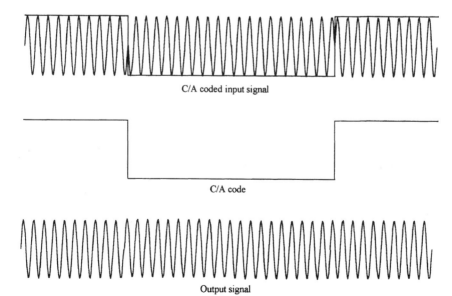

C/A coded input signal

C/A code

Output signal

FIGURE 7.1   C/A coded input signal multiplied by C/A code.

not 2.5 MHz. Therefore, one might calculate only 21 frequency components separated by 1 kHz using the discrete Fourier transform (DFT) to save calculation time. This decision depends on the speed of the two operations.

Since the beginning point of the C/A code in the input data is unknown, this point must be found. In order to find this point, a locally generated C/A code must be digitized into 5,000 points and multiply the input point by point with the input data. FFT or DFT is performed on the product to find the frequency. In order to search for 1 ms of data, the input data and the locally generated one must slide 5,000 times against each other. If the FFT is used, it requires 5,000 operations and each operation consists of a 5,000-point multiplication and a 5,000-point FFT. The outputs are 5,000 frames of data and each contains 2,500 frequency components because only 2,500 frequency components provide information and the other 2,500 components provide redundant information. There are a total of 1.25×10^7 ($5,000 \times 2,500$) outputs in the frequency domain. The highest amplitude among these 1.25×10^7 outputs can be considered as the desired result if it also crosses the threshold. Searching for the highest component among this amount of data is also time consuming. Since only 21 frequencies of the FFT outputs covering the desired 20 kHz are of interest, the total outputs can be reduced to 105,000 ($5,000 \times 21$). From this approach the beginning point of the C/A code can be found with a time resolution of 200 ns (1/5 MHz) and the frequency resolution of 1 kHz.

If 10 ms of data are used, it requires 5,000 operations because the signal only needs to be correlated for 1 ms. Each operation consists of a 50,000-point multiplication and a 50,000 FFT. There are a total of 1.25×10^8 ($5,000 \times 25,000$) outputs. If only the 201 frequency components covering the desired 20 kHz are considered, one must sort through 1,005,000 ($5,000 \times 201$) outputs. The increase in operation time from 1 ms to 10 ms is quite significant. The time resolution for the beginning of the C/A code is still 200 ns but the frequency resolution improves to 100 Hz.

7.6   TIME DOMAIN CORRELATION

The conventional acquisition in a GPS receiver is accomplished in hardware. The hardware is basically used to perform the process discussed in the previous section. Suppose that the input data is digitized at 5 MHz. One possible approach is to generate a 5,000-point digitized data of the C/A code and multiply them with the input signal point by point. The 5,000-point multiplication is performed every 200 ns. Frequency analysis such as a 5,000-point FFT is performed on the products every 200 ns. Figure 7.2 shows such an arrangement. If the C/A code and the input data are matched, the FFT output will have a strong component. As discussed in the previous section, this method will generate 1.25×10^7 ($5,000 \times 2,500$) outputs. However, only the outputs within the proper frequency range of ±10 kHz will be sorted. This limitation simplifies the sorting processing.

Another way to implement this operation is through DFT. The locally generated local code is modified to consist of a C/A code and an RF. The RF is

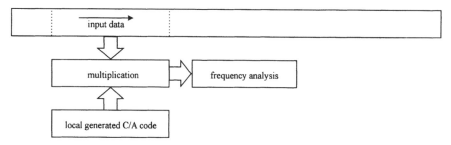

FIGURE 7.2 Acquisition with C/A code and frequency analysis.

complex and can be represented by $e^{j\omega t}$. The local code is obtained from the product of the complex RF and the C/A code, thus, it is also a complex quantity. Assume that the L1 frequency (1575.42 MHz) is converted to 21.25 MHz and digitized at 5 MHz; the output frequency is at 1.25 MHz as discussed in Section 6.8. Also assume that the acquisition programs search the frequency range of $1,250 \pm 10$ kHz in 1 kHz steps, and there are a total of 21 frequency components. The local code l_{si} can be represented as

$$l_{si} = C_s \exp(j2\pi f_i t) \tag{7.1}$$

where subscript s represents the number of satellites and subscript $i = 1, 2, 3 \ldots 21$, C_s is the C/A code of satellite S, $f_i = 1,250 - 10$, $1,250 - 9$, $1,250 - 8, \ldots 1250 + 10$ kHz. This local signal must also be digitized at 5 MHz and produces 5,000 data points. These 21 data sets represent the 21 frequencies separated by 1 kHz. These data are correlated with the input signal. If the locally generated signal contains the correct C/A code and the correct frequency component, the output will be high when the correct C/A phase is reached.

Figure 7.3 shows the concept of such an acquisition method. The operation of only one of these 21 sets of data will be discussed because the other 20 have the same operations. The digitized input signal and the locally generated one are multiplied point by point. Since the local signal is complex, the products obtained from the input and the local signals are also complex. The 5,000 real and imaginary values of the products are squared and added together and the square root of this value represents the amplitude of one of the output frequency bins. This process operates every 200 ns with every new incoming input data point. After the input data are shifted by 5,000 points, one ms of data are searched. In 1 ms there are 5,000 amplitude data points. Since there are 21 local signals, there are overall 105,000 ($5,000 \times 21$) amplitudes generated in 1 ms. A certain threshold can be set to measure the amplitude of the frequency outputs. The highest value among the 105,000 frequency bins that also crosses the threshold is the desired frequency bin. If the highest value occurs at the kth input data point, this point is the beginning of the C/A code. If the highest peak is generated by the f_i frequency component, this frequency component represents the carrier frequency of the input signal. Since the frequency resolution is 1 kHz, this resolution is not

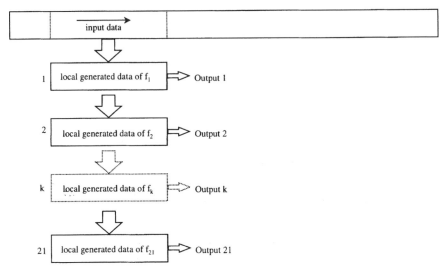

FIGURE 7.3 Acquisition through locally generated C/A and RF code.

accurate enough to be passed to the tracking program. More accurate frequency measurement is needed, and this subject will be discussed in Section 7.10.

The above discussion is for one satellite. If the receiver is designed to perform acquisition on 12 satellites in parallel, the above arrangements must be repeated 12 times.

7.7 CIRCULAR CONVOLUTION AND CIRCULAR CORRELATION[1]

This section provides the basic mathematics to understand a simpler way to perform correlation. If an input signal passes through a linear and time-invariant system, the output can be found in either the time domain through the convolution or in the frequency domain through the Fourier transform. If the impulse response of the system is $h(t)$, an input signal $x(t)$ can produce an output $y(t)$ through convolution as

$$y(t) = \int_{-\infty}^{\infty} x(t-\tau)h(\tau)d\tau = \int_{-\infty}^{\infty} x(\tau)h(t-\tau)d\tau \qquad (7.2)$$

The frequency domain response of $y(t)$ can be found from the Fourier transform as

$$Y(f) = \int_{-\infty}^{\infty} \int_{-\infty}^{\infty} x(\tau)h(t-\tau)d\tau e^{-j2\pi ft}dt$$

$$= \int_{-\infty}^{\infty} x(\tau) \left(\int_{-\infty}^{\infty} h(t-\tau)e^{-j2\pi ft}dt \right) d\tau \qquad (7.3)$$

Changing the variable by letting $t - \tau = u$, then

$$Y(f) = \int_{-\infty}^{\infty} x(\tau) \left(\int_{-\infty}^{\infty} h(u)e^{-j2\pi fu} du \right) e^{-2\pi f\tau} d\tau$$

$$= H(f) \int_{-\infty}^{\infty} x(\tau)e^{-j2\pi f\tau} d\tau = H(f)X(f) \tag{7.4}$$

In order to find the output in the time domain, an inverse Fourier transform on $Y(f)$ is required. The result can be written as

$$y(t) = x(t) * h(t) = \mathcal{F}^{-1}[X(f)H(f)] \tag{7.5}$$

where the $*$ represents convolution and $\mathcal{F}^{-1}$ represents inverse Fourier transform.

A similar relation can be found that a convolution in the frequency domain is equivalent to the multiplication in the time domain. These two relationships can be written as

$$x(t) * h(t) \leftrightarrow X(f)H(f)$$

$$X(f) * H(f) \leftrightarrow x(t)h(t) \tag{7.6}$$

This is often referred to as the duality of convolution in Fourier transform.

This concept can be applied in discrete time; however, the meaning is different from the continuous time domain expression. The response $y(n)$ can be expressed as

$$y(n) = \sum_{m=0}^{N-1} x(m)h(n-m) \tag{7.7}$$

where $x(m)$ is an input signal and $h(n-m)$ is system response in discrete time domain. It should be noted that in this equation the time shift in $h(n-m)$ is circular because the discrete operation is periodic. By taking the DFT of the above equation the result is

$$Y(k) = \sum_{n=0}^{N-1} \sum_{m=0}^{N-1} x(m)h(n-m)e^{(-j2\pi kn)/N}$$

$$= \sum_{m=0}^{N-1} x(m) \left[\sum_{n=0}^{N-1} h(n-m)e^{(-j2\pi(n-m)k)/N} \right] e^{(-j2\pi mk)/N}$$

$$= H(k) \sum_{m=0}^{N-1} x(m)e^{(-j2\pi mk)/N} = X(k)H(k) \tag{7.8}$$

Equations (7.7) and (7.8) are often referred to as the periodic convolution (or circular convolution). It does not produce the expected result of a linear convolution. A simple argument can illustrate this point. If the input signal and the impulse response of the linear system both have N data points, from a linear convolution, the output should be $2N - 1$ points. However, using Equation (7.8) one can easily see that the outputs have only N points. This is from the periodic nature of the DFT.

The acquisition algorithm does not use convolution; it uses correlation, which is different from convolution. A correlation between $x(n)$ and $h(n)$ can be written as

$$z(n) = \sum_{m=0}^{N-1} x(m)h(n+m) \tag{7.9}$$

The only difference between this equation and Equation (7.7) is the sign before index m in $h(n+m)$. The $h(n)$ is not the impulse response of a linear system but another signal. If the DFT is performed on $z(n)$ the result is

$$Z(k) = \sum_{n=0}^{N-1} \sum_{m=0}^{N-1} x(m)h(n+m)e^{(-j2\pi kn)/N}$$

$$= \sum_{m=0}^{N-1} x(m) \left[\sum_{n=0}^{N-1} h(n+m)e^{(-j2\pi(n+m)k)/N} \right] e^{(j2\pi mk)/N}$$

$$= H(k) \sum_{m=0}^{N-1} x(m)e^{(j2\pi mk)/N} = H(k)X^{-1}(k) \tag{7.10}$$

where $X^{-1}(k)$ represents the inverse DFT. The above equation can also be written as

$$Z(k) = \sum_{n=0}^{N-1} \sum_{m=0}^{N-1} x(n+m)h(m)e^{(-j2\pi kn)/N} = H^{-1}(k)X(k) \tag{7.11}$$

If the $x(n)$ is real, $x(n)^* = x(n)$ where $*$ is the complex conjugate. Using this relation, the magnitude of $Z(k)$ can be written as

$$|Z(k)| = |H^*(k)X(k)| = |H(k)X^*(k)| \tag{7.12}$$

This relationship can be used to find the correlation of the input signal and the locally generated signal. As discussed before, the above equation provides a periodic (or circular) correlation and this is the desired procedure.

7.8   ACQUISITION BY CIRCULAR CORRELATION[2]

The circular correlation method can be used for acquisition and the method is suitable for a software receiver approach. The basic idea is similar to the discussion in Section 7.6; however, the input data do not arrive in a continuous manner. This operation is suitable for a block of data. The input data are sampled with a 5 MHz ADC and stored in memory. Only 1 ms of the input data are used to find the beginning point of the C/A code and the searching frequency resolution is 1 kHz.

To perform the acquisition on the input data, the following steps are taken.

1. Perform the FFT on the 1 ms of input data $x(n)$ and convert the input into frequency domain as $X(k)$ where $n = k = 0$ to 4999 for 1 ms of data.
2. Take the complex conjugate $X(k)$ and the outputs become $X(k)^*$.
3. Generate 21 local codes $l_{si}(n)$ where $i = 1, 2, \ldots 21$, using Equation (7.1). The local code consists of the multiplication of the C/A code satellite s and a complex RF signal and it must be also sampled at 5 MHz. The frequency f_i of the local codes are separated by 1 kHz.
4. Perform FFT on $l_{si}(n)$ to transform them to the frequency domain as $L_{si}(k)$.
5. Multiply $X(k)^*$ and $L_{si}(k)$ point by point and call the result $R_{si}(k)$.
6. Take the inverse FFT of $R_{si}(k)$ to transform the result into time domain as $r_{si}(n)$ and find the absolute value of the $|r_{si}(n)|$. There are a total of 105,000 (21 × 5,000) of $|r_{si}(n)|$.
7. The maximum of $|r_{si}(n)|$ in the nth location and ith frequency bin gives the beginning point of C/A code in 200 ns resolution in the input data and the carrier frequency in 1 kHz resolution.

The above operation can be represented in Figure 7.4. The result shown in this figure is in the time domain and only one of the 21 local codes is shown.

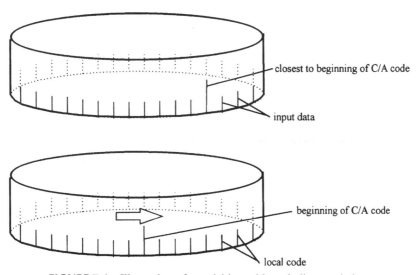

FIGURE 7.4 Illustration of acquisition with periodic correlation.

One can consider that the input data and the local data are on the surfaces of the two cylinders. The local data is rotated 5,000 times to match the input data. In other words, one cylinder rotates against the other one. At each step, all 5,000 input data are multiplied by the 5,000 local data point by point and results are summed together. It takes 5,000 steps to cover all the possible combinations of the input and the local code. The highest amplitude will be recorded. There are 21 pairs of cylinders (not shown). The highest amplitude from the 21 different frequency components is the desired value, if it also crosses a certain threshold.

Computer program (p7_1) will show the above operation with fine frequency information in Section 7.13. The determination and setting a specific threshold are not included in this program, but the results are plotted in time and frequency domain. One can determine the results from the plots.

7.9 MODIFIED ACQUISITION BY CIRCULAR CORRELATION[4]

This method is the same as the one above. The only difference is that the length of the FFT is reduced to half. In step 3 of the circular correlation method in the previous section the local code $l_{si}(n)$ is generated. Since the $l_{si}(n)$ is a complex quantity, the spectrum is asymmetrical as shown in Figure 7.5. From this figure it is obvious that the information is contained in the first-half spectrum lines. The

FIGURE 7.5 Spectrum of the locally generated signal.

second-half spectrum lines contain very little information. Thus the acquisition through the circular correlation method can be modified as follows:

1. Perform the FFT on the 1 ms of input data $x(n)$ and convert the input into frequency domain as $X(k)$ where $n = k = 0$ to 4,999 for 1 ms of data.
2. Use the first 2,500 $X(k)$ for $k = 0$ to 2,499. Take the complex conjugate and the outputs become $X(k)^*$.
3. Generate 21 local codes $l_{si}(n)$ where $i = 1, 2, \ldots 21$, using Equation (7.1) as discussed in the previous section. Each $l_{si}(n)$ has 5,000 points.
4. Perform the FFT on $l_{si}(n)$ to transform them to the frequency domain as $L_{si}(k)$.
5. Take the first half of $L_{si}(k)$, since the second half of $L_{si}(k)$ contains very little information. Multiply $L_{si}(k)$ and $X(k)^*$ point by point and call the result $R_{si}(k)$ where $k = 0 \sim 2499$.
6. Take the inverse FFT of $R_{si}(k)$ to transform the result back into time domain as $r_{si}(n)$ and find the absolute value of the $|r_{si}(n)|$. There are a total of 52,500 (21 × 2,500) of $|r_{si}(n)|$.
7. The maximum of $|r_{si}(n)|$ is the desired result, if it is also above a predetermined threshold. The ith frequency gives the carrier frequency with a resolution of 1 kHz and the nth location gives the beginning point of C/A code with a 400 ns time resolution.
8. Since the time resolution of the beginning of the C/A code with this method is 400 ns, the resolution can be improved to 200 ns by comparing the amplitudes of nth location with $(n-1)$ and $(n+1)$ locations.

In this approach from steps 5 through 7 only 2,500 point operations are performed instead of the 5,000 points. The sorting process in step 7 is simpler because only half the outputs are used. Step 8 is very simple. Therefore, this approach saves operation time. Simulated results show that this method has slightly lower signal-to-noise ratio, about 1.1 dB less than the regular circular correction method. This might be caused by the signal loss in the other half of the frequency domain.

7.10 DELAY AND MULTIPLY APPROACH[3-5]

The main purpose of this method is to eliminate the frequency information in the input signal. Without the frequency information one need only use the C/A code to find the initial point of the C/A code. Once the C/A is found, the frequency can be found from either FFT or DFT. This method is very interesting from a theoretical point of view; however, the actual application for processing the GPS signal still needs further study. This method is discussed as follows. First let us assume that the input signal $s(t)$ is complex, thus

$$s(t) = C_s(t)e^{j2\pi ft} \tag{7.13}$$

where $C_s(t)$ represents the C/A code of satellite s. The delayed version of this signal can be written as

$$s(t - \tau) = C_s(t - \tau)e^{j2\pi f(t-\tau)} \qquad (7.14)$$

where τ is the delay time. The product of $s(t)$ and the complex conjugate of the delayed version $s(t - \tau)$ is

$$s(t)s(t - \tau)^* = C_s(t)C_s(t - \tau)^* e^{j2\pi ft} e^{-j2\pi f(t-\tau)} \equiv C_n(t)e^{j2\pi f\tau} \qquad (7.15)$$

where

$$C_n(t) \equiv C_s(t)C_s(t - \tau) \qquad (7.16)$$

can be considered as a "new code," which is the product of a Gold code and its delayed version. This new "new code" belongs to the same family as the Gold code.[5] Simulated results show that its autocorrelation and the cross correlation can be used to find its beginning point of the "new code." The beginning point of the "new code" is the same as the beginning point of the C/A code. The interesting thing about Equation (7.15) is that it is frequency independent. The term $e^{j2\pi f\tau}$ is a constant, because f and τ are both constant. The amplitude of $e^{j2\pi f\tau}$ is unity. Thus, one only needs to search for the initial point of the "new code." Although this approach looks very attractive, the input signal must be complex. Since the input data collected are real, they must be converted to complex. This operation can be achieved through the Hilbert transform discussed in Section 6.13 or down converted into a complex signal; however, additional calculations are required.

A slight modification of the above method can be used for a real signal.[4] The approach is as follows. The input signal is

$$s(t) = C_s(t)\sin(2\pi ft) \qquad (7.17)$$

where $C_s(t)$ represents the C/A code of satellite s. The delayed version of the signal can be written as

$$s(t - \tau) = C_s(t - \tau)\sin[2\pi f(t - \tau)] \qquad (7.18)$$

The product of $s(t)$ and the delayed signal $s(t - \tau)$ is

$$\begin{aligned} s(t)s(t - \tau) &= C_s(t)C_s(t - \tau)\sin(2\pi ft)\sin[2\pi f(t - \tau)] \\ &\equiv \frac{C_n(t)}{2}\{\cos(2\pi f\tau) - \cos[2\pi f(2t - \tau)]\} \end{aligned} \qquad (7.19)$$

where $C_n(t)$ is defined in Equation (7.16). In the above equation there are two terms: a dc term and a high-frequency term. Usually the high frequency can be filtered out. In order to make this equation usable, the $|\cos(2\pi f\tau)|$ must be close to unity. Theoretically, this is difficult to achieve, because the frequency f

is unknown. However, since the frequency is within 1250 ± 10 kHz, it is possible to select a delay time to fulfill the requirement. For example, one can chose $2 \times \pi \times 1,250 \times 10^3 \tau = \pi$, thus, $\tau = 0.4 \times 10^{-6}$ s $= 400$ ns. Since the input data are digitized at 5 MHz, the sampling time is 200 ns (1/5 MHz). If the input signal is delayed by two samples, the delay time $\tau = 400$ ns. Under this condition $|\cos(2\pi f \tau)| = |\cos(\pi)| = 1$. If the frequency is off by 10 kHz, the corresponding value of $|\cos(2\pi f \tau)| = |\cos(2\pi \times 1,260 \times 10^3 \times 0.4 \times 10^{-6})| = 0.9997$, which is close to unity. Therefore, this approach can be applied to real data. The only restriction is that the delay time cannot be arbitrarily chosen as in Equation (7.15). Other delay times can also be used. For example, delay times of a multiple of 0.4 us can be used, if the delay line is not too long. For example, if $\tau = 1.6$ us, when the frequency is off by ± 10 kHz, the $|\cos(2\pi f \tau)| = 0.995$. One can see that $|\cos(2\pi f \tau)|$ decreases faster if a long delay time is used for a frequency off the center value of 1,250 kHz. If the delay time is too long the $|\cos(2\pi f \tau)|$ may no longer be close to unity.

The problem with this approach is that when two signals with noise are multiplied together the noise floor increases. Because of this problem one cannot search for 1 ms of data to acquire a satellite. Longer data are needed for acquiring a certain satellite.

One interesting point is that a navigation data change does not have a significant effect on the correlation result. Figure 7.6 shows this result. In Figure 7.6a there is no phase shift due to navigation data. The "new code" created by the multiplication of the C/A code and its delayed version will be repetitive every

(a) Without phase transition.

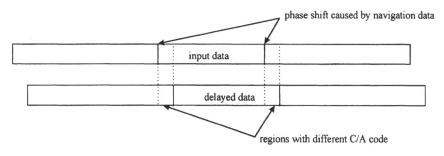

(b) With two phase transitions.

FIGURE 7.6 Effect of phase transition on the delay and multiplication method.

millisecond as the original C/A code. If there are two phase shifts by the navigation data, the only regions where the original C/A code and the delayed C/A code have a different code are shown in Figure 7.6b. The rest of the regions generate the same "new code." If there are 5,000 data points per C/A code, delaying two data points only degrades the performance by 2/5,000, caused by a navigation transition. Therefore, using this acquisition method one does not need to check for two consecutive data sets to guarantee that there is no phase shift by the navigation data in one of the sets. One can choose a longer data record and the result will improve the correlation output.

Experimental results indicate that 1 ms of data are not enough to find any satellite. The minimum data length appears to be 5 ms. Sometimes, one single delay time of 400 ns is not enough to find a desired signal. It may take several delay times, such as 0.4, 0.8, and 1.2 us, together to find the signals. A few weak signals that can be found by the circular correlation method cannot be found by this delay and multiplication method. With limited results, it appears that this method is not suitable to find weak signals.

7.11   NONCOHERENT INTEGRATION

Sometimes using 1 ms of data through the circular correlation method cannot detect a weak signal. Longer data records are needed to acquire a weak signal. As mentioned in Sections 7.4 and 7.5, an increase in the data length can require many more operations. One way to process more data is through noncoherent integration. For example, if 2 ms of data are used, the data can be divided into two 1 ms blocks. Each 1 ms of data are processed separately and the results are summed together. This operation basically doubles the number of operations, except for the addition in the last stage. The improvement obtained by this method is less than with the coherent method but there are fewer operations. The detailed discussion can be found in Sections 10.16 and 10.17.

Another advantage of this latter method is the ability to keep performing acquisition on successive 1 ms of data and summing the results. The final result can be compared with a certain threshold. Whenever the result crosses the threshold, the signal is found. A maximum data length can be chosen. If a signal cannot be found within the maximum data length, this process will automatically stop. Using this approach, strong signals can be found from 1 ms of data, but weak signals will take a longer data length. If there is phase shift caused by the navigation data, only the 1 ms of data with the phase shift will be affected. Therefore, the phase shift will have minimum impact on this approach. Conventional hardware receivers often use this approach.

7.12   COHERENT PROCESSING OF A LONG RECORD OF DATA[6]

This section presents the concept of processing a long record of data coherently with fewer operations. The details are easy to implement, thus they will not be included. The common approach to find a weak signal is to increase the acquisition data length. The advantage of this approach is the improvement in signal-to-noise ratio. One simple explanation is that an FFT with 2 ms of data produces a frequency resolution of 500 Hz in comparison with 1 kHz resolution of 1 ms of data. Since the signal is narrow band after the spectrum despread, the signal strength does not reduce by the narrower frequency resolution. Reducing the resolution bandwidth reduces the noise to half; therefore, the signal-to-noise ratio improves by 3 dB. Since the computational speed advances, data of longer than 10 ms are used for acquisition as discussed in Section 10.5.

The idea here is to perform FFT with fewer points. Let us use 10 ms of data (or 50,000 points) as an example to illustrate the idea. The center frequency of data is at 1.25 MHz. If one multiplies these data with a complex cw signal of 1.25 MHz, the input signal will be converted into a baseband and a high-frequency band at 2.5 MHz. If the high-frequency component is filtered out, only the baseband signal will be processed. Let us assume that the high frequency signal is filtered out. The baseband signal is a down-converted version of the input with the C/A code. One can multiply this signal by the C/A code point by point. If the correct phase of the C/A code is achieved, the output is a cw signal and the maximum frequency range is ± 10 kHz, caused by the Doppler frequency shift. Since the bandwidth of this signal is 20 kHz, one can sample this signal at 50 kHz, which is 2.5 times the bandwidth. With this sampling frequency there are only 500 data points in 10 ms. However, the signal is sampled at 5 MHz and there are 50,000 data points. One can average 100 points to make one data point. This averaging is equivalent to a low-pass filter; therefore, it eliminates the high-frequency components created by the multiplication of the 1.25 MHz cw signal as well as noise in the collected signal. Since the data are 10 ms long, the frequency resolution from the FFT is 100 Hz.

This approach is stated as follows by using 10 ms of data as an example:

1. Multiply 10 ms of the input signal by a locally generated complex cw signal at 1.25 MHz and digitized at 5 MHz. Let us refer to this output as the low-frequency output because the maximum frequency is 10 kHz. The high-frequency components at about 2.5 MHz will be filtered out later, thus, they will be neglected in this discussion. A total of 50,000 points of data will be obtained.

2. Multiply these output data by the desired 10 C/A codes point by point to obtain a total of 50,000 points.

3. Average 100 adjacent points into one point. This process filters out the high frequency at approximately 2.5 MHz.

4. Perform 500-point FFT to find a high output in the frequency domain. This operation generates only 250 useful frequency outputs.

5. Shift one data point of local code with respect to the low-frequency output and repeat steps 3 and 4. Since the C/A repeats every ms, one needs to perform this operation 5,000 times instead of 50,000 times.

6. There are overall 1.25×10^6 $(250 \times 5,000)$ outputs in the frequency domain. The highest amplitude that crosses a certain threshold will be the desired value. From this value the beginning of the C/A code and the Doppler frequency can be obtained. The frequency resolution obtained is 100 Hz.

Although the straightforward approach is presented above, circular correlation can be used to achieve the same purpose with fewer operations.

7.13 BASIC CONCEPT OF FINE FREQUENCY ESTIMATION[7]

The frequency resolution obtained from the 1 ms of data is about 1 kHz, which is too coarse for the tracking loop. The desired frequency resolution should be within a few tens of Hertz. Usually, the tracking loop has a width of only a few Hertz. Using the DFT (or FFT) to find fine frequency is not an appropriate approach, because in order to find 10 Hz resolution, a data record of 100 ms is required. If there are 5,000 data points/ms, 100 ms contains 500,000 data points, which is very time consuming for FFT operation. Besides, the probability of having phase shift in 100 ms of data is relatively high.

The approach to find the fine frequency resolution is through phase relation. Once the C/A code is stripped from the input signal, the input becomes a cw signal. If the highest frequency component in 1 ms of data at time m is $X_m(k)$, k represents the frequency component of the input signal. The initial phase $\theta_m(k)$ of the input can be found from the DFT outputs as

$$\theta_m(k) = \tan^{-1}\left(\frac{\text{Im}(X_m(k))}{\text{Re}(X_m(k))}\right) \tag{7.20}$$

where Im and Re represent the imaginary and real parts, respectively. Let us assume that at time n, a short time after m, the DFT component $X_n(k)$ of 1 ms of data is also the strongest component, because the input frequency will not change that rapidly during a short time. The initial phase angle of the input signal at time n and frequency component k is

$$\theta_n(k) = \tan^{-1}\left(\frac{\text{Im}(X_n(k))}{\text{Re}(X_n(k))}\right) \tag{7.21}$$

These two phase angles can be used to find the fine frequency as

$$f = \frac{\theta_n(k) - \theta_m(k)}{2\pi(n - m)} \tag{7.22}$$

This equation provides a much finer frequency resolution than the result obtained from DFT. In order to keep the frequency unambiguous, the phase difference $\theta_n - \theta_m$ must be less than 2π. If the phase difference is at the maximum value of 2π, the unambiguous bandwidth is $1/(n - m)$ where $n - m$ is the delay time between two consecutive data sets.

7.14   RESOLVING AMBIGUITY IN FINE FREQUENCY MEASUREMENTS

Although the basic approach to find the fine frequency is based on Equation (7.22), there are several slightly different ways to apply it. If one takes the kth frequency component of the DFT every millisecond, the frequency resolution is 1 kHz and the unambiguous bandwidth is also 1 kHz. In Figure 7.7a five frequency components are shown and they are separated by 1 kHz. If the input signal falls into the region between two frequency components as shown in Figure 7.7b, the phase may have uncertainty due to noise in the system.

One approach to eliminate the uncertainty is to speed up the DFT operation. If the DFT is performed every 0.5 ms, the unambiguous bandwidth is 2 kHz. With a frequency resolution of 1 kHz and an unambiguous bandwidth of 2 kHz, there will be no ambiguity problem in determining the fine frequency. However, this approach will double the DFT operations.

The second approach is to use an amplitude comparison scheme without doubling the speed of the DFT operations, if the input is a cw signal. As shown in

(a) Frequency separation.

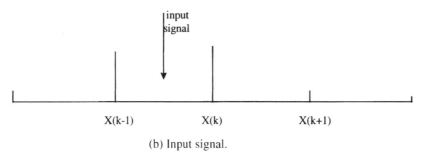

(b) Input signal.

FIGURE 7.7   Ambiguous ranges in frequency domain.

Figure 7.7b, the input signal falls in between two frequency bins. Suppose that the amplitude of $X(k)$ is slightly higher than $X(k-1)$; then $X(k)$ will be used in Equations (7.21) and (7.22) to find the fine frequency resolution. The difference frequency should be close to 500 Hz. The correct result is that the input frequency is about 500 Hz lower than $X(k)$. Due to noise the 500 Hz could be assessed as higher than $X(k)$; therefore, a wrong answer can be reached. However, for this input frequency the amplitudes of $X(k)$ and $X(k-1)$ are close together and they are much stronger than $X(k+1)$. Thus, if the highest-frequency bin is $X(k)$ and the phase calculated is in the ambiguous range, which is close to the centers between $X(k-1)$ and $X(k)$ or between $X(k)$ and $X(k+1)$, the amplitude of $X(k-1)$ and $X(k+1)$ will be compared. If $X(k-1)$ is stronger than $X(k+1)$, the input frequency is lower than $X(k)$; otherwise, the input frequency is higher than $X(k)$. Under this condition, the accuracy of fine frequency is determined by the phase but the sign of the difference frequency is determined by the amplitudes of two frequency components adjacent to the highest one.

However, the problem is a little more complicated than this, because it is possible that there is a 180-degree phase shift between two consecutive data sets due to navigation data. If this condition occurs, the input can no longer be treated as a cw signal. This possibility limits the ambiguous bandwidth to 250 Hz for 1 ms time delay. If the frequency is off by ± 250 Hz, the corresponding angle is $\pm \pi/2$. If the frequency is off by $+250$ Hz, the angle should be $+\pi/2$. However, a π phase shift due to the navigation will change the angle to $-\pi/2(+\pi/2 - \pi)$, which corresponds to a -250 kHz change. If the phase transition is not taken account of in finding the fine frequency, the result will be off by 500 kHz.

In order to avoid this problem, the maximum frequency uncertainty must be less than 250 Hz. If the maximum frequency difference is ± 200 Hz, which is selected experimentally, the corresponding phase angle difference is $\pm 2\pi/5$ as shown in Figure 7.8. If there is a π phase shift, the magnitude of the phase difference is $3\pi/5[| \pm (2\pi/5) \mp \pi|]$, which is greater than $2\pi/5$. From this arrangement, the phase difference can be used to determine the fine frequency without creating erroneous frequency shift. If the phase difference is greater than $2\pi/5$, π can be subtracted from the result to keep the phase difference less than $2\pi/5$. In order to keep the frequency within 200 kHz, the maximum separation between the k values in $X(k)$ will be 400 kHz. If the input is in the middle of two adjacent k values, the input signal is 200 kHz from both of the k values. In the following discussion, let us keep the maximum phase difference at $2\pi/5$, which corresponds to a frequency difference of 200 kHz.

The last point to be discussed is converting the real and imaginary parts of $X(k)$ into an angle. Usually, the phase angle measured is between $\pm \pi$. The two angles in Equations (7.20) and (7.21) will be obtained in this manner. The difference angle between the two angles can be any value between 0 and 2π as shown in Figure 7.9. Since the maximum allowable difference angle is $2\pi/5$ for 200 kHz, the difference angle must be equal or less than $2\pi/5$. If the result is greater than $2\pi/5$, 2π can be either added or subtracted from the result, and the absolute value of the angle must be less than $2\pi/5$. If noise is taken into

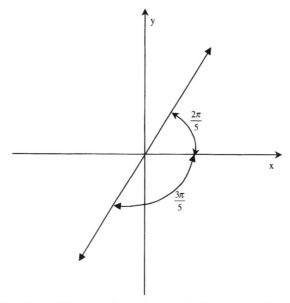

FIGURE 7.8   Phase difference of less than $2\pi/5$ will not cause frequency error.

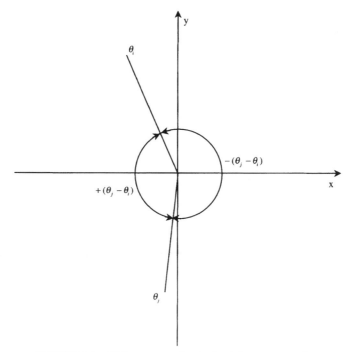

FIGURE 7.9   Difference angle obtained from two angles.

consideration, the $2\pi/5$ threshold can be extended slightly, such as using $2.3\pi/5$, which means the difference must be equal to or less than this value. If the final value of the adjusted phase difference is still greater than this threshold, it means that there is a phase shift between the two milliseconds of data and π should be subtracted from the result. Of course, the final angle should also be adjusted by adding or subtracting 2π to obtain the final result of less than the threshold.

From the above discussion, the following steps are required to find the beginning of the C/A code and the carrier frequency of a certain satellite:

1. Perform circular correlation on 1 ms of data; the starting point of a certain C/A code can be found in these data and the carrier frequency can be found in 1 kHz resolution.

2. From the highest-frequency component $X(k)$, perform two DFT operations on the same 1 ms of data: one is 400 kHz lower and the other one is 400 kHz higher than k in $X(k)$. The highest output from the three outputs $[X(k-1), X(k), X(k+1)]$ will be designated to be the new $X(k)$ and used as the DFT component to find the fine frequency.

3. Arbitrarily choose five milliseconds of consecutive data starting from the beginning of the C/A code. Multiply these data with 5 consecutive C/A codes; the result should be a cw signal of 5 ms long. However, it might contain one π phase shift between any of the 1 ms of data.

4. Find $X_n(k)$ on all the input data, where $n = 1, 2, 3, 4,$ and 5. Then find the phase angle from Equation (7.20). The difference angle can be defined as

$$\Delta\theta = \theta_{n+1} - \theta_n \tag{7.23}$$

5. The absolute value of the difference angle must be less than the threshold $(2.3\pi/5)$. If this condition is not fulfilled, 2π can be added or subtracted from $\Delta\theta$. If the result is still above the threshold, π can be added or subtracted from $\Delta\theta$ to adjust for the π phase shift. This result will also be tested against the $2.3\pi/5$ threshold. If the angle is higher than the threshold, 2π can be added or subtracted to obtain the desired result. After these adjustments, the final angle is the desired value.

6. Equation (7.22) can be used to find the fine frequency. Since there are 5 ms of data, there will be 4 sets of fine frequencies. The average value of these four fine frequencies will be used as the desired fine frequency value to improve accuracy.

Program (p7_1) listed at the end of this chapter can be used to find the initial point of the C/A code as well as the fine frequency. This program calls the digitizg.m, which generates digitized C/A code. The digitizg.m in turn calls codegen.m, which is a modified version of program (p5_2), and generates the C/A code of the satellites. These programs just provide the basic idea. They can be modified to solve certain problems. For example, one can add a threshold to the detection of a certain satellite. If the signal is weak, one can use several milliseconds of data and add them incoherently.

7.15 AN EXAMPLE OF ACQUISITION

In this section the acquisition computer program (p7_1) is used to find the initial point of a C/A code and the fine frequency. The computer program will operate on actual data collected. The experimental setup to collect the data is similar to Figure 6.5b. The data were digitized at 5 MHz. The data contain 7 satellites, numbers 6, 10, 17, 23, 24, 26, and 28. Most of the satellites in the data are reasonably strong and they can be found from 1 ms of data. However, this is a qualitative discussion, because no threshold is used to determine the probability of detection. Satellite 24 is weak; in order to confirm this signal several milliseconds of data need to be added incoherently.

The input data in the time domain are shown in Figure 7.10. As expected, the data look like noise. The frequency plot of the input can be found through the FFT operation as shown in Figure 7.11. The bandwidth is 2.5 MHz as expected. The spectrum shape resembles the shape of the filter in the RF chain. After the circular correlation, the beginning of the C/A code of satellite 6 is shown in Figure 7.12. The beginning of the C/A code is at 2884. The amplitudes of the 21 frequency components separated by 1 kHz are shown in Figure 7.13. The highest

FIGURE 7.10 Input data.

FIGURE 7.11 FFT of input data.

FIGURE 7.12 Beginning of C/A code of satellite 6.

FIGURE 7.13 Frequency component of the despread signal of satellite 6.

component occurs at $k = 7$. From Figures 7.12 and 7.13, one can see that the initial point of the C/A code and the frequency are clearly shown. Since the data are actually collected, the accuracy of the fine frequency is difficult to determine because the Doppler frequency is unknown. The fine frequency also depends on the frequency accuracy of the local oscillator used in the down conversion and the accuracy of the sampling frequency.

One way to get a feeling of the calculated fine frequency accuracy is to use different portions of the data. Six fine frequencies are calculated from different portions of the input data. The data used are 1–25,000, 5001–30,001, 10,001–35,001, 15,001–40,001, 20,001–45,001, 25,001–50,001. These data are five milliseconds long and the starting points are shifted by 1 ms. Between two adjacent data sets four out of the five milliseconds of data are the same. Therefore, the calculated fine frequency should be close. The fine frequency differences between these six sets are −2.4, 9.0, −8.2, 5.4, and 2.3 Hz. These data are collected at a stationary set. The frequency change per millisecond should be very small as discussed in Chapter 3. Thus, the frequency difference can be considered as the inaccuracy of the acquisition method. When the signal strength changes, the difference of the fine frequency also changes. For a weak signal the frequency difference can be in tens of Hertz, if the same calculation method is used.

The acquisition performed on a weak signal (satellite 24) is shown in Figures 7.14 and 7.15. From these figures, it is difficult to assess whether the

FIGURE 7.14 Beginning of C/A code of satellite 24.

FIGURE 7.15 Frequency component of the despread signal of satellite 24.

beginning point of the C/A code and the frequency are the correct values. The beginning of the C/A code and the fine frequency can be found by adding more calculations incoherently. The correct results are different from the maximum shown in Figures 7.14 and 7.15.

7.16 SUMMARY

The concept of signal acquisition is discussed in this section. The circular correlation method can provide fast acquisition without sacrificing detection sensitivity. Although there are methods that can operate at a faster speed than the circular correlation method, they usually have lower detection sensitivity. It appears that the circular correction method using 1 ms of data is able to acquire signals from most satellites. The delay and multiplication method must use more than 1 ms of data to acquire a signal because of the high noise level created. The acquisition provides the beginning of the C/A code and a coarse frequency. The fine frequency can be found through phase comparison to within a few tens of Hertz. To speed up the operation of the acquisition process is always desirable in software GPS receivers.

REFERENCES

1. Oppenheim, A. V., Schafer, R. W., *Digital Signal Processing*, Prentice-Hall, Englewood Cliffs, NJ, 1975.
2. Van Nee, D. J. R., Coenen, A. J. R. M., "New fast GPS code acquisition technique using FFT," *Electronics Letters*, vol. 27, pp. 158–160, January 17, 1991.
3. Tomlinson, M., School of Electronic, Communication and Electrical Engineering, University of Plymouth, United Kingdom, private communication.
4. Lin, D., Tsui, J., "Acquisition schemes for software GPS receiver," *ION GPS-98*, pp. 317–325, Nashville, TN, September 15–18, 1998.
5. Wolfert, R., Chen, S., Kohli, S., Leimer, D., Lascody, J., "Rapid direct P(Y)-code acquisition in a hostile environment," *ION GPS-98*, pp. 353–360, Nashville, TN, September 15–18, 1998.
6. Spilker, J. J., "GPS signal structure and performance characteristics," *Navigation*, Institute of Navigation, vol. 25, no. 2, pp. 121–146, Summer 1978.
7. Tsui, J. B. Y., *Digital Techniques for Wideband Receivers*, Artech House, Boston, 1995.

```
% p7_1.m performs acquisition on collected data

clear
% ****** initial condition ******
svnum=input('enter satellite number = ');
intodat=10001; %input('enter initial pt into data (multiple
of n) = ');
fs=5e6; % *** sampling freq
ts=1/fs; % *** sampling time
n=fs/1000; % *** data pt in 1 ms
nn=[0:n-1]; % *** total no. of pts
fc=1.25e6; % *** center freq without Doppler
nsat=length(svnum); % *** total number of satellites to be
processed

% ****** input data file *********
fid=fopen('d:\gps\Big_data\srvy1sf1.dat','r');
fseek(fid,intodat-1,'bof');
x2=fread(fid,6*n,'schar');

yy = zeros(21,n);

% ****** start acquisition ******
code=digitizg(n,fs,0,svnum); % digitize C/A code
xf = fft(x2(1:n)');
for i = [1:21]; %**** find coarse freq 1 kHz resolution
   fr=fc-10000+(i-1)*1000;
   lc=code.* exp(j*2*pi*fr*ts*nn); % generate local code
   lcf=fft(lc);
   yy(i,:)=ifft(xf .* conj(lcf)); % circular correlation
end
[amp crw]=max(max(abs(yy'))); % find highest peak
[amp ccn]=max(max(abs(yy)));

pt_init=ccn; % initial point
cfrq=fc+1000*(crw-11); % coarse freq

% ****** gerenate 5 ms data by stripping C/A code ******
z5=x2(pt_init:pt_init+5*n-1); % take 5 ms data starting with
C/A code
za5=z5' .* [code code code code code];% create cw from 5 sets
of data

% ****** find medium freq resolution 400 kHz apart ******
for i = [1:3];
   fr=cfrq-400+(i-1)*400;
   mfrq0(i)=sum(za5(1:5000) .* exp(j*2*pi*fr*ts*nn));
   mfrq1(i)=abs(mfrq0(i));
end
```

```
[mamp mrw] = max(mfrq1); % find highest peak
mfrq=cfrq+200*(mrw-2);
fr=mfrq; % medium freq

% ****** find fine freq ******
zb5=za5 .* exp(j*2*pi*fr*ts*[0:5*n-1]); % one DFT component
zc5=diff(-angle(sum(reshape(zb5,n,5)))); % find difference
angle
zc5fix=zc5;

%****** Adjust phase and take out possible phase shift
******
threshold=2.3*pi\5;
for i=1:4;
 if abs(zc5(i))>threshold;% for angle adjustment
     zc5(i)=zc5fix(i)-2*pi;
     if abs(zc5(i))>threshold;
        zc5(i)=zc5fix(i)+2*pi; % end
        if abs(zc5(i))>2.2*pi\5; %for pi phase shift
        correction
           zc5(i)=zc5fix(i)-pi;
              if abs(zc5(i))>threshold;
                 zc5(i)=zc5fix(i)-3*pi;
                 if abs(zc5(i))>threshold;
                    zc5(i)=zc5fix(i)+pi; %end
                 end
              end
        end
     end
 end
end

dfrq=mean(zc5)*1000/(2*pi);
frr=fr+dfrq;% fine freq

plot(abs(yy(crw,1:n)))
title(['GPS = ' num2str(svnum) '      max at '
num2str(pt_init)])
figure
plot(abs(yy(:,ccn)),'*')
% title(['GPS = ' num2str(svnum) '     Freq = ' num2str(frr)])

format
pt_init
format long e
frr
```

```
% digitizg.m  This prog generate the C/A code and digitize it
function code2 = digitizg(n,fs,offset,svnum);

% code   - gold code
% n      - number of samples
% fs     - sample frequency in Hz;
% offset - delay time in second must be less than 1/fs can
not shift left
% svnum   - satellite number;

gold_rate = 1.023e6;     %gold code clock rate in Hz.
ts=1/fs;
tc=1/gold_rate;

cmd1 = codegen(svnum); % generate C/A code
code_in=cmd1;

% ******* creating 16 C/A code for digitizing *******
code_a = [code_in code_in code_in code_in];
code_a=[code_a code_a];
code_a=[code_a code_a];

% ******* digitizing *******
b = [1:n];
c = ceil((ts*b+offset)/tc);
code = code_a(c);

% ******* adjusting first data point *******
if offset>=0;
    code2=[code(1) code(1:n-1)];
else
    code2=[code(n) code(1:n-1)];
end
```

```
% codegen.m generate one of the 32 C/A codes written by
D.Akos modified by J. Tsui

function [ca_used]=codegen(svnum);

% ca_used : a vector containing the desired output sequence
% the g2s vector holds the appropriate shift of the g2 code
to generate
% the C/A code (ex. for SV#19 - use a G2 shift of
g2s(19)=471)
% svnum: Satellite number

g2s = [5;6;7;8;17;18;139;140;141;251;252;254;255;256;257; ...
       258;469;470;471;
       472;473;474;509;512;513;514;515;516;859;860;861;862];

g2shift=g2s(svnum,1);

% ******* Generate G1 code *******
      %   load shift register
          reg = -1*ones(1,10);
      for i = 1:1023,
          g1(i) = reg(10);
          save1 = reg(3)*reg(10);
          reg(1,2:10) = reg(1:1:9);
          reg(1) = save1;
      end,

% ******* Generate G2 code *******
      %   load shift register
           reg = -1*ones(1,10);
      for i = 1:1023,
          g2(i) = reg(10);
          save2 = reg(2)*reg(3)*reg(6)*reg(8)*reg(9)*reg(10);
          reg(1,2:10) = reg(1:1:9);
          reg(1) = save2;
      end,

% ******* Shift G2 code *******
      g2tmp(1,1:g2shift)=g2(1,1023-g2shift+1:1023);
      g2tmp(1,g2shift+1:1023)=g2(1,1:1023-g2shift);
      g2 = g2tmp;

% ******* Form single sample C/A code by multiplying G1 and
G2

ss_ca = g1.*g2;
ca_used=-ss_ca;
```

CHAPTER EIGHT

Tracking GPS Signals

8.1 INTRODUCTION

One might think that the basic method of tracking a signal is to build a narrow-band filter around an input signal and follow it. In other words, while the frequency of the input signal varies over time, the center frequency of the filter must follow the signal. In the actual tracking process, the center frequency of the narrow-band filter is fixed, but a locally generated signal follows the frequency of the input signal. The phases of the input and locally generated signals are compared through a phase comparator. The output from the phase comparator passes through a narrow-band filter. Since the tracking circuit has a very narrow bandwidth, the sensitivity is relatively high in comparison with the acquisition method.

When there are phase shifts in the carrier due to the C/A code, as in a GPS signal, the code must be stripped off first as discussed in Section 7.5. The tracking process will follow the signal and obtain the information of the navigation data. If a GPS receiver is stationary, the expected frequency change due to satellite movement is very slow as discussed in Chapter 3. Under this condition, the frequency change of the locally generated signal is also slow; therefore, the update rate of the tracking loop can be low. In order to strip off the C/A code another loop is needed. Thus, to track a GPS signal two tracking loops are required. One loop is used to track the carrier frequency and is referred to as the carrier loop. The other one is used to track the C/A code and is referred to as the code loop.

In this chapter the basic loop concept will be discussed first. Two tracking methods will be discussed. The first one is the conventional tracking loop. The

Fundamentals of Global Positioning System Receivers: A Software Approach, Second Edition,
by James Bao-Yen Tsui
ISBN 0-471-70647-7 Copyright © 2005 John Wiley & Sons, Inc.

only unique point of this method is that the tracking loop will be presented in digital form and the tracking will be accomplished in software. The second method is referred to as the block adjustment of synchronizing signal (BASS) method. The BASS method is also implemented in software and the performance might be slightly sensitive to noise. The details of the two methods will be presented.

8.2   BASIC PHASE-LOCKED LOOPS[1-4]

In this section the basic concept of the phase-locked loop will be described, which includes the transfer function, the error transfer function, the noise bandwidth, and two types of input signals.

The main purpose of a phase-locked loop is to adjust the frequency of a local oscillator to match the frequency of an input signal, which is sometimes referred to as the reference signal. A basic phase-locked loop is shown in Figure 8.1.

Figure 8.1a shows the time domain configuration and Figure 8.1b shows the s-domain configuration, which is obtained from the Laplace transform. The input signal is $\theta_i(t)$ and the output from the voltage-controlled oscillator (VCO) is $\theta_f(t)$. The phase comparator $\sum$ measures the phase difference of these two signals. The amplifier k_0 represents the gain of the phase comparator and the low-pass filter limits the noise in the loop. The input voltage V_o to the VCO controls its output frequency, which can be expressed as

$$\omega_2(t) = \omega_0 + k_1 u(t) V_o \tag{8.1}$$

where ω_0 is the center angular frequency of the VCO, k_1 is the gain of the VCO, and $u(t)$ is a unit step function, which is defined as

$$u(t) = \begin{cases} 0 & \text{for} \quad t < 0 \\ 1 & \text{for} \quad t > 0 \end{cases} \tag{8.2}$$

The phase angle of the VCO can be obtained by integrating Equation (8.1) as

$$\int_0^t \omega_2(t)dt = \omega_0 t + \theta_f(t) = \omega_0 t + \int_0^t k_1 u(t) V_o dt$$

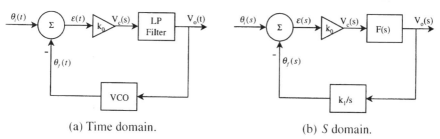

(a) Time domain.   (b) S domain.

FIGURE 8.1   A basic phase-locked loop.

where

$$\theta_f(t) = \int_0^t k_1 u(t) V_o dt \tag{8.3}$$

The Laplace transform of $\theta_f(t)$ is

$$\theta_f(s) = V_o(s) \frac{k_1}{s} \tag{8.4}$$

From Figure 8.1b the following equations can be written.

$$V_c(s) = k_0 \epsilon(s) = k_0[\theta_i(s) - \theta_f(s)] \tag{8.5}$$

$$V_o(s) = V_c(s) F(s) \tag{8.6}$$

$$\theta_f(s) = V_o(s) \frac{k_1}{s} \tag{8.7}$$

From these three equations one can obtain

$$\epsilon(s) = \theta_i(s) - \theta_f(s) = \frac{V_c(s)}{k_0} = \frac{V_o(s)}{k_0 F(s)} = \frac{s\theta_f(s)}{k_0 k_1 F(s)} \quad \text{or}$$

$$\theta_i(s) = \theta_f(s) \left(1 + \frac{s}{k_0 k_1 F(s)} \right) \tag{8.8}$$

where $\epsilon(s)$ is the error function. The transfer function $H(s)$ of the loop is defined as

$$H(s) \equiv \frac{\theta_f(s)}{\theta_i(s)} = \frac{k_0 k_1 F(s)}{s + k_0 k_1 F(s)} \tag{8.9}$$

The error transfer function is defined as

$$H_e(s) = \frac{\epsilon(s)}{\theta_i(s)} = \frac{\theta_i(s) - \theta_f(s)}{\theta_i(s)} = 1 - H(s) = \frac{s}{s + k_0 k_1 F(s)} \tag{8.10}$$

The equivalent noise bandwidth is defined as[1]

$$B_n = \int_0^\infty |H(j\omega)|^2 df \tag{8.11}$$

where ω is the angular frequency and it relates to the frequency f by $\omega = 2\pi f$.
   In order to study the properties of the phase-locked loops, two types of input signals are usually studied. The first type is a unit step function as

$$\theta_i(t) = u(t) \quad \text{or} \quad \theta_i(s) = \frac{1}{s} \tag{8.12}$$

The second type is a frequency-modulated signal

$$\theta_i(t) = \Delta\omega t \quad \text{or} \quad \theta_i(s) = \frac{\Delta\omega}{s^2} \tag{8.13}$$

These two types of signals will be discussed in the next two sections.

8.3 FIRST-ORDER PHASE-LOCKED LOOP[1-4]

In this section, the first-order phase-locked loop will be discussed. A first-order phase-locked loop implies the denominator of the transfer function $H(s)$ is a first-order function of s. The order of the phase-locked loop depends on the order of the filter in the loop. For this kind of phase-locked loop, the filter function is

$$F(s) = 1 \tag{8.14}$$

This is the simplest phase-locked loop. For a unit step function input, the corresponding transfer function from Equation (8.9) becomes

$$H(s) = \frac{k_0 k_1}{s + k_0 k_1} \tag{8.15}$$

The denominator of $H(s)$ is a first order of s.
The noise bandwidth can be found as

$$
\begin{aligned}
B_n &= \int_0^\infty \frac{(k_0 k_1)^2 df}{\omega^2 + (k_0 k_1)^2} = \frac{(k_0 k_1)^2}{2\pi} \int_0^\infty \frac{d\omega}{\omega^2 + (k_0 k_1)^2} \\
&= \frac{(k_0 k_1)^2}{2\pi k_0 k_1} \tan^{-1}\left(\frac{\omega}{k_0 k_1}\right)_0^\infty = \frac{k_0 k_1}{4}
\end{aligned} \tag{8.16}
$$

With the input signal $\theta_i(s) = 1/s$, the error function can be found from Equation (8.10) as

$$\epsilon(s) = \theta_i(s) H_e(s) = \frac{1}{s + k_0 k_1} \tag{8.17}$$

The steady-state error can be found from the final value theorem of the Laplace transform, which can be stated as

$$\lim_{t \to \infty} y(t) = \lim_{s \to 0} s Y(s) \tag{8.18}$$

Using this relation, the final value of $\epsilon(t)$ can be found as

$$\lim_{t \to \infty} \epsilon(t) = \lim_{s \to 0} s\epsilon(s) = \lim_{s \to 0} \frac{s}{s + k_0 k_1} = 0 \tag{8.19}$$

With the input signal $\theta_i(s) = \Delta\omega/s^2$, the error function is

$$\epsilon(s) = \theta_i(s)H_e(s) = \frac{\Delta\omega}{s}\frac{1}{s + k_0 k_1} \tag{8.20}$$

The steady-state error is

$$\lim_{t \to \infty} \epsilon(t) = \lim_{s \to 0} s\epsilon(s) = \lim_{s \to 0} \frac{\Delta\omega}{s + k_0 k_1} = \frac{\Delta\omega}{k_0 k_1} \tag{8.21}$$

This steady-state phase error is not equal to zero. A large value of $k_0 k_1$ can make the error small. However, from Equation (8.15) the 3 dB bandwidth occurs at $s = k_0 k_1$. Thus, a small final value of $\epsilon(t)$ also means large bandwidth, which contains more noise.

8.4   SECOND-ORDER PHASE-LOCKED LOOP[1-4]

A second-order phase-locked loop means the denominator of the transfer function $H(s)$ is a second-order function of s. One of the filters to make such a second-order phase-locked loop is

$$F(s) = \frac{s\tau_2 + 1}{s\tau_1} \tag{8.22}$$

Substituting this relation into Equation (8.9), the transfer function becomes

$$H(s) = \frac{\dfrac{k_0 k_1 \tau_2 s}{\tau_1} + \dfrac{k_0 k_1}{\tau_1}}{s^2 + \dfrac{k_0 k_1 \tau_2 s}{\tau_1} + \dfrac{k_0 k_1}{\tau_1}} \equiv \frac{2\zeta\omega_n s + \omega_n^2}{s^2 + 2\zeta\omega_n s + \omega_n^2} \tag{8.23}$$

where ω_n is the natural frequency, which can be expressed as

$$\omega_n = \sqrt{\frac{k_0 k_1}{\tau_1}} \tag{8.24}$$

and ζ is the damping factor, which can be shown as

$$2\zeta\omega_n = \frac{k_0 k_1 \tau_2}{\tau_1} \quad \text{or} \quad \zeta = \frac{\omega_n \tau_2}{2} \tag{8.25}$$

The denominator of $H(s)$ is a second order of s.

The noise bandwidth can be found as[1]

$$B_n = \int_0^\infty |H(\omega)|^2 df = \frac{\omega_n}{2\pi} \int_0^\infty \frac{1 + \left(2\zeta \dfrac{\omega}{\omega_n}\right)^2}{\left[1 - \left(\dfrac{\omega}{\omega_n}\right)^2\right]^2 + \left(2\zeta \dfrac{\omega}{\omega_n}\right)^2} d\omega$$

$$= \frac{\omega_n}{2\pi} \int_0^\infty \frac{1 + 4\zeta^2 \left(\dfrac{\omega}{\omega_n}\right)^2}{\left(\dfrac{\omega}{\omega}\right)^4 + 2(2\zeta^2 - 1)\left(\dfrac{\omega}{\omega_n}\right)^2 + 1} d\omega = \frac{\omega_n}{2}\left(\zeta + \frac{1}{4\zeta}\right) \quad (8.26)$$

This integration can be found in the appendix at the end of this chapter. The error transfer function can be obtained from Equation (8.10) as

$$H_e(s) = 1 - H(s) = \frac{s^2}{s^2 + 2\zeta\omega_n s + \omega_n^2} \quad (8.27)$$

When the input is $\theta_i(s) = 1/s$, the error function is

$$\epsilon(s) = \frac{s}{s^2 + 2\zeta\omega_n s + \omega_n^2} \quad (8.28)$$

The steady-state error is

$$\lim_{t\to\infty} \epsilon(t) = \lim_{s\to 0} s\epsilon(s) = 0 \quad (8.29)$$

When the input is $\theta_i(s) = 1/s^2$, the error function is

$$\epsilon(s) = \frac{1}{s^2 + 2\zeta\omega_n s + \omega_n^2} \quad (8.30)$$

The steady-state error is

$$\lim_{t\to\infty} \epsilon(t) = \lim_{s\to 0} s\epsilon(s) = 0 \quad (8.31)$$

In contrast to the first-order loop, the steady-state error is zero for the frequency-modulated signal. This means the second-order loop tracks a frequency-modulated signal and returns the phase comparator characteristic to the null point. The conventional phase-locked loop in a GPS receiver is usually a second-order one.

8.5   TRANSFORM FROM CONTINUOUS TO DISCRETE SYSTEMS[5,6]

In the previous sections, the discussion is based on continuous systems. In order to build a phase-locked loop in software for digitized data, the continuous system must be changed into a discrete system. This discussion is based on reference 5. The transfer from the continuous s-domain into the discrete z-domain is through bilinear transform as

$$s = \frac{2}{t_s} \frac{1 - z^{-1}}{1 + z^{-1}} \qquad (8.32)$$

where t_s is the sampling interval. Substituting this relation in Equation (8.22) the filter is transformed to

$$F(z) = C_1 + \frac{C_2}{1 - z^{-1}} = \frac{(C_1 + C_2) - C_1 z^{-1}}{1 - z^{-1}} \qquad (8.33)$$

where

$$C_1 = \frac{2\tau_2 - t_s}{2\tau_1}$$

$$C_2 = \frac{t_s}{\tau_1} \qquad (8.34)$$

This filter is shown in Figure 8.2.

The VCO in the phase-locked loop is replaced by a direct digital frequency synthesizer and its transfer function $N(z)$ can be used to replace the result in Equation (8.7) with

$$N(z) = \frac{\theta_f(z)}{V_o(z)} \equiv \frac{k_1 z^{-1}}{1 - z^{-1}} \qquad (8.35)$$

Using the same approach as Equation (8.8), the transfer function $H(z)$ can be written as

$$H(z) = \frac{\theta_f(z)}{\theta_i(z)} = \frac{k_0 F(z) N(z)}{1 + k_0 F(z) N(z)} \qquad (8.36)$$

Substituting the results of Equations (8.33) and (8.35) into the above equation, the result is

$$H(z) = \frac{k_0 k_1 (C_1 + C_2) z^{-1} - k_0 k_1 C_1 z^{-2}}{1 + [k_0 k_1 (C_1 + C_2) - 2] z^{-1} + (1 - k_0 k_1 C_1) z^{-2}} \qquad (8.37)$$

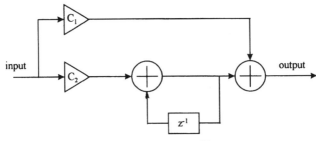

FIGURE 8.2 Loop filter.

By applying bilinear transform in Equation (8.32) to Equation (8.23), the result can be written as,

$$H(z) = \frac{[4\zeta\omega_n + (\omega_n t_s)^2] + 2(\omega_n t_s)^2 z^{-1} + [(\omega_n t_s)^2 - r\zeta\omega_n t_s]z^{-2}}{[4 + 4\zeta\omega_n t_s + (\omega_n t_s)^2] + [2(\omega_n t_s)^2 - 8]z^{-1} + [4 - 4\zeta\omega_n t_s + (\omega_n t_s)^2]z^{-2}}$$
(8.38)

By equating the denominator polynomials in the above two equations, C_1 and C_2 can be found as

$$C_1 = \frac{1}{k_0 k_1} \frac{8\zeta\omega_n t_s}{4 + 4\zeta\omega_n t_s + (\omega_n t_s)^2}$$

$$C_2 = \frac{1}{k_0 k_1} \frac{4(\omega_n t_s)^2}{4 + 4\zeta\omega_n t_s + (\omega_n t_s)^2}$$
(8.39)

The applications of these equations will be discussed in the next two sections.

In reference 6 a third-order phase-locked loop is also implemented. The filter is implemented in digital format and the result can be used for phase-locked loop designs, but it is not included in this book.

8.6 CARRIER AND CODE TRACKING[4]

Before discussing the usage of the above equations, let us concentrate on the tracking of GPS signals. The input to a conventional phase-locked loop is usually a continuous wave (cw) or frequency-modulated signal and the frequency of the VCO is controlled to follow the frequency of the input signal. In a GPS receiver the input is the GPS signal and a phase-locked loop must follow (or track) this signal. However, the GPS signal is a bi-phase coded signal. The carrier and code frequencies change due to the Doppler effect, which is caused by the motion of the GPS satellite as well as from the motion of the GPS receiver as discussed in Chapter 3. In order to track the GPS signal, the C/A code information must be removed. As a result, it requires two phase-locked loops to track a GPS signal. One loop is to track the C/A code and the other one is to track the carrier frequency. These two loops must be coupled together as shown in Figure 8.3.

In Figure 8.3, the C/A code loop generates three outputs: an early code, a late code, and a prompt code. The prompt code is applied to the digitized input signal and strips the C/A code from the input signal. Stripping the C/A code means to multiply the C/A code to the input signal with the proper phase as shown in Figure 7.1. The output will be a cw signal with phase transition caused only by the navigation data. This signal is applied to the input of the carrier loop. The output from the carrier loop is a cw with the carrier frequency of the input signal. This signal is used to strip the carrier from the digitized input signal, which means using this signal to multiply the input signal. The output is a signal with only a C/A code and no carrier frequency, which is applied to the input of the code loop.

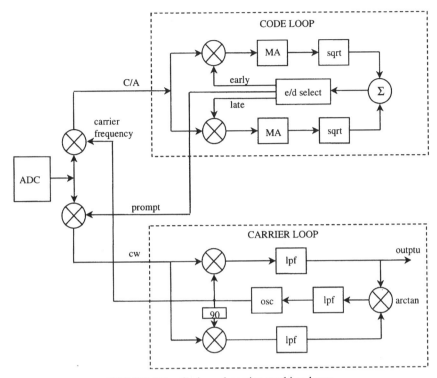

FIGURE 8.3 Code and carrier tracking loops.

The acquisition program determines the beginning of the C/A code. The code loop generates early and late C/A codes and these two codes are the C/A code time shifted typically by approximately one-half-chip time of 0.489 us (1/2 × 1.023 × 10⁶) or less. The early and late codes correlate with the input C/A codes to produce two outputs. Each output passes through a moving average filter and the output of the filter is squared. The two squared outputs are compared to generate a control signal to adjust the rate of the locally generated C/A code to match the C/A code of the input signal. The locally generated C/A code is the prompt C/A code and this signal is used to strip the C/A code from the digitized input signal.

The carrier frequency loop receives a cw signal phase modulated only by the navigation data as the C/A code is stripped off from the input signal. The acquisition program determines the initial value of the carrier frequency. The voltage-controlled oscillator (VCO) generates a carrier frequency according to the value obtained from the acquisition program. This signal is divided into two paths: a direct one and one with a 90-degree phase shift. These two signals are correlated with the input signal. The outputs of the correlators are filtered and their phases are compared against each other through an arctangent comparator. The arctangent operation is insensitive to the phase transition caused by the navigation data and it can be considered as one type of a Costas loop. A Costas loop is a phase-locked loop, which is insensitive to phase transition. The output of the comparator is filtered again and generates a control signal. This control signal is used to tune the oscillator to generate a carrier frequency to follow the input cw signal. This carrier frequency is also used to strip the carrier from the input signal.

8.7 USING THE PHASE-LOCKED LOOP TO TRACK GPS SIGNALS[6,7]

In this section, the application of the equations derived in Sections 8.3 through 8.5 will be discussed. A tracking program using the phase-locked loop will be discussed. The input data to the tracking loop are collected from actual satellites. In this discussion second-order phase-locked loops will be used. Several constants must be determined such as the noise bandwidth, the gain factors of the phase detector, and the VCO (or the digital frequency synthesizer). These constants are determined through trial and error and are by no means optimized. This tracking program is applied only on limited data length. Although it generates satisfactory results, further study might be needed if it is used in a software GPS receiver designed to track long records of data. The following steps can be applied to both the code loop and the carrier loop:

1. Set the bandwidths and the gain of the code and carrier loops. The loop gain includes the gains of the phase detector and the VCO. The bandwidth of the code loop is narrower than the carrier loop because it tracks the signal for a longer period of time. Choose the noise bandwidth of the code loop to be 1 Hz and the carrier loop to be 20 Hz. This is one set of several possible selections that the tracking program can operate or function.
2. Select the damping factor in Equation (8.25) to be $\zeta = .707$. This ζ value is often considered close to optimum.
3. The natural frequency can be found from Equation (8.26).
4. Choose the code loop gain $(k_0 k_1)$ to be 50 and the carrier loop gain to be $4\pi \times 100$. These values are also one set of several possible selections. The constants C_1 and C_2 of the filter can be found from Equation (8.39).

These four steps provide the necessary information for the two loops. Once the constants of the loops are known, the phase of the code loop and the phase of the carrier frequency can be adjusted to follow the input signals. In this approach, the loops usually update every millisecond, because the C/A code is one millisecond long. At every millisecond the C/A code must be regenerated and the initial phase of the C/A code must be continuous from the previous one. This initial code phase can be related to fine time resolution. The phase of carrier frequency is updated from the output of the arctangent phase comparator. The output is obtained from the in-phase channel of the carrier loop as shown in Figure 8.3. A typical output data set is shown in Figure 8.4. In this figure, the amplitude

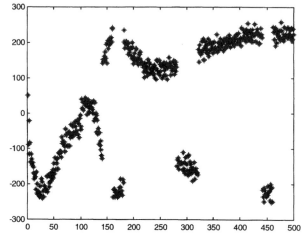

FIGURE 8.4   Outputs from conventional tracking method of sv17.

changes with time; this is the transit effect of the tracking loop. Finally, the amplitude reaches a steady state.

8.8   CARRIER FREQUENCY UPDATE FOR THE BLOCK ADJUSTMENT OF SYNCHRONIZING SIGNAL (BASS) APPROACH[8,9]

The purpose of introducing the BASS method is to present a different tracking program from the conventional method. This program is used for the software GPS receiver discussion in the next chapter. In this program, once the C/A code is generated it is used all the time. No initial phase adjustment is required such as in the conventional phase-locked loop discussed in previous sections. The fine time resolution can be obtained from the early and the late outputs of the code loop, which will be discussed in Sections 8.11 and 8.12.

This discussion is based on reference 8. The operation is on 1 ms of data for simplicity; however, other data lengths can be used. The concept is based on discrete Fourier transform (DFT). If the digitized input signal is $x(n)$, the DFT output $X(k)$ can be written as

$$X(k) = \sum_{n=0}^{N-1} x(n)e^{-j2\pi nk/N} \tag{8.40}$$

where k represents a certain frequency component and N is the total number of data points. If $x(n)$ is obtained from digitizing a sinusoidal wave, the highest amplitude $|X(k_i)|$ represents the frequency of the input signal. The real (Re) and imaginary (Im) parts of $X(k_1)$ can be used to obtain phase angle θ as

$$\theta = \tan^{-1}\left(\frac{\text{Im}[X(k_i)]}{\text{Re}[X(k_i)]}\right) \tag{8.41}$$

K——— processing time ———✳———————————➤|

θ_t θ_{t+m}

|◄——— time between processing ———➤|

FIGURE 8.5   Phase angle from two consecutive data sets.

where θ presents the initial phase of the sine wave with respect to the Kernel function. If k is an integer, the initial phase of the Kernel function is zero. In general, if the frequency of the input signal is an unknown quantity, all the components of k ($k = 0 \sim N - 1$) must be calculated. However, only half of the k values ($k = 0 \sim N/2 - 1$) provide useful information as discussed in Section 6.13. The highest component $X(k_i)$ can be found by comparing all the $X(k)$ values. For this operation, the fast Fourier transform (FFT) is often used to save calculation time.

If the frequency of the input signal can be found within a frequency resolution cell, which is equal to $1/Nt_s$ (where t_s is the sampling time), the desired $X(k)$ can be found from one component of the DFT. It should be noted that to calculate one component of $X(k)$, the k value need not be an integer as in the case of FFT. Since the input frequency is estimated from the acquisition method, the $X(k)$ can be found from one k value of Equation (8.40). The purpose of this operation is to find the fine frequency of the input signal.

The phase angle θ can be used to find the fine frequency of the input signal as discussed in Section 7.13. Figure 8.5 shows that the data points are divided into two different time domains. In each time domain, the same $X(k_i)$ are calculated. The corresponding phase angles are θ_n and θ_{n+m} and they are separated by time m. The fine frequency can be obtained as

$$f = \frac{\delta\theta}{m} \equiv \frac{\theta_{n+m} - \theta_n}{m} \tag{8.42}$$

This relation can provide much finer frequency resolution than the DFT result. The frequency resolution depends on the angle resolution measured.[9] The difference angle $\delta\theta$ must be less than $2\pi/5$ as discussed in Section 7.14 and this requirement limits the time between the two consecutive DFT calculations. Since the frequency k is very close to the input frequency, which changes slowly with time, the unambiguous frequency range is not a problem. This approach is used to find the correct frequency and update it accordingly.

8.9   DISCONTINUITY IN KERNEL FUNCTION

In conventional DFT operation the k value in Equation 8.40 is an integer. However, in applying Equation (8.40) to the tracking program the k value is usually a noninteger, because in using an integer value of k, the frequency generated from the kernel function $e^{-j2\pi nk/N}$ can be too far from the input frequency. If the k value is far from the input signal, the amplitude of $X(k)$ obtained from Equation (8.40) will be small, which implies that the sensitivity of the processing is low. In order to avoid this problem, the k value should be kept as close to the input frequency as possible. A k value close to the input frequency can also reduce the frequency ambiguity. Under this condition, the k value is usually no longer an integer.

When the k value is an integer, the initial phase of the Kernel function $e^{-j2\pi nk/N}$ is zero and the values obtained from two consecutive sets are continuous. The beginning point of the first set is $n = 0$ and the beginning point of the second set is $n = N$. It is easily shown that

$$e^{-j2\pi nk/N}\big|_{n=0} = e^{-j2\pi nk/N}\big|_{n=N} \quad \text{if} \quad k = \text{integer} \tag{8.43}$$

If k is not an integer this relation no longer holds. The following example is used to illustrate this point. Assume that $N = 256$, and $n = 0 \sim 255$. For any integer value of k, 256 data points can be generated from $e^{-j2\pi nk/N}$ for $n = 0 \sim 255$. Two sets of the same 256 data points are placed in cascade to generate a total of 512 data points. There is no discontinuity from data point 256, the last data point of the first set, to data point 257, the beginning of the second set. Since the values generated from $e^{-j2\pi nk/N}$ are complex, the continuity can be shown graphically only in real and imaginary parts of $e^{-j2\pi nk/N}$. Figure 8.6a shows the real and imaginary results of $k = 20$. In this figure only the points from 240 to 270 are plotted and there is no discontinuity. Figure 8.6b shows the results of $k = 20.5$ and there is a discontinuity between point 256 and 257 in both the real and imaginary portions of $e^{-j2\pi nk/N}$. The discontinuity will affect the application of Equation (8.42).

Figure 8.7 shows a cw input signal and two sections of the real part of $e^{-j2\pi nk/N}$. If $e^{-j2\pi nk/N}$ is continuous, the two sets of DFT can be considered as the correlation of the input signal with one complex cw signal. Under this condition, Equations (8.41) and (8.42) can be used to find the fine frequency. If the kernel function has a discontinuity, the two sets of DFT are the input signal correlated with two sets of kernel functions. Under this condition there is a phase discontinuity in the phase relation. In order to use Equation (8.42), the phase discontinuity must be taken into consideration.

This discontinuity can be found by calculating the phase angle at $n = N$. In the previous example, $N = 256$ and $n = 0 \sim 255$ are used to generate the values of the kernel function. In order to generate a continuous kernel function, the value of $e^{-j2\pi nk/N} = e^{-j2\pi k}$ ($n = 256$) must equal e^{-0} (or zero degree). If k is an integer, this relation is true. If k is not an integer, this relation does not hold and the phase difference between $e^{-j2\pi k}$ and e^{-0} is the phase discontinuity. This phase must be subtracted from the phase angle before Equation (8.42) can be properly used.

If the difference phase from the Kernel function is subtracted at the end of each millisecond, two situations can occur between the two adjacent milliseconds. One is that there is no phase change and the other one is that there is a π phase

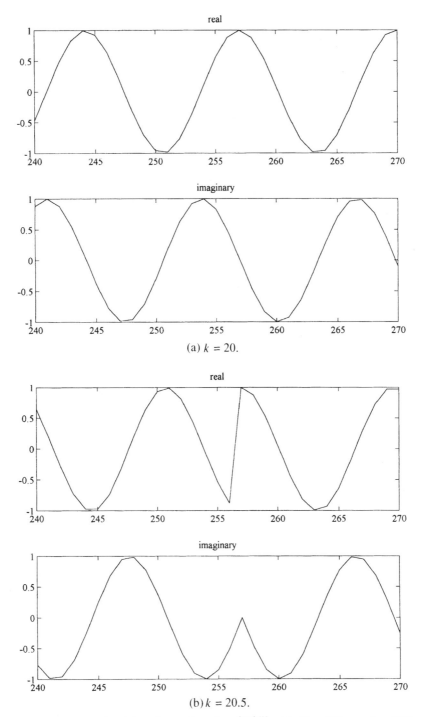

FIGURE 8.6   Real and imaginary plot of $e^{-j2\pi nk/N}$ with $N = 256$, $n = 0 \sim 255$.

(a) Input signal.

Two sets of real part of
$e^{-j2\pi nk/N}$ continuous

(b) Without discontinuity.

Two sets of real part of
$e^{-j2\pi nk/N}$ discontinuous

(c) With discontinuity.

FIGURE 8.7 An input signal and two sections of the real part of $e^{-j2\pi nk/N}$.

change due to the navigation data. Since there is noise in the input data, the phase change will not exactly equal the desired values of 0 to $\pm\pi$. For example, if the phase shift is close to 0 or 2π, it is considered that there is no phase shift. If the phase shift is close to $\pm\pi$, it is considered that there is a π phase shift. In general, a threshold can be set at $\pm\pi/2$. In Figure 8.8, the thresholds are set at $\pi/2$ and $3\pi/2$. If the absolute value of the difference angle is within the range $\pi/2$ and $3\pi/2$, it can be classified as a π phase shift. Otherwise, there is no phase shift. The π phase shift cannot occur within 20 ms and it occurs only at a multiple of 20 ms.

8.10 ACCURACY OF THE BEGINNING OF C/A CODE MEASUREMENT

The input signal is digitized at 5 MHz, or every data point is separated by 200 ns. With this time resolution, the corresponding distance resolution is about 60 m ($3 \times 10^8 \times 200 \times 10^{-9}$), which is not accurate enough to solve for a user position. Since the GPS signal and digitizing clock of the receiver cannot be synchronized, it is not likely to match a data point with the true beginning of the C/A code. Under the worst condition, the digitized beginning of the C/A code can be 100 ns away from the true value, when the true beginning of the C/A code falls at the middle of two digitizing points. The acquisition program can only measure the accuracy of the beginning of the C/A code to the digitized resolution. It is desirable to measure the beginning of the C/A code very accurately.

In the conventional tracking loop discussed in Section 8.7, the locally generated C/A code is updated every millisecond. The purpose of the updating is to generate a C/A code to match the C/A code in the input signal and generate a carrier frequency to match the carrier frequency in the input signal. Only

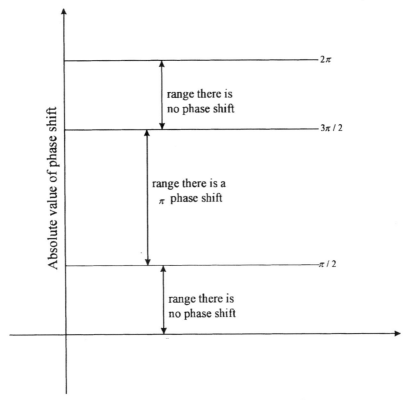

FIGURE 8.8   Thresholds of a phase shift.

the matching of the C/A code will be discussed here. For example, if the true beginning of the C/A code is exactly at the middle of two digitizing points, it is desirable to generate the local C/A code to match exactly at the same point. However, noise in the signal will prevent this from occurring. The locally generated code can only be close to the desired value. One can measure the beginning of the locally generated C/A code to find fine time resolution.

In the BASS tracking program, the locally generated C/A code is a fixed one. The first data point always starts from the true beginning of the C/A code and this same code is used all the time. Under the worst condition, this locally generated C/A code and the digitized input can be 100 ns apart. Two approaches can be used to find the true starting point of the C/A code in the input signal with better time resolution.

One method uses three signals: a prompt, an early signal, and a late signal. The locally generated C/A code can be considered as the prompt code. From this signal early and late signals will be generated at a fixed chip spacing. Digitizing the C/A code at 5 MHz generates the prompt code. Since the C/A code is 1 ms long, it produces 5,000 data points. The early and late codes can be obtained by shifting the prompt data.

FIGURE 8.9 Input signal, prompt, early, and late codes.

Figure 8.9 shows the prompt C/A code, which starts from data point 1 to 5,000. The early code is generated by shifting two points that are arbitrarily chosen, 4,999 and 5,000, to the front; thus, the early code has the data points in the sequence of 4,999, 5,000, 1, 2 ... 4,998. Shifting points 1 and 2 to the end generates the late code; thus, the late code has the sequence of 3, 4, ... 5,000, 1, 2. All three codes are correlated with the input signal.

The other approach uses five signals: one prompt, two early, and two late signals. Shifting the prompt signal by 4 points generates the additional early and late signals. Both approaches use approximations. A detailed discussion will be presented in the following sections.

8.11 FINE TIME RESOLUTION THROUGH IDEAL CORRELATION OUTPUTS[8]

As discussed in Section 5.7, the correlation peak of the C/A code is 1,023. If the C/A code is off by more than one chip, the correlation has three values: 63, −1, 65. The first method assumes that the correlation value is 0 if the two signals are off by more than one chip for simplicity. Figure 8.10 shows the correlation within one chip.

Each chip is 977.5 ns (1,000 us/1023) long without considering the Doppler effect; let us refer this as the chip time. The three correlation values are: y_p from the prompt C/A code, y_e from the early code, and y_l from the late code. The time d is measured from the prompt to the early or to the late in units of chip time. For this special case, the time between the early and prompt or late and prompt is 400 ns, which can be written in units of chip time as $d = .4092$ (400/977.5). Suppose that y_p is x seconds from the ideal peak. This method uses the values of y_e and y_l to find the value of x, which can be either positive or negative.

The value of x can be found from the following equations as

$$y_p = 1023(1 - |x|)$$

$$y_l = 1023(1 - x - d)$$

$$y_e = 1023(1 + x - d) \tag{8.44}$$

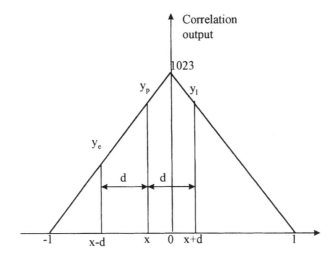

Correlation lag in unit of chip

FIGURE 8.10   Correlation output of a C/A code.

The units of x and d are measured in chip time at 977.5 ns. The ratio of y_l and y_e can be written as

$$r \equiv \frac{y_l}{y_e} = \frac{1 - x - d}{1 + x - d} \quad \text{or}$$

$$x = \frac{(1 - r)(1 - d)}{1 + r} \tag{8.45}$$

The value x can be found from this equation once r and d are calculated. The value of r can be found from y_l and y_e, as shown in the first portion of Equation 8.45. The above calculations are performed every 5,000 data points (1 ms), or approximately 1 C/A code when Doppler shift is taken into consideration.

Now let us find the amplitude of y_p under the worst mismatched condition. This information can show the sensitivity degradation of this approach. Under the worst mismatched condition, the closest digitized point is 100 ns from the beginning of the C/A code as shown in Figure 8.11. Since $x = \pm 100$ ns and $d = 400$ ns, after normalizing x to the chip time, one can find that $y_p \approx 1023 \times (1 - 100/977.5) = 918.35$ from Equation (8.44), which is about -0.94 dB ($20 \times \log(918.35/1023)$) below the ideal correlation peak. Therefore, the worst situation is that the correlation peak is about 1 dB less than the ideal case. Occasionally, the C/A loop will be off more than $d/4$, because the update is not performed every millisecond and the noise in the data may cause error. In the conventional tracking loop the C/A code is generated every millisecond with the initial phase properly adjusted. Therefore, the locally generated C/A code can better match the input signal with slightly higher peak than the BASS method. In general,

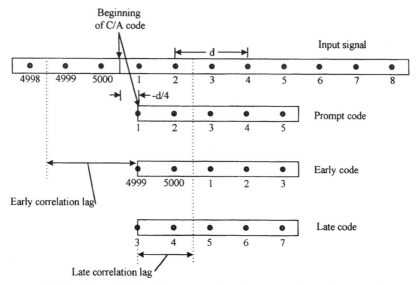

FIGURE 8.11 Digitized input signal and prompt code are 100 ns off.

the acquisition program has less sensitivity than the tracking program. From experimental results it appears that as long as the acquisition program can find a signal, the tracking program can track it without any difficulty.

In Equation (8.45), when $x > 100$ ns or $x < -100$ ns, the prompt code and the input data are misaligned by more than half the sampling time. Under this condition, the input data should be shifted one sample to align better with the code as shown in Figure 8.12. In this figure, before the data shifting the x is larger than $d/4$ (or half the sampling time). All three codes, the prompt, early, and late, move to the left, which is equivalent to shifting the input data to the right as shown in Figure 8.12. After the shifting of input data, x is smaller than $d/4$.

When $x = 100$ ns, $r = 0.705$ calculated from Equation (8.45). When $x = -100$ ns, $r = 1.419(1/0.705)$. These two r values (0.705 and 1.419) can be considered as a threshold. If the r value is less than 0.705 or larger than 1.419, this means $x > 100$ ns of $x < -100$ ns. Under both conditions, the input data should be shifted one sample to better align with the locally generated code. It is impractical to use the r value from 1 ms of data to determine whether to make a shift of the input data because of the noise. It usually takes 10 or 20 ms of data to make a decision. From Section 3.6 and considering a 10 kHz Doppler frequency shift on the carrier, it takes approximately 16 ms to shift the C/A code by half the sampling time of 100 ns, which can justify this update rate.

In the actual tracking program used, the update is performed every 10 ms. The r value is calculated every millisecond, but ten r values are averaged and the averaged value is compared with the threshold to determine whether a shift in input data is needed. The x value calculated from the averaged r value through Equation (8.45) is considered as the fine time resolution.

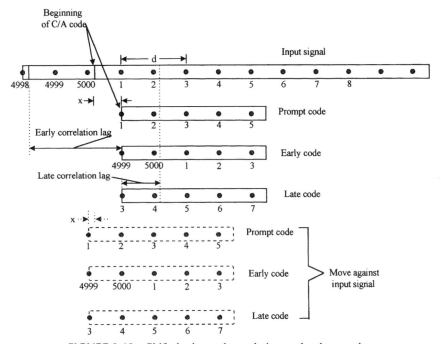

FIGURE 8.12   Shift the input data relative to the three codes.

8.12   FINE TIME RESOLUTION THROUGH CURVE FITTING

The discussion in the previous section is based on an ideal correlation of the C/A code. The correlation output is triangle shaped with values varying from 0 to 1,023. The actual correlation of the C/A signal does not start from zero and the shape is not a triangle. In an actual receiver, since the bandwidth is limited, the correlation does not have a sharp peak as shown in Figure 8.10, but is a smooth one as shown in Figure 8.13. The top of the correlation output is rounded due to the limited bandwidth. To take this shape of correlation function into consideration, a quadratic equation can be used to model it. In order to perform the curve fitting, the correlation data must contain the highest value and the two values on either side of it. If this situation does not occur, a wrong result can be drawn. In order to guarantee that this situation occurs, usually more than three correlation values are needed. In general, five correlation peaks with two early and two late codes should be sufficient. The two early and late codes are obtained by shifting the prompt code by $\pm d$ and $\pm 2d$. The highest y value and its two adjacent neighbors are used in the following equation. The quadratic equation to model the correlation peak can be written as

$$y = ax^2 + bx + c \qquad (8.46)$$

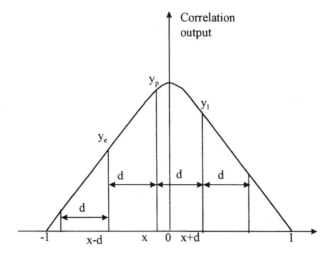

Correlation lag in unit of chip

FIGURE 8.13   Correlation Output with Limited Bandwidth.

where y represents the correlation value and x represents the x-axis, which uses time d as a unit. This equation can be solved to obtain a, b, and c with three sets of x and y values as

$$
\begin{bmatrix} y_1 \\ y_2 \\ y_3 \end{bmatrix} = \begin{bmatrix} x_1^2 & x_1 & 1 \\ x_2^2 & x_2 & 1 \\ x_3^2 & x_3 & 1 \end{bmatrix} \begin{bmatrix} a \\ b \\ c \end{bmatrix} \quad \text{or}
$$

$$
Y = XA \quad \text{with}
$$

$$
Y = \begin{bmatrix} y_1 \\ y_2 \\ y_3 \end{bmatrix}
$$

$$
X = \begin{bmatrix} x_1^2 & x_1 & 1 \\ x_2^2 & x_2 & 1 \\ x_3^2 & x_3 & 1 \end{bmatrix}
$$

$$
A = \begin{bmatrix} a \\ b \\ c \end{bmatrix} \tag{8.47}
$$

The solution can be written as

$$
A = X^{-1}Y \tag{8.48}
$$

where X^{-1} is the inverse of a matrix. Once a, b, c from Equation (8.48) are found, the maximum value of y can be found by taking the derivative of y with respect to x and setting the result to zero. The result is

$$\frac{dy}{dx} = 2ax + b = 0 \quad \text{or}$$

$$x = \frac{-b}{2a} \tag{8.49}$$

Thus, x is the desired result. If $x > 100$ ns ($d/2$) or $x < -100$ ns ($-d/2$), it is required to shift the input data with respect to the locally generated C/A code as discussed in the previous section.

In actual application, every millisecond five y values are generated. Every 10 milliseconds the y values are averaged to generate five averaged y values. The highest averaged y value and its two adjacent neighbors are used to find the desired x value. This x value is considered as the fine time resolution and used to determine whether a shift in input data is needed.

This method and the method discussed in Section 8.11 are both used in the BASS tracking program. The differences between these two programs are insignificant. The results from the method discussed in Section 8.11 are used in Chapter 9.

8.13   OUTPUTS FROM THE BASS TRACKING PROGRAM

Besides the phase angles measured through the BASS program, two additional outputs are important for calculating user position. One is the beginning of the C/A code and the other is the fine time resolution. Since the update is performed every 10 ms, the shift of the beginning of the C/A code is checked at this rate. If the misalignment between the locally generated C/A code and the input data is more than 100 ns, the input data must be shifted either to the right or to the left one data point to better match the local generated code. The actual starting points (reference to the input data points) of the C/A code for every 10 ms must be kept. They will be used in the next chapter to find the beginning of the subframes.

The C/A code beginning points of 6 satellites for 81 ms of data are listed in Table 8.1. These values are the input data points that represent the beginning of a C/A code. For each satellite the data points are separated by 10 ms. The

TABLE 8.1   C/A Code Beginning Points

SV #	Starting Point	Beginning of C/A Code							
6	2884	52885	102885	152885	202885	252885	302885	352885	402885
10	3814	53815	103815	153815	203815	253815	303815	353815	403815
17	0469	50470	100470	150470	200470	250470	300470	350470	400469
23	2200	52202	102201	152201	202201	252201	302201	352201	402201
26	2664	52664	102664	152664	202664	252664	302664	352664	402664
28	3269	53269	103270	153270	203270	253269	303269	353269	403269

first starting point is obtained from the acquisition method. If there is no shifting of input data points, the difference between numbers of the same row should be 50,000, which represents 10 ms of data. Sometimes the starting points are slightly different from the values obtained from the tracking program.

In satellites 6, 10, and 26 there are no input data point shifts. However, the first starting points of satellites 6 and 10 are off by 1 point from the rest of the data. For satellite 17 the last data point has a shift. For satellite 23 the second data point shifts 2 points then shifts back one point and for satellite 28 the data points shift back and forth and these effects are caused by noise.

The values of the beginning of the C/A code keep increasing every 10 ms as shown in Table 8.1. Each value is about 50,000 points more than the previous one. In a long record of data these values can become extremely large. It is inconvenient to store these values. However, these values need not be so large. The reason for keeping these large values is easier to explain in the next chapter. In actual programming these values are kept between 1 and 5,000. For example, the beginning of the C/A code for satellite 6 will have the same value of 2,885 instead of the values listed in the table. The beginning of the C/A code is used to find the beginning of the subframes, which can be located within 1 ms of input data. This topic will be discussed in the next chapter. Since within 1 ms the beginning of the C/A code is from 1 to 5,000, a value within this range is sufficient to locate the beginning of the first subframe.

The time resolution in the above data is 200 ns, determined by the sampling frequency of 5 MHz, thus, the beginning of the C/A code can be measured with this time resolution. With each value of the beginning of the C/A code there is a fine time x calculated by Equations (8.45) or (8.49), which is not included in Table 8.1. These fine times can be used to improve the overall time resolution.

8.14   COMBINING RF AND C/A CODE

In Sections 8.8 to 8.12 the BASS tracking method is discussed. In order to track the input signals for each satellite, two quantities are locally generated: a complex RF frequency and the C/A code. Once the C/A is generated, it is used all the time. The locally generated complex RF signal is updated at most every 10 ms because the carrier frequency changes very slowly for a stationary receiver as discussed in Chapter 3. The locally generated C/A code and RF signal can correlate with the input signal simultaneously. One convenient approach is to combine the C/A code and the Rf signal through point-by-point multiplication to generate a new code. This new code can be used as the prompt code. This prompt code is shifted two data points to the left and right to obtain the early and late codes. These three codes consisting of the C/A code and RF signals are used to correlate with the input signal. The phase of the RF in the prompt signal is important because it is used to find the fine frequency and the phase transition in the navigation data. The amplitudes of the early and late codes are used only to determine the fine time resolution; therefore, the phase of RF signals in the early and late codes is not important. The fine frequency is calculated from Equation (8.42) and the new frequency is used in generating the local RF signal for the next 10 ms of data. The phase discontinuity in the Kernel function must be calculated and necessary adjustments must be made as discussed in Section 8.9.

The outputs from the BASS method are the phase angles obtained from the prompt code. The phase angles are calculated every millisecond. After the phase adjustment discussed in Section 8.9, the phase should either be continuous between two milliseconds of data or changing by π. The absolute values of the phase angle are not very important because it depends on the initial sampling point. The difference angle is defined as the phase difference between two adjacent milliseconds. If the difference angle is within $\pm\pi/2$, it is considered as no phase shift. If the difference is outside the range of $\pm\pi/2$, there is a π phase shift. The π phase shift should happen in multiples of 20 ms.

8.15 TRACKING OF LONGER DATA AND FIRST PHASE TRANSITION

The BASS tracking program discussed in this chapter is based on tracking 1 ms of input data. If the signal is weak, it is possible to track more than 1 ms of data to improve sensitivity. It seems that the maximum data length that can be tracked coherently is 20 ms without very sophisticated processing because the navigation data is 20 ms long. The input data must be properly selected and there should not be a phase transition within the selected 20 ms of data. Under this condition, the sensitivity is improved by 13 dB (10 log (20/1)) over tracking 1 ms of data. If the data length is longer than 20 ms, it may contain a phase transition due to the navigation data. The phase transition will disturb the operation of the tracking program. If multiples of 20 ms of data are used, the tracking process must cover all the possible combinations of phase transition. The tracking program can be rather complicated and calculation intensive.

In order to track 20 ms of data without a navigation data phase transition, one must find a phase transition in the data. This requirement puts an additional restraint on the acquisition program. The acquisition is not only required to find the beginning of the C/A code and the carrier frequency, it must also find a phase transition. If the phase transition cannot be found, it is impractical to track 20 ms of data. Therefore, finding the phase transition becomes an important requirement to process weak signals. In Chapter 10, the first navigation data transition is determined in the acquisition program rather than the tracking.

8.16 SUMMARY

In this chapter the concept of tracking a GPS signal is discussed. Two approaches are presented: the conventional and the BASS methods. A general discussion on the conventional phase-locked loop is presented and its application to GPS receiver is discussed. In addition, a BASS method is presented in detail. This method needs to generate the C/A code only once; thus it may save calculation time for software receiver design. Theoretically, the BASS method at worst case may lose about 1 dB of sensitivity with 5 MHz sampling rate due to the potential misalignment between the input signal and the locally generated signal. Two methods of generating fine time resolution are discussed. The outputs from the tracking are also discussed. These outputs will be used in the next chapter to find the user position.

APPENDIX[(10)]

Equation (8.26) can be written as

$$B_n = \frac{\omega_n}{2\pi} \int_0^\infty \frac{1 + 4\zeta^2 x}{x^4 + 2(2\zeta^2 - 1)x + 1} dx = \frac{\omega_n}{2\pi} I_1 + \frac{\omega_n}{2\pi} 4\zeta^2 I_2 \tag{8A.1}$$

where

$$I_1 = \int_0^\infty \frac{dx}{x^4 + 2(2\zeta^2 - 1)x^2 + 1} \tag{8A.2}$$

$$I_2 = \int_0^\infty \frac{x^2 dx}{x^4 + 2(2\zeta^2 - 1)x^2 + 1} \tag{8A.3}$$

These integrals can be found from[(11)]

$$\int_0^\infty \frac{x^{\mu-1} dx}{(\beta + x^2)(\gamma + x^2)} = \frac{\pi}{2} \frac{\gamma^{\mu/2-1} - \beta^{\mu/2-1}}{\beta - \gamma} \csc \frac{\mu\pi}{2} \tag{8A.4}$$

with the condition that

$$|\arg \beta| < \pi, |\arg \gamma| < \pi \quad 0 < \text{real } \mu < 4$$

The values of β and γ in Equation (8A.4) can be found by comparing this equation with Equation (8A.3) or (8A.2) as

$$\beta\gamma = 1$$
$$\beta + \gamma = 2(\zeta^2 - 1) \tag{8A.5}$$

Solving for β and γ, the results are

$$\beta = (\zeta + \sqrt{\zeta^2 - 1})^2$$
$$\gamma = (\zeta - \sqrt{\zeta^2 - 1})^2 \tag{8A.6}$$

To obtain the result of I_1 where $\mu = 1$ is

$$I_1 = \frac{\pi}{2} \frac{\gamma^{1/2-1} - \beta^{1/2-1}}{\beta - \gamma} \csc \frac{\pi}{2} = \frac{\pi}{4\zeta} \tag{8A.7}$$

The result of I_2 can be obtained with $\mu = 3$ as

$$I_2 = \frac{\pi}{2} \frac{\gamma^{3/2-1} - \beta^{3/2-1}}{\beta - \gamma} \csc \frac{3\pi}{2} = \frac{\pi}{4\zeta} \tag{8A.8}$$

Interestingly, the two integrals provide the same results. Substituting these values into Equation (8A.1), the result is

$$B_n = \frac{\omega_n}{2} \left(\zeta + \frac{1}{4\zeta} \right) \tag{8A.9}$$

which is the desired result.

REFERENCES

1. Gardner, F. M., *Phaselock Techniques*, 2nd ed., Wiley, New York, 1979.
2. Best, R. E., *Phase-locked Loops, Theory, Design, and Applications*, McGraw-Hill, New York, 1984.
3. Stremler, F. G., *Introduction to Communications Systems*, 2nd ed., Addison-Wesley, Reading, MA, 1982.
4. Ziemer, R. E., Peterson, R. L., *Digital Communications and Spread Spectrum System*, p. 265, Macmillan, New York, 1985.
5. Chung, B. Y., Chien, C., Samueli, H., Jain, R., "Performance analysis of an all-digital BPSK direct-sequence spread-spectrum IF receiver architecture," *IEEE Journal of Selected Areas in Communications*, vol. 11, no. 7, pp. 1096–1107, September 1993.
6. Van Dierendonck, A. J., "GPS receivers," Chapter 8 in Parkinson, B. W., Spilker, J. J. Jr., *Global Positioning System: Theory and Applications*, vols. 1 and 2, American Institute of Aeronautics and Astronautics, 370 L'Enfant Promenade, SW, Washington, DC, 1996.
7. Stockmaster, M. H., Rockwell Collins, Cedar Rapids, IA, private communication. 1997.
8. Tsui, J. B. Y., Stockmaster, M. H., Akos, D. M., "Block adjustment of synchronizing signal (BASS) for global positioning system (GPS) receiver signal processing," *ION GPS 1997 Symposium*, pp. 637–643, Kansas City, MO, September 15–19, 1997.
9. Tsui, J. B. Y., *Digital Techniques for Wideband Receivers*, Chapter 10 of this book, Artech House, Norwood, MA, 1995.
10. Yu, J. S., West Virginia Institute of Technology, private communication.
11. Gradshteyn, I. S., Ryzhik, I. M., *Table of Integrals, Series and Products*, Equation 3.264-2, p. 300, Academic Press, New York, 1980.

CHAPTER NINE

GPS Software Receivers

9.1 INTRODUCTION

This chapter can be considered as a summary of all the previous chapters. As mentioned in Chapter 1, this book does not follow the signal path of a GPS receiver but rather follows the design concept of the GPS. Therefore, the previous chapters are presented in the following order. The user position can be found from the known positions of at least four satellites and the distances to them. In Chapters 2 to 4 the satellite constellation and the earth-centered, earth-fixed coordinate system are introduced and the equations to calculate the user position are presented. The structure of the GPS signal and the acquisition and tracking of the signal are discussed in Chapters 5, 7, and 8, respectively. Although this approach should give a better understanding of the overall concepts associated with GPS, the discussion may not flow smoothly from a receiver design perspective.

In this chapter the GPS receiver is discussed following the actual signal flow through the receiver. The input signal will be digitized first followed by acquisition and tracking. Once the tracking is achieved, the output will be converted into navigation data through subframe matching and parity checking. From the subframes the ephemeris data such as the week number can be found. The position of the satellite can be determined from the ephemeris data. The pseudoranges between the receiver and the satellites can also be determined. Once all the necessary information is obtained, satellite positions and the user position can be calculated. Finally, the user position is put in the desired coordinate system. The presentation in this chapter follows this order, which is shown in Figure 1.1. The block adjustment of synchronized signal (BASS) is used for the tracking program. The conventional approach will provide very similar results.

Fundamentals of Global Positioning System Receivers: A Software Approach, Second Edition,
by James Bao-Yen Tsui
ISBN 0-471-70647-7 Copyright © 2005 John Wiley & Sons, Inc.

9.2   INFORMATION OBTAINED FROM TRACKING RESULTS

As previously discussed, assume that the input GPS signal is down converted to 21.25 MHz, filtered to about 2 MHz bandwidth, and digitized at 5 MHz. The minimum requirement for the collected data is that they must contain the first three subframes. The information in these three subframes is used to find the satellite position and the user position as discussed in Chapter 5. The results from the conventional tracking program are shown in Figure 9.1, which plots the amplitude of the output signal. Each data point is obtained from 1 millisecond of digitized data. The signal from satellite 6 is relatively strong and the signal from satellite 28 is relatively weak, which can be observed from the two amplitude scales.

The same signal is tracked by the BASS method and the results are shown in Figure 9.2. The vertical scale in Figure 9.2 shows the angle separation rather than the signal amplitude. Since the difference in angle is always π for a phase change, the scales stay the same for both plots. In this figure, the stronger signal shows a tighter groups than the weak one.

In addition to the output signal, the conventional approach will output the initial phase of the locally generated C/A code. This initial phase of the C/A represents the fine time resolution in the tracking loop. In the BASS method, the

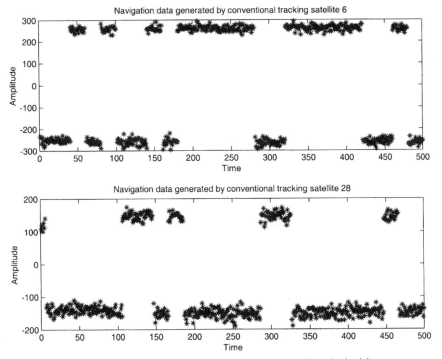

FIGURE 9.1   Tracking results from conventional phase-locked loop.

FIGURE 9.2 Tracking results from the BASS method.

C/A code is repeatedly used and the initial phase stays constant. As mentioned in Chapter 8, fine time resolution x can be obtained from the ratio of the correlations of the early and late C/A codes. These x values are calculated every 10 ms and the results are shown in Figure 9.3. The data should be close to a straight line. One can see that the results are fairly noisy, even though every point is generated from averaging 10 ms of data. Once the x value is greater than 100 ns or less than -100 ns, the next set of input data point is shifted by 200 ns. This effect causes the discontinuities in the plots. In Figure 9.3a, both the 18th and 20th points are greater than 100 ns but point 19 is smaller than -100 ns. This indicates that the input data point shifts back and forth from 18 to 19 and back to 20, then to point 21 again. Among these four points of data the input shifts three times. The cause of this shifting back and forth is noise. In Figure 9.3b, the first data point is much less than -200 ns. The reason is that the initial point is obtained from the acquisition program and the result might not be as accurate.

In Figure 9.3a the slope of the plot is positive and in Figure 9.3b the slope is negative. These represent the positive and negative Doppler frequency shift. The slopes of the plots represent the magnitude of the Doppler frequency shift.

From these plots one can decide that 1 ms of data should not be used to determine the input data point shift because it is very noisy. The tracking program

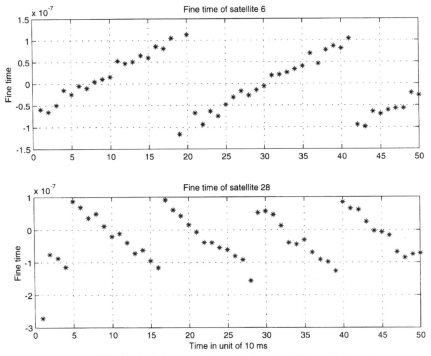

FIGURE 9.3   Fine time obtained from BASS method.

reports these fine times every 10 ms and also reports the beginning of the C/A code as discussed in Section 8.12. These fine times are used to estimate the pseudorange. Even the fine time calculated from 10 ms of data is somewhat noisy and if an individual point is used to calculate the user position, the accuracy may not be very good. A smoothing method such as the least mean square fit should be used to find the fine time from a relatively long record of data such as many tens of milliseconds of data points. These methods should improve the accuracy of the fine time, which should provide better accuracy in the calculated user position.

9.3   CONVERTING TRACKING OUTPUTS TO NAVIGATION DATA

The next step is to change the output data (every 20 ms) as shown in Figures 9.1 and 9.2 into $+1$ and -1 (or 0) values. There are several ways to accomplish this. One common way is to find the difference between adjacent millisecond outputs. If the difference is beyond a certain threshold, there is a data transition. For the conventional tracking program, the threshold is usually obtained from the minimum anticipated amplitude of the output. Since strong and weak signals produce different amplitudes as shown in Figure 9.1, the minimum values should be used as the threshold. For the BASS method the threshold is at $\pm\pi/2$ as discussed in Section 8.9.

From these transitions, it is easy to change the tracking results into navigation data. The navigation data transition points must correspond to individual points in the collected input data, which have a time resolution of 200 ns. This time resolution can be used to find the relative time difference between different satellites. The following steps can be applied to accomplish this goal. This method represents only one way to solve the problem and is by no means the optimum one. This method is presented because it might be easier to understand. The following steps are used to convert phase transition to navigation data:

1. Find all the navigation data transitions. The beginning of the first navigation data should be within the first 20 ms of output data because the navigation data are 20 ms long. However, there might not be a phase transition within 20 ms of data. The first phase transition can be used to find the beginning of the first navigation data. The first phase transition detected in the output data is the beginning of the navigation data. If the first phase transition is within the first 20 ms of data, this point is also the beginning of the first navigation data. If the first phase transition occurs at a later time, a multiple number of 20 ms should be subtracted from it. The remainder is the beginning of the first navigation data. For simplicity let us just call it the first navigation data point instead of the beginning of the first navigation data. This information will be stored and used to find the coarse pseudorange discussed in Section 9.6. The first navigation data point can be padded with data points of the same sign to make the first navigation data point always occur at 21 ms. This approach creates one navigation data point at the beginning of the data from partially obtained information. For example, if the first phase transition occurs at 97 ms, by subtracting 80 ms from this value, the first navigation data point occurs at 17 ms. These 17 ms of data are padded with 4 ms of data of the same sign to make the first navigation data 20 ms long. This process makes the first navigation data point at 21 ms. This operation also changes the rest of the beginnings of the navigation data by 4 ms. Thus, the navigation data points occur at 21, 41, 61, and so on.

Figure 9.4 illustrates the above example. The upper part of Figure 9.4 shows the output data from the tracking program and the bottom part shows the result padded with additional data. The adjusted first navigation data point at 21 ms is stored. If the first phase transition occurs at 40 ms, by subtracting 40, the adjusted first navigation data point occurs at 0 ms. Twenty-one ms of data with either + or − can be added in front of the first navigation data point to make it occur at 21 ms.

2. Once the navigation data points are determined, the validity of these transitions must be checked. These navigation data points must be separated by multiples of 20 ms. If these navigation data points do not occur at a multiple of 20 ms, the data contain errors and should be discarded.

3. After the navigation data points pass the validity check, these outputs are converted into navigation data. Every 20 outputs (or 20 ms) convert into one

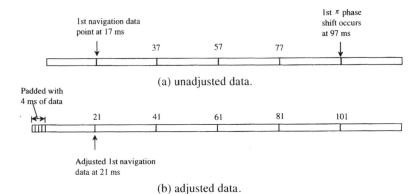

FIGURE 9.4   Adjustment of the first navigation data point.

navigation data bit. The signs of these navigation data are arbitrarily chosen. The navigation data are designated as $+1$ and -1. The parity check process can put the navigation data in the correct polarity.

9.4 SUBFRAME MATCHING AND PARITY CHECK

After the outputs from the tracking are converted into navigation data, the next step is to find the subframes in these data. As discussed in Section 5.9 and Figure 5.7, a subframe will start with the preamble of pattern (10001011) in the first word (the telemetry). In the second word HOW (the hand over word), bits 20–22 are the subframe ID and last two bits (29,30) are the parity code (00). However, simply searching for these data does not guarantee that the beginning of a subframe will be found. One can search for more than one subframe at a time. If matches are found for more than one subframe, it has a better probability of being correct.

It is important to notice that the polarities of the words in a subframe may change. Therefore, one should perform correlation on only one word (30 navigation data bits) at a time. In other words, each word should be separately correlated. The code to match the preamble can be written as $(1 -1 -1 -1 \, 1 -1 \, 1 \, 1)$. Since the polarity of the word is not known, the matched result can be ± 8. Once a match is found, 300 data points (1 subframe) later there should be another preamble match. If a match is not found, the first match is not a preamble. One can repeat this method to find the beginning of several subframes. More matches can improve the confidence level. The last two bits in the HOW can also be used for subframe matching. Once a subframe is found, the subframe number can be found from bits 20–22 of the HOW. The subframe numbers must be from 1 to 5 and they must be properly ordered from 1, 2, 3, 4, 5, 1, and so forth.

The parity check has been discussed in detail in Section 5.11. The procedure will not be repeated here. Two programs (p9_1) and (p9_2) are listed at the end of this chapter. They are used to match subframes and check parity.

The subframe-finding program matches three consecutive preambles. If all three are correctly matched, this declares that the beginning of a subframe is found. The search of three consecutive preambles is arbitrarily chosen. First the preamble from 360 data points is searched. This data length is one subframe plus two words, which should have at least one preamble match (there could be more than one). If multiple matches are found only one of them will be the preamble. If a match is found, two more preambles 300 data points after the beginning of the first match are sought. If the search of the two preambles fails, the first match is not a preamble but some other data with that pattern. If both preambles are matched, all three preambles are considered as the beginnings of three consecutive subframes.

The next step is to check the polarity of the last two bits in the HOW. These two bits should both be negative, thus, the sum of these two bits should be -2. However, the sum of these two bits could be either $+2$ or -2. If the sum is zero, there is a mistake and the beginnings of the three subframes must be wrong. This can be considered as an additional check. If the sum is -2 the sign of the HOW word is correct and the subframe number can be found from bits 20–22 of the HOW. If the sum is $+2$, the polarity of the HOW must be inverted first; then find the subframe number. From the subframe number one can search for the beginning of subframes 1, 2, and 3, because they contain the information to calculate the user position.

9.5 OBTAINING EPHEMERIS DATA FROM SUBFRAME 1

Once the beginning of subframe 1 is found, the following information can be obtained. The navigation data are in two forms: the binary and the two's complement as discussed in Section 5.12 and Table 5.8. For the convenience of calculations, most of these data are converted into decimal form:

1. WN: The week number has 10 bits from 61–70 in binary form. These data are converted into decimal form. As discussed in Section 5.12, this number starts from midnight January 5/morning January 6, 1980 and a rollover of 1,023 weeks must be taken care of. The decoded time must match the data collection time.

2. T_{GD}: The estimated group delay differential has 8 bits from 197–204 in two's complementary form. These data are converted into decimal form.

3. t_{oc}: the satellite clock corrections have 16 bits from 219–234 in binary form. These data are converted into decimal form.

4. a_{f2}: The satellite clock corrections have 8 bits from 241–248 in two's complementary form. These data are converted into decimal form.

5. a_{f1}: The satellite clock corrections have 16 bits from 249–264 in two's complementary form. These data are converted into decimal form.

6. a_{f0}: The satellite clock corrections have 22 bits from 271–292 in two's complementary form. These data are converted into decimal form.

7. IODC: The issue of data, clock has 10 bits. Bits 83–84 are the most significant bits (MSB) and bits 211–218 are the least significant bits (LSB). As discussed in Section 5.13, the LSB of the IODC will be compared with the issue of data, ephemeris (IODE) of subframes 2 and 3. Whenever these three data sets are not matched, a data set cutover has occurred and new data must be collected.

8. TOW: The time of the week has 17 bits from 31–47 in binary form. These data are converted into decimal form and the time resolution is 6 seconds as shown in Figures 5–7 and 5–8. In order to convert to seconds the data are multiplied by 6. Another important factor is that the TOW is the next subframe, not the present subframe as discussed in Section 5.10. In order to obtain the time of the present subframe 6 seconds must be subtracted from the results.

9.6 OBTAINING EPHEMERIS DATA FROM SUBFRAME 2

The data from subframe 2 can be obtained and converted into decimal form in a similar manner as discussed in the previous section. Referencing Figure 5.9b, the following sets of navigation data can be obtained from subframe 2:

1. IODE: The issue of data, ephemeris has 8 bits from 61–68. This bit pattern is compared with the 8 least significant bits (LSB) of the issue of data, clock (IODC) in subframe 1, and the IODE in subframe 3. If they are different, a data set cutover has occurred and these data cannot be used and new data should be collected.

2. C_{rs}: The amplitude of the sine harmonic correction term to the orbit radius has 16 bits from 69–84 in two's complementary form. These data are converted into decimal form.

3. Δn: The mean motion difference from computed value has 16 bits from 91–106 in two's complementary form. These data are converted into decimal form. The unit is in semicircles/sec, thus, the data are multiplied by π to change to radians.

4. M_o: The mean anomaly at reference time has 32 bits in two's complementary form. These data are divided into two parts, the 8-bit MSB from 107–114 and 24-bit LSB from 121–144, and are converted into decimal form. The unit is in semicircles, thus, the data are multiplied by π to change to radians.

5. C_{uc}: The amplitude of the cosine harmonic correction term to the argument of latitude has 16 bits from 151–166 in two's complementary form. These data are converted into decimal form.

6. e_s: The eccentricity of satellite orbit has 32 bits in binary form. These data are divided into two parts, the 8-bit MSB from 167–174 and 24-bit LSB from 181–204, and converted into decimal form.

7. C_{us}: The amplitude of the sine harmonic correction term of the argument of latitude has 16 bits from 211–226 in two's complementary form. These data are converted into decimal form.

8. $\sqrt{a_s}$: The square root of the semi-major axis of the satellite orbit has 32 bits in binary form. These data are divided into two parts, the 8-bit MSB from 227–234 and 24-bit LSB from 241–264, and converted into decimal form.

9. t_{oe}: The reference time ephemeris has 16 bits from 271–286 in binary form. These data are converted into decimal form.

9.7 OBTAINING EPHEMERIS DATA FROM SUBFRAME 3

The data from subframe 3 will be obtained in a similar way. Referencing Figure 5.9c, the following data can be obtained from subframe 3:

1. C_{ic}: The amplitude of the cosine harmonic correction term to the angle of inclination has 16 bits from 61–76 in two's complementary form. These data are converted into decimal form.

2. Ω_0: The longitude of the ascending node of orbit plane at weekly epoch has 32 bits in two's complementary form. These data are divided into two parts, the 8-bit MSB from 77–84 and 24-bit LSB from 91–114, and converted into decimal form. The unit is in semicircles, thus, the data are multiplied by π to change to radians.

3. C_{is}: The amplitude of the sine harmonic correction term to the angle of inclination has 16 bits from 121–126 in two's complementary form. These data are converted into decimal form.

4. i_0: The inclination angle at reference time has 32 bits in two's complementary form. These data are divided into two parts, the 8-bit MSB from 137–144 and 24-bit LSB from 151–174, and converted into decimal form. The unit is in semicircles, thus, the data are multiplied by π to change to radians.

5. C_{rc}: The amplitude of the cosine harmonic correction term to the orbit radius has 16 bits from 181–196 in two's complementary form. These data are converted into decimal form.

6. ω: The argument of perigee has 32 bits in two's complementary form. These data are divided into two parts, the 8-bit MSB from 197–204 and 24-bit LSB from 211–234, and converted into decimal form. The unit is in semicircles, thus, the data are multiplied by π to change to radians.

7. $\dot{\Omega}$: The rate of right ascension has 24 bits from 241–264 in two's complementary form. These data are converted into decimal form. The unit is in semicircles, thus, the data are multiplied by π to change to radians.

8. IODE: The issue of data, ephemeris has 8 bits from 271–278. This bit pattern is compared with the 8 least significant bits (LSB) of the issue of data, clock (IODC) in subframe 1, and the IODE in subframe 2. If they

are different, a data set cutover has occurred and these data cannot be used and new data should be collected.

9. idot: The rate of inclination angle has 14 bits from 279–292 in two's complementary form. These data are converted into decimal form. The unit is in semicircles, thus, the data are multiplied by π to change to radians.

It should be noted that the TOWs from subframes 2 and 3 are not decoded because the TOW from subframe 1 will provide the necessary information. All the data from subframes 1, 2, and 3 are decoded and converted to decimal form and have the desired units. The following steps are to calculate the satellite positions and user position.

Three computer program (p9_3), (p9_4), and (p9_5) are listed at the end of this chapter and they are used to obtain the navigation data from subframes 1, 2, and 3.

9.8 TYPICAL VALUES OF EPHEMERIS DATA

Some of the ephemeris data are user located dependent. Others are somewhat user location independent, such as the inclination angle. Some of the ephemeris data that are user location independent are listed in Table 9.1 as a reference. These data are from three different satellites and most of the values have about the same order of magnitude.

9.9 FINDING PSEUDORANGE

In collecting the digitized data there is no absolute time reference and the only time reference is the sampling frequency. As a result, the pseudorange can be

TABLE 9.1   Typical Ephemeris Data

Δn	5.117713173295686e − 009	5.055924885279763e − 009	4.529831542808932e − 009
$\sqrt{a_s}$	5.153714639663696e + 003	5.153681760787964e + 003	5.153659612655640e + 003
i_0	9.560779626333219e − 001	9.478276108359106e − 001	9.581404971861649e − 001
$\dot{\Omega}$	−8.277844805462636e − 009	−8.656432003710486e − 009	−7.999618930523888e − 009
idot	−6.393123442109902e − 011	7.857470152313846e − 011	−6.468126566291079e − 010
e_s	4.112668684683740e − 003	1.435045502148569e − 003	9.771452634595335e − 003
crs	−1.375000000000000e + 000	1.093750000000000e + 001	−1.453125000000000e + 001
cuc	−1.583248376846314e − 007	5.904585123062134e − 007	−7.711350917816162e − 007
cus	4.576519131660461e − 006	1.367181539535523e − 006	1.043826341629028e − 005
cic	2.048909664154053e − 008	2.793967723846436e − 008	−1.471489667892456e − 007
cis	−8.940696716308594e − 008	3.725290298461914e − 009	1.862645149230957e − 007
crc	2.872500000000000e + 002	3.436875000000000e + 002	1.749375000000000e + 002
a_{f0}	6.267381832003593e − 005	5.940999835729599e − 005	3.222310915589333e − 004
a_{f1}	1.136868377216160e − 012	3.410605131648481e − 012	6.662048690486699e − 011
a_{f2}	0	0	0
t_{oc}	4.032000000000000e + 005	4.031840000000000e + 005	4.032000000000000e + 005
t_{gd}	4.656612873077393e − 010	1.396983861923218e − 009	1.396983861923218e − 009

measured only in a relative way as shown in Figure 9.5, because the clock bias of the receiver is an unknown quantity. In this figure the points represent individual input digitized data and they are separated by 200 ns because the sampling rate is 5 MHz. The relative pseudorange is the distance (or time) between two reference points. In this figure the beginning point of subframe 1 is used as the reference point. All the beginning points of subframe 1 from different satellites are transmitted at the same time except for the clock correction terms of each satellite. As a result one can consider that the subframes from different satellites are transmitted at the same time. Since the beginnings of subframe 1 from different satellites are received at different times, this difference time represents the time (or distance) difference from the satellite to the receiver. Therefore, it represents the relative pseudorange. Subframe 1 occurs every 30 seconds and the maximum time difference between two satellites is about 19 ms as discussed in Chapter 3. From this information it is guaranteed that the subframe 1's transmitted at the same time from different satellites are compared. In other words, if the difference time between two satellites is in the tens of milliseconds range, the two subframe 1's must be transmitted at the same time and they cannot be separated by 30 seconds.

Now the digitized input data point corresponding to the beginning of subframe 1 must be found. This information can be obtained from three input data: (1) the beginning points of the C/A code obtained from the tracking program with a time resolution of 10 ms and accuracy of 200 ns as discussed in Section 8.12; (2) the beginning of the first navigation data (simplified as the first navigation data point) obtained from Section 9.3; (3) the beginning of subframe 1 obtained through subframe matching as discussed in Section 9.4 with a time resolution of 20 ms.

Figure 9.6 illustrates the relations among these three quantities. In Figures 9.6a, b, and d the results are obtained from actual collected data, but Figure 9.6c is artificially created for illustration. The upper part of the figure represents the original input data with the first beginning of the C/A code in the first 5,000 digitized input data points obtained from the acquisition program. The rest of the beginnings of the C/A code are obtained from the tracking program. In the bottom of the figure, the beginning of subframe 1 is given. These figures are used to help find the beginning of subframe 1 in terms of the beginning of the C/A code.

FIGURE 9.5   Relative pseudorange.

The following four examples are used to illustrate this operation. These four cases include the first phase transitions occurring at 0, 7(<10), 10, and 17(>10), which covers all the possible cases.

In Figure 9.6a the first phase transition occurs at 0 point and the data are padded with 21 points. The first π phase shift determines the first navigation data point. The beginning of subframe 1 is at the 100th navigation data point

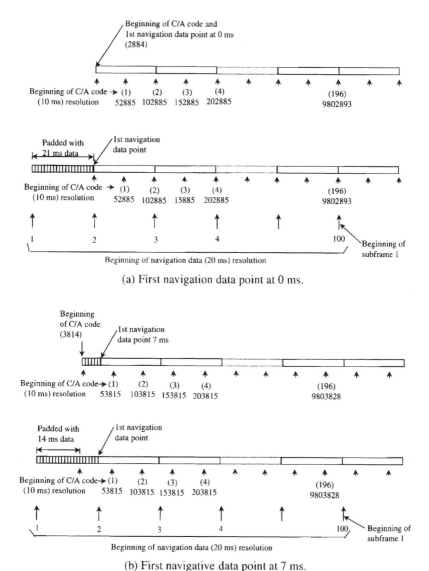

(a) First navigation data point at 0 ms.

(b) First navigative data point at 7 ms.

FIGURE 9.6 Relations among the beginning of the C/A code, first navigation data point, and beginning of subframe 1.

(c) First navigation data point at 10 ms.

(d) First navigation data point at 17 ms.

FIGURE 9.6   (*continued*)

(20 ms resolution). The corresponding beginning of the C/A code is at 196 (10 ms resolution). Thus, the navigation data point at 196 is lined up with the beginning of subframe 1. This relation can be obtained from the beginnings of the C/A code labeled (1), (2), (3), ... and the beginnings of navigation data labeled 1, 2, 3, ... at the bottom of the figure. The corresponding data point can be found from beginning of the C/A code as 9802893. This number 9802893 is obtained from the tracking program. This is the beginning of the C/A code with index of 196 and obtained from the tracking program.

In Figure 9.6b the first navigation data point occurs at 7 ms and the data are padded with 14 points at the beginning. The beginning of the subframe 1 is at the 100th navigation data point. The beginning of the C/A code in front of the beginning of subframe 1 is 196. The corresponding beginning of the C/A code is 9803828. However, this point does not align with the beginning of subframe 1. In order to align with the beginning of subframe 1, 7 ms will be added. These 7 ms come from the first navigation data point at 7 ms shown at the top of the figure. Since each millisecond contains 5,000 digitized data, 5,000 must be multiplied by this 7 ms to obtain the beginning of subframe 1 in terms of the digitized input data points. Thus, the beginning of subframe 1 is at $9803828 + 7 \times 5000 = 9838828$.

In Figure 9.6c the first phase transition is at 10 ms and the data are padded with 11 points at the beginning. The beginning of subframe 1 is at the 100th navigation data point. The beginning of the C/A code is aligned with the beginning of subframe 1 at 197. The corresponding beginning of the C/A code is 9850115.

In Figure 9.6d the first navigation data point is at 17 ms and the data are padded with 4 points at the beginning. The beginning of the subframe 1 is at the 99th navigation data point. The beginning of the C/A code in front of subframe 1 is at 195. The corresponding beginning of the C/A code with index of 195 is 9752661. However, in order to align with the beginning of subframe 1, 7 ms will be added. This 7 ms comes from the first navigation data point at 17 ms. Since the beginning of the C/A code has a time resolution of 10 ms, 10 ms are subtracted from the 17 ms to obtain 7 ms. The final value is $9752661 + 7 \times 5000 = 9787661$.

From the above discussion, one can see that it takes two steps to find the beginning of subframe 1 in terms of the actual digitized input data points. The first step is to find the index of the beginning of the C/A code just before subframe 1. The second step is to find the time between the desired beginning of the C/A code to the beginning of subframe 1. The first step can be accomplished through the following equation:

$$ind = 2(sfb1 - 2) + integer(nav1/10) \tag{9.1}$$

where *ind* is the index of the desired beginning of the C/A code; *sfb*1 is the beginning of subframe 1; *nav*1 is the first navigation data point and *integer* means takes the integer part of the result.

The second step is to find the difference in milliseconds (*difms*), which can be written as

$$difms = rem(nav1/10) \tag{9.2}$$

where *rem* means to take the remainder of the value in the parenthesis. The desired input point corresponding to the beginning of subframe 1 can be written as

$$dat = bca(ind) + difms \times 5000 \tag{9.3}$$

where *dat* is the digitized input data point; *bca* is the beginning of the C/A code.

Let us use these three equations to find the desired values in Figure 9.6. The results are listed in Table 9.2.

The satellite are designated as a, b, c, and d instead of a real satellite number because the information in satellite c is artificially created to illustrate a special

TABLE 9.2   Coarse Relative Pseudorange (time)

Sat	nav 1*	sfb 1**	ind	difms	bca(ind)***	dat	diff of dat
a	0	100	196	0	9802893	9802893	0
b	7	100	196	7	9803828	9838828	35935
c	10	100	197	0	9850115	9850115	47222
d	17	99	195	7	9752661	9787661	−15232

*Obtained from tracking program and adjusted to a value less than 20.
**Obtained from subframe 1 matching program.
***Obtained from tracking program.

case. The values in the second and third columns are obtained from the tracking and subframe matching programs. The *ind* and *difms* are calculated from Equations (9.1) and (9.2). The values of *bca(ind)* are also obtained from the tracking program. The final values of *dat* can be found from Equation (9.3). The last column is the relative time difference with respect to satellite a, which can be found by subtracting 9802893 from the *dat* values. In order to obtain time resolution better than 200 ns, the fine time obtained from the tracking program must be used. This time is calculated every 10 ms and used to determine the beginning of the C/A code. For simplicity, the fine time associated with index (196, 196, 197, 195) will be used to find the fine pseudorange (time). This operation is just to add the fine time to the difference time. For better results the fine time value can be obtained from manipulating more data points such as a least squares data fitting.

One can use the relative times 0, 35935, 47222, and −15232 to calculate the pseudoranges. In this calculation, some of the pseudoranges will be negative. A constant might be added to the relative times to make them positive; however, this is not necessary but only a convenient step. Since the time delay from the satellites to the user is in the range of 67 to 86 ms as shown in Section 3.3, a value between these two numbers is a reasonable choice. Although the discussion is about pseudorange, the actual units are in time which can be changed into distance by multiplying the speed of light. The pseudoranges ρ can be found as

$$\rho = c(const + diff\ of\ dat + finetime) \qquad (9.4)$$

where $c = 299792458$ m/s is speed of light; *const* is an arbitrarily chosen constant to make all the pseudoranges positive; the relative transit time (*diff of dat*) is listed in the last column of Table 9.1; and the fine time is obtained from the tracking program.

Let us choose the *const* = 75 ms. For the above example, the four pseudoranges (pr) can be calculated as

$$\rho_1 = 299792458 \times (75 \times 10^{-3})$$
$$\rho_2 = 299792458 \times (75 \times 10^{-3} + 35935 \times 200 \times 10^{-9})$$

$$\rho_3 = 299792458 \times (75 \times 10^{-3} + 47222 \times 200 \times 10^{-9})$$

$$\rho_4 = 299792458 \times (75 \times 10^{-3} - 15232 \times 200 \times 10^{-9})$$

In this equation the fine time is not included. For the actual calculation the fine time must be included in the above equation. In addition, the ionospheric correction term from Equation (5.8) or (5.9) and the tropospheric correction term from Equation (5.19) must also be included. The result obtained from Equation (5.19) is in meters and it must be divided by the speed of light to change into time. It should be noted that the ionospheric correction constants are in subframe 4. Using only the information in the first three subframes cannot make ionospheric correction.

In the above discussion, the beginning of the C/A code has a time resolution of 10 milliseconds. The beginning of subframe 1 has a time resolution of 20 milliseconds. The first navigation data point and the difference in milliseconds have a time resolution of 1 millisecond. These quantities can be used to determine the beginning of subframe 1 to within 1 millisecond. Thus, the value of the beginning of the C/A code can be limited up to 5,000 (1 ms). The large values of the beginnings of the C/A code 9802893, 9803828, 9850115, and 9752661 shown in Figures 9.6a, 9.6b, 9.6c, and 9.6d are not necessary. Instead four new values can be the remainder of these four values minus multiples of 5,000. The results are 2893, 3828, 115, and 2661. These new data are listed in Table 9.3.

The only difference between Tables 9.3 and 9.2 is in the sixth column. The beginnings of subframe 1 can be obtained as

$$dat = 10 \times 5000 \times ind + 5000 \times difms + bca(ind) \tag{9.5}$$

because dat has a time resolution of 10 ms, $difms$ has a time resolution of 1 ms, and each millisecond has 5,000 data points. Using this equation the same values of dat can be obtained and listed in the above table. In the actual software receiver program, beginnings of the C/A code with values equal to or less than 5,000 are used.

In the above discussion the main task is to find an input data point corresponding to the beginning of subframe 1. The above discussion is only one of many possible approaches to accomplish this goal.

TABLE 9.3 **Coarse Relative Pseudorange (time) with New Beginnings of the C/A Code**

Sat	nav 1	sfb 1	inds	difms	bca(ind)	dat	diff of dat
a	0	100	196	0	2893	9802893	0
b	7	100	196	7	3828	9838828	35935
c	10	100	197	0	115	9850115	47222
d	17	99	195	7	2661	9787661	-15232

9.10   GPS SYSTEM TIME AT TIME OF TRANSMISSION CORRECTED BY TRANSIT TIME (t_c)

Since time is a variable in the earth-centered, earth-fixed coordinate system, to determine the user position a time must be given. All the GPS signals from different satellites are transmitted at the same time except for the satellite clock error. However, all the signals arrive at the receiver at different times because of the different pseudoranges. The receiving time is equal to the transmission time plus the transit time. The transit time is the time the signal travels from the satellite to the user, which is equal to the pseudorange divided by the speed of light. It is reasonable to select one time (time of receiving) to measure the user position. Once a time of receiving is selected as a reference, the time of transmission can be obtained by subtracting the transit time from the time of receiving. Since the transit time for different satellites is different, the time of transmission for different satellites is also different. This seems unreasonable because all the times of transmission for different satellites are very close together. These differences can be explained as that selecting a receiving time as a reference causes the time of transmission to be different. This time of transmission can be referred to as time of transmission corrected by transit time and represented by t_c. As discussed in the previous section, the transit time cannot be measured because the user clock bias is unknown. Only the relative transit times among different satellites can be measured. The time t_c can be found from the relative transit time and the time of the week (TOW), which has a time resolution of seconds. The TOWs obtained from different satellites should have the same values because the time resolution is 6 seconds and the transit time is only 67–86 milliseconds. The time t_c can be obtained by subtracting the relative transit time from the TOW as shown in the following equation:

$$t_c = \text{TOW} - \textit{relative transit time} = \text{TOW} - \textit{diff of dat} \times 200 \times 10^{-9} \quad (9.6)$$

In this equation the subtracting implies that the time of transmission is before the receiving time. The relative transit time is the same as the difference in digitized data points (*diff of dat*). The factor 200×10^{-9} is the time between digitized points.

9.11   CALCULATION OF SATELLITE POSITION

Most of the equations used to find the satellite positions are from Chapter 4. In order to refer to these equations easily, they will be listed here again. However, the explanations of these equations will not be included here.

Let us use the obtained data to calculate the mean motion as shown in Equation (4.33):

$$n = \sqrt{\frac{\mu}{a_s^3}} + \Delta n \quad (9.7)$$

where $\mu = 9386005 \times 10^8$ m^3/s^2 is the earth's universal gravitational parameter, a_s is the semi-major axis of the satellite orbit obtained from subframe 2 in bits 227–234 and 241–264, and Δn is the mean motion difference obtained from subframe 2 in bits 91–106.

As discussed in Section 4.8, the correction of GPS time at time of transmission (t_c) must be performed first. The correction can be made from Equation (4.32) as follows:

$$\text{if } t_c - t_{oe} > 302400 \quad \text{then} \quad t_c \Rightarrow t_c - 604800$$

$$\text{if } t_c - t_{oe} < -302400 \quad \text{then} \quad t_c \Rightarrow t_c + 604800 \tag{9.8}$$

where t_c is obtained from Equation (9.6) and t_{oe} (subframe 2, bits 271–286) is the reference time ephemeris obtained from navigation data.

Once the GPS system time at time of transmission (t_c) is found, the mean anomaly can be found from Equation (4.34)

$$M = M_0 + n(t_c - t_{oe}) \tag{9.9}$$

where M_0 is the mean anomaly at reference time obtained from subframe 2 bits 107–114, 121–144. The eccentric anomaly E can be found from Equation (4.35) as

$$E = M + e_s \sin E \tag{9.10}$$

where e_s is the eccentricity of satellite orbit obtained from subframe 2 bits 167–174 and 181–204. Since this equation is nonlinear, the iteration method will be used to obtain E.

The relativistic correction term can be obtained from Equation (4.37)

$$\Delta t_r = F e_s \sqrt{a_s} \sin E \tag{9.11}$$

where $F = -4.442807633 \times 10^{-10}$ sec/m$^{1/2}$ is a constant, e_s, a_s, and E are mentioned in Equations (9.7) and (9.10). The overall time correction term is shown in Equation (4.38) as

$$\Delta t = a_{f0} + a_{f1}(t_c - t_{oc}) + a_{f2}(t_c - t_{oc}) + \Delta t_r - T_{GD} \tag{9.12}$$

where a_{f0} (271–292), a_{f1} (249–264), a_{f2} (241–248), t_{oc} (219–234) are satellite clock corrections, t_{GD} (197–204) is the estimated group delay differential, and all are obtained from subframe 1.

The GPS time at time of transmission can be corrected again from Equation (4.39) as

$$t = t_c - \Delta t \tag{9.13}$$

The true anomaly can be found from Equation (4.41) as

$$v_1 = \cos^{-1}\left(\frac{\cos E - e_s}{1 - e_s^2 \cos E}\right)$$

$$v_2 = \sin^{-1}\left(\frac{\sqrt{1 - e_s^2}\, \sin E}{1 - e_s \cos E}\right)$$

$$v = v_1 \, \text{sign}(v_2) \tag{9.14}$$

and angle ϕ can be found from Equation (4.42) as

$$\pi \equiv v + \omega \tag{9.15}$$

where ω is the argument of perigee (subframe 3, bits 197–204 and 211–234) obtained from navigation data.

The following correction terms are needed as shown in Equation (4.43):

$$\delta\phi = C_{us} \sin 2\phi + C_{uc} \cos 2\phi$$

$$\delta r = C_{rs} \sin 2\phi + C_{rc} \cos 2\phi$$

$$\delta i = C_{is} \sin 2\phi + C_{ic} \cos 2\phi \tag{9.16}$$

where C_{us} (subframe 2, bits 211–226), C_{ue} (subframe 2, bits 151–166), C_{rs} (subframe 2, bits 69–84), C_{rc} (subframe 3, bits 181–196), C_{is} (subframe 3, bits 121–126), and C_{ic} (subframe 3, bits 61–76) are obtained from navigation data. These three terms are used to correct the following terms as shown in Equations (4.44) and (4.45):

$$\phi \Rightarrow \phi + \delta\phi$$

$$r \Rightarrow r + \delta r$$

$$i \Rightarrow i + \delta i + \text{idot}(t - t_{oe}) \tag{9.17}$$

where idot (subframe 3, bits 279–292) is the rate of inclination angle and is obtained from the navigation data, t is obtained from Equation (9.13).

The angle between the accenting node and the Greenwich meridian Ω_{er} can be found from Equation (4.46) as

$$\Omega_{er} = \Omega_e + \dot{\Omega}(t - t_{oe}) - \dot{\Omega}_i e t \tag{9.18}$$

The final two steps are to find the position of the satellite from Equation (4.47) and adjust the pseudorange by the overall clock correction term as

$$\begin{bmatrix} x \\ y \\ z \end{bmatrix} = \begin{bmatrix} r \cos\Omega_{er} \cos\phi - r \sin\Omega_{er} \cos i \sin\phi \\ r \sin\Omega_{er} \cos\phi + r \cos\Omega_{er} \cos i \sin\phi \\ r \sin i \sin\phi \end{bmatrix}$$

$$\rho_i \Rightarrow \rho_i + c\Delta t \tag{9.19}$$

where Δt is obtained from Equation (9.12) and ρ_i is the pseudorange to satellite i.

A computer program (p9_6) is listed at the end of this chapter to illustrate the calculation of the satellite positions.

9.12   CALCULATION OF USER POSITION IN CARTESIAN COORDINATE SYSTEM

The calculation of user position is discussed in Chapter 2. The inputs are the positions of the satellites and the pseudoranges. Theoretically, the user position can be solved from Equation (2.5) as

$$\rho_1 = \sqrt{(x_1 - x_u)^2 + (y_1 - y_u)^2 + (z_1 - z_u)^2} + b_u$$
$$\rho_2 = \sqrt{(x_2 - x_u)^2 + (y_2 - y_u)^2 + (z_2 - z_u)^2} + b_u$$
$$\rho_3 = \sqrt{(x_3 - x_u)^2 + (y_3 - y_u)^2 + (z_3 - z_u)^2} + b_u$$
$$\rho_4 = \sqrt{(x_4 - x_u)^2 + (y_4 - y_u)^2 + (z_4 - z_u)^2} + b_u \qquad (9.20)$$

However, this equation is difficult to solve. A linearized version of Equation (2.7) can be used to solve the user position through iteration as

$$\delta\rho_i = \frac{(x_i - x_u)\delta x_u + (y_i - y_u)\delta y_u + (z_i - z_u)\delta z_u}{\sqrt{(x_i - x_u)^2 + (y_1 - y_u)^2 + (z_i - z_u)^2}} + \delta b_u$$
$$= \frac{(x_i - x_u)\delta x_u + (y_i - y_u)\delta y_u + (z_i - z_u)\delta z_u}{\rho_i - b_n} + \delta b_n \qquad (9.21)$$

Following the steps in Section 2.6 and using program (p2_1) in Chapter 2, the user position x_u, y_u, z_u can be found in the Cartesian coordinate system.

9.13   ADJUSTMENT OF COORDINATE SYSTEM OF SATELLITES

As discussed in Section 4.10, the earth-centered, earth-fixed coordinate system is a function of time. The time used to calculate the position of a satellite and the time used to calculate user position are different. The time used to calculate the satellite position should be adjusted to be the same time for calculating user position. The following three equations are used in an iterative way to perform the adjustment.

First the pseudorange and the transit time can be found from Equation (4.48) as

$$\rho = \sqrt{(x - x_u)^2 + (y - y_u)^2 + (z - z_u)^2}$$
$$t_t = \sqrt{(x - x_u)^2 + (y - y_u)^2 + (z - z_u)^2}/c \qquad (9.22)$$

where x, y, z and x_u, y_u, z_u are the coordinates of the satellite and the user, respectively, c is the speed of light. Use this transit time to modify the angle Ω_{er} in Equation (4.49) as

$$\Omega_{er} \Rightarrow \Omega_{er} - \dot{\Omega}_{ie} t_t \tag{9.23}$$

Use this new Ω_{er} in the first portion of Equation (9.19) to calculate the satellite position x, y, z in the new coordinate system. From these satellite positions, the user position x_u, y_u, z_u will be calculated again from Equation (9.21).

These four equations (9.19), (9.21), (9.22), and (9.23) can be used in an iterative way until the changes in x, y, z (or x_u, y_u, z_u) are below a predetermined value. The final position will be the desired user position x_u, y_u, z_u.

9.14   CHANGING USER POSITION TO COORDINATE SYSTEM OF THE EARTH

Once the user position x_u, y_u, z_u in Cartesian coordinate system is found, it should be converted into a spherical coordinate system, because the user position on the surface of the earth is given in geodetic latitude L, longitude l, and altitude h as shown in Equations (2.17)–(2.19):

$$r = \sqrt{x_u^2 + y_u^2 + z_u^2}$$

$$L_c = \tan^{-1}\left(\frac{z_u}{\sqrt{x_u^2 + y_u^2}}\right)$$

$$l = \tan^{-1}\left(\frac{y_u}{x_u}\right) \tag{9.24}$$

where L_c is the geocentric latitude. However, the surface of the earth is not a perfect sphere; the shape of the earth must be taken into consideration. The geodetic latitude L is used in maps and should be calculated from L_c through Equations (2.50) or (2.51) as

$$L = L_c + e_p \sin 2L \quad \text{or}$$

$$L_{i+1} = L_c + e_p \sin 2L_i \tag{9.25}$$

where e_p is the ellipticity. The second portion of the above equation is written in iterative form. The altitude can be found from Equation (2.57) as

$$h = \sqrt{x_u^2 + y_u^2 + z_u^2} - a_e(1 - e_p \sin^2 L) \tag{9.26}$$

These last three values, latitude L, longitude l, and altitude h, are the desired user position. The latitude and longitude are often expressed in degrees, minutes, and seconds or in degrees and minutes.

A computer program (p9_7) is listed at the end of the chapter to illustrate the calculation of the user position. This program integrates the user position and the satellite position correction together.

9.15   TRANSITION FROM ACQUISITION TO TRACKING PROGRAM

The above discussion is based on data digitized and stored in memory. However, in building a real-time receiver, the information obtained from the acquisition must be passed to the tracking program in a timely manner. In a software approach, the acquisition performs on a set of data already collected and the tracking will process the incoming data. For ordinary signal strength it takes the acquisition program slightly less than 1 second to process 1 ms of digitized data to find the signal using a 400 MHz pentium computer. Therefore, the tracking program will process data collected about 1 second later than that used for acquisition. Figure 9.7 shows this operation.

The question to be answered is whether the maximum acquisition time is short enough such that the tracking program still can process the new data. This section will present the maximum allowed time separation between the data for acquisition and the data for tracking. These results are obtained experimentally.

Two parameters, the carrier frequency and the beginning of the C/A code obtained from the acquisition program, will pass to the tracking program. If these two parameters are known, one can start to track the data. The carrier obtained from the acquisition will be used in the tracking program but a different beginning of the C/A code from the acquisition must be used in the incoming data. Theoretically, the beginning of the C/A code can be predicted from the carrier frequency. However, as mentioned in Section 6.15, the sampling frequency may be off from the desired value. Therefore, a set of digitized data must first be experimentally tested. The testing procedure is to process the collected data and find the relation between the carrier frequency and the beginning of the C/A code. This result can be obtained by tracking several satellites in the data. Once this result is obtained, it can be used for all satellites.

In this data collection system, the oscillator used for down conversion and the clock used for digitization are not phase locked. As a result, the following discussion may be different, if the down conversion oscillator and the digitization clock are phase locked.

The data used for this illustration are obtained from an I-Q channel digitizer. The nominal sampling frequency is 3.2 MHz as discussed in Section 6.13.

FIGURE 9.7   Transition time required from acquisition to tracking.

The data are converted into real form with an equivalent sampling frequency of 6.4 MHz and a center frequency of 1.6 MHz.

Figure 9.8 shows the accumulative beginning point shift of the C/A code versus time. The results are straight lines; let us refer to them as the accumulative shift lines. Each line contains 12,400 input points of data and is for a certain satellite. If the sampling frequency is accurate, the lines should have both positive and negative slopes to represent the positive and negative Doppler frequency shift. The results in Figure 9.8 indicate that the sampling frequency is not at 1.6 MHz, because the slopes of the lines are all positive.

The slopes of the lines are plotted against the measured Doppler frequency shift and the result is shown in Figure 9.9. The measured Doppler frequency is the difference between the measured frequency through the tracking program and the assumed center frequency of 1.6 MHz. The results are close to a straight line. From these two figures one can determine the desired shift of the beginning of the C/A code. For example, if a certain value of Doppler frequency is measured, from Figure 9.9 the slope of the accumulative line can be obtained. This slope can be used to draw a line as shown in Figure 9.8. The accumulative beginning point of the C/A code can be found as a function of time.

This method is used to find the maximum time from the data used for acquisition to the data to be tracked. For all of the satellites the minimum time obtained is slightly over 30 seconds. It takes only 1 second to perform the acquisition; therefore, there is plenty of time to pass the necessary information to the tracking

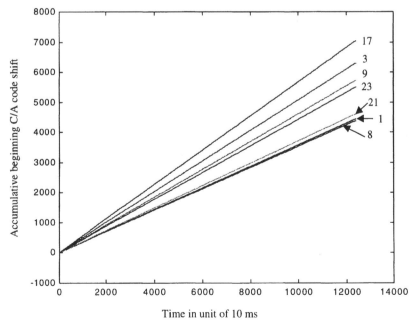

FIGURE 9.8 Accumulative beginning point shift of C/A code versus time.

FIGURE 9.9   Doppler frequency versus slopes of the accumulative lines.

program. Therefore, the two parameters, carrier frequency and the beginning of the C/A code, can be used for a real-time receiver.

9.16   SUMMARY

This chapter describes a GPS receiver by following the signal flow. Digitized data are used as input to perform acquisition and find the signals of all the satellites. Once the signals are found they will be tracked. The results of the tracking program can be converted into navigation data. In general, 30 seconds of data should contain the information of subframes 1, 2, and 3. The preamble words and the subframe numbers can be used to find the subframes. Parity checking must be performed to ensure that the data are without errors. Ephemeris data can be found from subframes 1, 2, and 3. The GPS time at time of transmission can be found from the beginning of subframe 1 and the time of the week. The pseudoranges for different satellites can be found from the relative beginning points of subframe 1. The use of subframe 1 as a reference is arbitrary. From the data obtained, the positions of the satellites and the user position can be obtained. Finally, it is illustrated that the transition from acquisition to tracking in a software receiver can be achieved for real-time applications.

```
% p9_1.m This program finds subframes

function start_sf1 = findsf1(navd)

[nsat, n] = size(navd);
marker = [1 -1 -1 -1 1 -1 1 1];
start_sf1 = zeros(1, nsat);

for m = 1:nsat,
   c = xcorr(marker, navd(m, 1:360)); % find data bit
sequences that match preambles
    indn = find(abs(c)>7.99);
    pts = indn - 360 + 1;
   flag = 1;
   n = 1;
   while flag & n <= length(pts),
      pt = pts(n);
      k = 1;
      err = 0;
      while k < 3 & ~err,                    % check for
preambles in next two subframes
         sfmark = navd(m, pt + (k * 300):pt + (k * 300) + 7);
         if abs(sum(sfmark .* marker)) < 8,
            err = 1;
         end
         k = k + 1;
      end

      end_HOW = sum(navd(m, pt + 58:pt + 59));  % Check
parity bits at end of HOW
      if ~end_HOW | err,                     % If anything
wrong, go to next point
         n = n + 1;
      else
         id = navd(m, pt + 49:pt + 51);            % Find
subframe #
         if   end_HOW == 2,                   % Find
polarity of HOW word
            id = -id;
         end
            sf(m) = 0;
            id = fliplr(id);

            for k = 1:length(id),
            if id(k) == 1,
               sf = sf + 2 ^ (k - 1);
            end
            end
         if sf(m) < 6 & sf(m) > 0,                % Ensure
```

```
subframe # is 1-5
            flag = 0;
        else
            n = n + 1;
        end
      end
    end
  if ~flag,                                % Find beginning
of subframe #1
      if sf(m) == 1,
          start_sf1(m) = pt;
      else
          start_sf1(m) = pt + ((6 - sf(m)) * 300);
          id = navd(m, start_sf1(m) + 49:start_sf1(m) + 51);
          if id ~= [-1 -1 1] & id ~= [1 1 -1],
              disp(['Error in finding sat ' num2str(m) ' - sf1
id did not match.'])
              start_sf1(m) = 0;
          end
        end
    else
       disp(['Error in finding sat ' num2str(m)])
    end
end
```

```
% p9_2.m This program checks the parity code

function [pt, navd]=matchsubf(navd);

nsat = size(navd,1);

pt = zeros(nsat);

pt = findsf1(navd)';
dlen = (floor(size(navd,2) / 30) - 1) * 30;

h1 = [1 1 1 0 1 1 0 0 0 1 1 1 1 1 0 0 1 1 0 1 0 0 1 0];
% from GPS Theory & App p.131
h2 = [0 1 1 1 0 1 1 0 0 0 1 1 1 1 1 0 0 1 1 0 1 0 0 1]; % by
B.W. Parkinson & J.J. spilker
h3 = [1 0 1 1 1 0 1 1 0 0 0 1 1 1 1 1 0 0 1 1 0 1 0 0];
h4 = [0 1 0 1 1 1 0 1 1 0 0 0 1 1 1 1 1 0 0 1 1 0 1 0];
h5 = [1 0 1 0 1 1 1 0 1 1 0 0 0 1 1 1 1 1 0 0 1 1 0 1];
h6 = [0 0 1 0 1 1 0 1 1 1 1 0 1 0 1 0 0 0 1 0 0 1 1 1];
H = [h1; h2; h3; h4; h5; h6];

for m=1:nsat;
   if pt(m)>0
        for pnt=pt(m):30:dlen, %for different initial pts
        D29 = navd(m, pnt-2);
        D30 = navd(m, pnt-1);
        navd(m, pnt:pnt+23) = D30 * navd(m, pnt:pnt+23);
            d = navd(m, pnt:pnt+23);
        Df = [D29 D30 D29 D30 D30 D29];
        for k = 1:6,
           temp = H(k,:) .* d;
           p(k) = prod([Df(k) temp(find(temp))]);
        end
        if p ~= navd(m, pnt+24:pnt+29),
           disp('Parity check failed!')
        end
        end
     end
end
ind=find(pt~=0);
pt=pt(ind);
navd = -navd;
navd=(navd+1)./2;
```

```
% p9_3.m DECODE1.M decode navigation data in subfram 1 into
ephermeris data

function[week_no, tgd, toc, af2, af1, af0,
iode1,tow1]=decode1(points,navdata);

nsat=length(points);
for m=1:nsat;
  week_no(m)=bi2de(navdata(m,points(m)-1+70:-1:points(m)-
  1+61));
  tgd(m)=comp2dec(navdata(m,points(m)-1+204:-1:points(m)-
  1+197),-31);
  toc(m)=bi2de(navdata(m,points(m)-1+234:-1:points(m)-
  1+219))*2^4;
  af2(m)=comp2dec(navdata(m,points(m)-1+248:-1:points(m)-
  1+241),-55);
  af1(m)=comp2dec(navdata(m,points(m)-1+264:-1:points(m)-
  1+249),-43);
  af0(m)=comp2dec(navdata(m,points(m)-1+292:-1:points(m)-
  1+271),-31);
  iode1(m,:)=[navdata(m,points(m)-1+211:points(m)-1+218)];
  tow1(m)=bi2de(navdata(m,points(m)-1+47:-1:points(m)-
  1+31))*6;
end
```

```
% p9_4.m DECODE2.M decode navigation data in subfram 2 into
ephermeris data

clear angmat navd rymkmat
load d:\gps\big_data\srvy8210;
angmat=ang;
ptnumat=transpt;
rymkmat=fintime;

[st navd]=navdat(angmat); %*find naviga data (20ms) & first
angle transition
[points, navdata]=matsubf(navd); %*find start pt of subfram 2
prod 0 1s

nsat=length(points);
for m=1:nsat;
  iode2(m,:)=navdata(m,points(m)-1+61:points(m)-1+68);
  crs(m)=comp2dec(navdata(m,points(m)-1+84:-1:points(m)-
  1+69),-5);
  deln(m)=comp2dec(navdata(m,points(m)-1+106:-1:points(m)-
  1+91),-43)*pi;

  m01=navdata(m,points(m)-1+114:-1:points(m)-1+107);
  m02=navdata(m,points(m)-1+144:-1:points(m)-1+121);
  m03=[m02 m01];
  m0(m)=comp2dec(m03,-31)*pi;
  clear m01 m02 m03;
  cuc(m)=comp2dec(navdata(m,points(m)-1+166:-1:points(m)-
  1+151),-29);

  e1=navdata(m,points(m)-1+174:-1:points(m)-1+167);
  e2=navdata(m,points(m)-1+204:-1:points(m)-1+181);
  e3=[e2 e1];
  e(m)=bi2de(e3)*2^(-33);
  clear e1 e2 e3;
  cus(m)=comp2dec(navdata(m,points(m)-1+226:-1:points(m)-
  1+211),-29);

  sa1=navdata(m,points(m)-1+234:-1:points(m)-1+227);
  sa2=navdata(m,points(m)-1+264:-1:points(m)-1+241);
  sa3=[sa2 sa1];
  sa(m)=bi2de(sa3)*2^(-19);
  clear sa1 sa2 sa3;

  toe(m)=bi2de(navdata(m,points(m)-1+286:-1:points(m)-
  1+271))*2^4;
end
```

```
% p9_5.m DECODE3.M decode navigation data in subfram 3 into
ephermeris data

clear angmat navd rymkmat
load d:\gps\big_data\srvy8310;
angmat=ang;
ptnumat=transpt;
rymkmat=fintime;

[st navd]=navdat(angmat); %*find naviga data (20ms) & first
angle transition
[points, navdata]=matsubf(navd); %*find start pt of subfrm 3
pro 0 1s

nsat=length(points);
for m=1:nsat;
  cic(m)=comp2dec(navdata(m,points(m)-1+76:-1:points(m)-
  1+61),-29);

  comega1=navdata(m,points(m)-1+84:-1:points(m)-1+77);
  comega2=navdata(m,points(m)-1+114:-1:points(m)-1+91);
  comega3=[comega2 comega1];
  comega0(m)=comp2dec(comega3,-31)*pi;
  clear comega1 comega2 comega3;
  cis(m)=comp2dec(navdata(m,points(m)-1+136:-1:points(m)-
  1+121),-29);

  i01=navdata(m,points(m)-1+144:-1:points(m)-1+137);
  i02=navdata(m,points(m)-1+174:-1:points(m)-1+151);
  i03=[i02 i01];
  i0(m)=comp2dec(i03,-31)*pi;
  clear i01 i02 i03;
  crc(m)=comp2dec(navdata(m,points(m)-1+196:-1:points(m)-
  1+181),-5);

  omega1=navdata(m,points(m)-1+204:-1:points(m)-1+197);
  omega2=navdata(m,points(m)-1+234:-1:points(m)-1+211);
  omega3=[omega2 omega1];
  omega(m)=comp2dec(omega3,-31)*pi;
  clear omega1 omega2 omega3;
  comegadot(m)=comp2dec(navdata(m,points(m)-1+264:-
  1:points(m)-1+241),-43)*pi;
  iode3(m,:)=navdata(m,points(m)-1+271:points(m)-1+278);
  idot(m)=comp2dec(navdata(m,points(m)-1+292:-1:points(m)-
  1+279),-43)*pi;
end
```

```
% p9_6.m SATPOSM.M Use ephemeris data to calculate satellite
position
% modified for data generated from OUR OWN COLLECTED DATA
receiver
% generat find PR from coarse PR
% JT 24 Sept 96

function[outp]=p(inpdat);
tc=inpdat(1);
toe=inpdat(2);
deln=inpdat(3);
asq=inpdat(4);
ra=inpdat(5);
i=inpdat(6);
w=inpdat(7);
delra=inpdat(8);
M=inpdat(9);
idot=inpdat(10);
e=inpdat(11);
crs=inpdat(12);
cuc=inpdat(13);
cus=inpdat(14);
cic=inpdat(15);
cis=inpdat(16);
crc=inpdat(17);
af0=inpdat(18);
af1=inpdat(19);
af2=inpdat(20);
toc=inpdat(21);
tgd=inpdat(22);
PRc=inpdat(23);

%**** define data

wer=7.2921151467e-5;
GM=3986005e8;
a=asq^2;
mu=GM;
n=(mu/(a^3))^.5+deln;

if tc-toe>302400;
  tc=tc-604800;
elseif tc-toe<-302400;
 tc=tc+604800;
end

M=M+n*(tc-toe);

Eold=M;
```

```
error=1;
while error>1e-12;
  E=M+e*sin(Eold);
  error=abs(E-Eold);
  Eold=E;
end

F=-4.442807633e-10;
deltr=F*e*asq*sin(E);
delt=af0+af1*(tc-toc)+af2*(tc-toc)^2+deltr-tgd;
t=tc-delt;

v1=acos((cos(E)-e)/(1-e*cos(E)));
v2=asin(((1-e^2)^.5)*sin(E)/(1-e*cos(E)));
v=v1*sign(v2);
%r=a*(1-e^2)/(1+e*cos(v));

phi=v+w;
delphi=cus*sin(2*phi)+cuc*cos(2*phi);
delr=crc*cos(2*phi)+crs*sin(2*phi);
deli=cic*cos(2*phi)+cis*sin(2*phi);

phi=phi+delphi;
r=a*(1-e*cos(E));
r=r+delr;
i=i+deli+idot*(t-toe);
omeger=ra+delra*(t-toe)-wer*t;

x=r*cos(phi)*cos(omeger)-r*sin(phi)*cos(i)*sin(omeger);
y=r*cos(phi)*sin(omeger)+r*sin(phi)*cos(i)*cos(omeger);
z=r*sin(phi)*sin(i);
PRf=PRc+delt*299792458;
tf=t;

outp=[x y z PRf tf];
```

```
% p9_7.m Userpos.m use pseudorange and satellite positions to
calculate user position
% JT 30 April 96
function[upos] = userpos(inp);

[mm nn]=size(inp);
nsat=nn;

xquess = 0; yquess = 0; zquess = 0; tc = 0;

sp = inp(1:3,1:nn);
pr = inp(4,:);
qu(1) = xquess; qu(2) = yquess; qu(3) = zquess;
for j = 1:nsat
    rn(j)=((qu(1)-sp(1,j))^2+(qu(2)-sp(2,j))^2+(qu(3)-
    sp(3,j))^2)^.5;
end

rn0 = rn;
h(:,4) = ones(nsat,1);

erro = 1;
while erro > 0.01;
    for j = 1:nsat;
        for k = 1:3;
            h(j,k) = (qu(k)-sp(k,j))/(rn(j)); %** find h
        end
    end
    dr = pr - (rn + ones(1,nsat)*tc);   %** find del pr
    dl = pinv(h)*dr';
    tc = tc + dl(4);
    for k = 1:3;
        qu(k) = qu(k) + dl(k);  %** find new position
    end
    erro = dl(1)^2 + dl(2)^2 + dl(3)^2;
    for j = 1:nsat;
        rn(j)=((qu(1)-sp(1,j))^2+(qu(2)-
sp(2,j))^2+(qu(3)-sp(3,j))^2)^.5; %**new pr
    end
    inp = sat_corr(inp', qu)';   %Correct satellite position
    sp = inp(1:3,1:nn);
end
xuser = qu(1); yuser = qu(2); zuser = qu(3); bias = tc;
format long
format short
rsp = (xuser^2+yuser^2+zuser^2)^.5;
Lc = atan(zuser/(xuser^2+yuser^2)^.5);
lsp = atan(yuser/xuser)*180/pi;
```

```
e=1/298.257223563;
Ltemp=Lc;
erro1=1;
while erro1>1e-6;
    L=Lc+e*sin(2*Ltemp);
    erro1=abs(Ltemp-L);
    Ltemp=L;
end
Lflp=L*180/pi;
re=6378137;
h=rsp-re*(1-e*(sin(L)^2));
upos = [xuser yuser zuser bias rsp Lflp lsp h]';
```

```
% p9_7_1.m This program corrects satellite position and called
by program 9_7

function outp = sat_corr(satp1, userp)

[m, n] = size(satp1);
outp = satp1;
Omegadot_e = 7.2921151467e-5;
c = 299792458;

for k = 1:m,
    x = satp1(k, 1);
    y = satp1(k, 2);
    z = satp1(k, 3);
    r = satp1(k, 6);                    %Corrected Radius
    i = satp1(k, 7);                    %Corrected Inclination
    phi = satp1(k, 8);                  %Argument of Latitude
    omeger = satp1(k, 9);
    xp = r * cos(phi);                  %In-plane x position
    yp = r * sin(phi);                  %In-plane y position
    xr = userp(1); yr = userp(2); zr = userp(3);
    err = 1000;

    while err > 1,
        xold = x; yold = y; zold = z;
        tprop = ((x - xr) ^ 2 + (y - yr) ^ 2 + (z - zr) ^ 2) ^
        0.5 / c;
        Omega_p = omeger - Omegadot_e * tprop;
        x = xp * cos(Omega_p) - yp * cos(i) * sin(Omega_p);
        y = xp * sin(Omega_p) + yp * cos(i) * cos(Omega_p);
        z = yp * sin(i);
    err = ((x - xold) ^ 2 + (y - yold) ^ 2 + (z - zold) ^ 2) ^
    0.5;
    end
    outp(k, 1:3) = [x y z];
end
```

```
% DECODE1.M decode navigation data in subfram 1 into
ephermeris data

function[week_no, tgd, toc, af2, af1, af0,
iode1,tow1]=decode1(points,navdata);

nsat=length(points);
for m=1:nsat;
  week_no(m)=bi2de(navdata(m,points(m)-1+70:-1:points(m)-
  1+61));
  tgd(m)=comp2dec(navdata(m,points(m)-1+204:-1:points(m)-
  1+197),-31);
  toc(m)=bi2de(navdata(m,points(m)-1+234:-1:points(m)-
  1+219))*2^4;
  af2(m)=comp2dec(navdata(m,points(m)-1+248:-1:points(m)-
  1+241),-55);
  af1(m)=comp2dec(navdata(m,points(m)-1+264:-1:points(m)-
  1+249),-43);
  af0(m)=comp2dec(navdata(m,points(m)-1+292:-1:points(m)-
  1+271),-31);
  iode1(m,:)=[navdata(m,points(m)-1+211:points(m)-1+218)];
  tow1(m)=bi2de(navdata(m,points(m)-1+47:-1:points(m)-
  1+31))*6;
end
```

```
% DECODE2.M decode navigation data in subfram 2 into
ephermeris data

clear angmat navd rymkmat
load d:\gps\big_data\srvy8210;
angmat=ang;
ptnumat=transpt;
rymkmat=fintime;

[st navd]=navdat(angmat); %*find naviga data (20ms) & first
angle transition
[points, navdata]=matsubf(navd); %*find start pt of subfram 2
prod 0 1s

nsat=length(points);
for m=1:nsat;
  iode2(m,:)=navdata(m,points(m)-1+61:points(m)-1+68);
  crs(m)=comp2dec(navdata(m,points(m)-1+84:-
1:points(m)-1+69),-5);
  deln(m)=comp2dec(navdata(m,points(m)-1+106:-1:points(m)-
1+91),-43)*pi;

  m01=navdata(m,points(m)-1+114:-1:points(m)-1+107);
  m02=navdata(m,points(m)-1+144:-1:points(m)-1+121);
  m03=[m02 m01];
  m0(m)=comp2dec(m03,-31)*pi;
  clear m01 m02 m03;
  cuc(m)=comp2dec(navdata(m,points(m)-1+166:-1:points(m)-
1+151),-29);

  e1=navdata(m,points(m)-1+174:-1:points(m)-1+167);
  e2=navdata(m,points(m)-1+204:-1:points(m)-1+181);
  e3=[e2 e1];
  e(m)=bi2de(e3)*2^(-33);
  clear e1 e2 e3;
  cus(m)=comp2dec(navdata(m,points(m)-1+226:-1:points(m)-
1+211),-29);

  sa1=navdata(m,points(m)-1+234:-1:points(m)-1+227);
  sa2=navdata(m,points(m)-1+264:-1:points(m)-1+241);
  sa3=[sa2 sa1];
  sa(m)=bi2de(sa3)*2^(-19);
  clear sa1 sa2 sa3;

  toe(m)=bi2de(navdata(m,points(m)-1+286:-1:points(m)-
1+271))*2^4;
end
```

```
% DECODE3.M decode navigation data in subfram 3 into
ephermeris data

clear angmat navd rymkmat
load d:\gps\big_data\srvy8310;
angmat=ang;
ptnumat=transpt;
rymkmat=fintime;

[st navd]=navdat(angmat); %*find naviga data (20ms) & first
angle transition
[points, navdata]=matsubf(navd); %*find start pt of subfrm 3
pro 0 1s

nsat=length(points);
for m=1:nsat;
  cic(m)=comp2dec(navdata(m,points(m)-1+76:-1:points(m)-
  1+61),-29);

  comega1=navdata(m,points(m)-1+84:-1:points(m)-1+77);
  comega2=navdata(m,points(m)-1+114:-1:points(m)-1+91);
  comega3=[comega2 comega1];
  comega0(m)=comp2dec(comega3,-31)*pi;
  clear comega1 comega2 comega3;
  cis(m)=comp2dec(navdata(m,points(m)-1+136:-1:points(m)-
  1+121),-29);

  i01=navdata(m,points(m)-1+144:-1:points(m)-1+137);
  i02=navdata(m,points(m)-1+174:-1:points(m)-1+151);
  i03=[i02 i01];
  i0(m)=comp2dec(i03,-31)*pi;
  clear i01 i02 i03;
  crc(m)=comp2dec(navdata(m,points(m)-1+196:-1:points(m)-
  1+181),-5);

  omega1=navdata(m,points(m)-1+204:-1:points(m)-1+197);
  omega2=navdata(m,points(m)-1+234:-1:points(m)-1+211);
  omega3=[omega2 omega1];
  omega(m)=comp2dec(omega3,-31)*pi;
  clear omega1 omega2 omega3;
  comegadot(m)=comp2dec(navdata(m,points(m)-1+264:-
  1:points(m)-1+241),-43)*pi;
  iode3(m,:)=navdata(m,points(m)-1+271:points(m)-1+278);
  idot(m)=comp2dec(navdata(m,points(m)-1+292:-1:points(m)-
  1+279),-43)*pi;
end
```

CHAPTER TEN

Acquisition of Weak Signals

10.1 INTRODUCTION

In this and the following chapters the reception of weak signals will be studied. This chapter deals with acquisition of weak signals, and the next chapter discusses signal tracking. Because the software approach is very flexible, there are many different ways to deal with this problem. Only one approach will be included here.

First the sensitivity of a receiver will be discussed. Because the GPS signal is a spread spectrum, the processing gain must be used to obtain a decent signal-to-noise ratio (S/N). There are two ways to improve the S/N: coherent and noncoherent integrations. This problem has been well studied by the radar designers, and the results will be used here. With the combination of coherent and noncoherent integrations, the sensitivity of a receiver can be determined. From the sensitivity, the probability of detection and probability of false alarm will be obtained.

In general, there are three categories of acquiring a GPS signal. The first category is performed without a priori information. The second category is performed with a priori information, and the third category is with aided information. Performance without a priori information means that the user does not know his approximate location nor the time of day. Thus the acquisition must search all the satellites with all possible Doppler frequencies. Obviously this method is time-consuming. Acquisition with a priori information means that one knows approximately the location and time of the day. From the almanac data the number of available satellites and the approximate Doppler frequency can be obtained. Thus with a priori information the acquisition can be simplified, saving time. Most situations are between these two categories. For example, one knows the available satellites but not the approximate Doppler frequency.

Fundamentals of Global Positioning System Receivers: A Software Approach, Second Edition, by James Bao-Yen Tsui
ISBN 0-471-70647-7   Copyright © 2005 John Wiley & Sons, Inc.

The category with aided information is that a separate station receives the GPS signals of normal strength, decodes them, and transmits the information, such as carrier frequency and navigation data, to the users in the neighborhood. With this information the user can acquire very weak signals. The acquisition method used in this approach can be very close to the tracking method, and it is will be discussed briefly in Section 12.13.

In order to achieve high sensitivity of a receiver, the dynamic of the receiver must be low, which means the receiver cannot have high acceleration. The receiver is either stationary or moving at relatively low constant velocity. Under this condition most of the Doppler frequency change is caused by the motion of the satellites. From Section 3.8 we know that a maximum Doppler frequency shift of about 1 Hz/sec is caused by the satellite motion.

The actual procedure of acquiring a weak signal contains the following five steps:

1. Perform coherent integration.
2. Perform noncoherent integration.
3. From the noise distribution, set a threshold to detect a signal. The detected signal contains the initial phase of the C/A code and the coarse carrier frequency.
4. Find fine carrier frequency for determining an accurate navigation data transition.
5. Find the location of the navigation data transition.

Before the presentation of the actual acquisition method the sensitivity of a receiver will be discussed.

10.2   SIGNAL-TO-NOISE RATIO (*S/N*)[1-3]

In order to determine the *S/N*, the strength of the input signal and noise floor must be known. The GPS signal strength is at a nominal level of −130 dBm, as discussed in Section 5.2. This noise floor is discussed in Section 6.3, and the noise figure of the receiver is discussed in Section 6.5. Let us arbitrarily assume that the antenna temperature is 180 K, and the corresponding noise floor is at about −176 dBm/Hz. Let us also assume that the receiver front end has a noise figure of 2 dB; thus the overall noise floor is at −174 dBm/Hz or −114 dBm/MHz. For simplicity, the discussion is based on this noise floor.

For a GPS receiver there is no unique way to present the *S/N* because a unique bandwidth is not universally acceptable. Sometimes the signal power level is used for sensitivity; other times *S/N* is used. Here four terms will be introduced: the input power level, the input *S/N* referenced to the input bandwidth, the *S/N* referenced to 1 kHz and the *S/N* referenced to 1 Hz, which is often expressed as C/N_0.

TABLE 10.1 Input Power and S/N with Different Bandwidths

Input Power (dBm)	Input S/N (dB/2 MHz)	S/N (dB/kHz)	C/N₀ (dB/Hz)
−130	−19 dB	14 dB	44 dB
−140	−29 dB	4 dB	34 dB
−150	−39 dB	−6 dB	24 dB

The nominal input signal strength is at −130 dBm (sometimes referred as −160 dBw.) If the gain of the receiving antenna is assumed to be unity, the signal strength is −130 dBm. The input bandwidth of a C/A code receiver is approximately 2 MHz (2.046 MHz knoll to knoll,) and the corresponding noise floor is at about −111 dBm (−114 + 3). Referenced to this 2 MHz input bandwidth, the input $S/N = -19$ dB (−130 + 111). The noise floor for 1 kHz is at −144 dBm; thus the corresponding $S/N = 14$ dB (−130 + 144). The noise floor for 1 Hz is at −174 dBm; thus $C/N_0 = 44$ dB (−130 + 174). Table 10.1 shows these relations.

Under normal conditions the S/N at the input of the receiver is about −19 dB, which is too weak to be detected. In order to detect the signal, processing must be provided to increase the S/N. The two general processing methods are referred to as coherent and noncoherent integrations. Before the discussion of these two methods, the limit of receiver sensitivity, the desired probability of detection, and the probability of false alarm will be presented because they are related to the S/N.

10.3 LIMITATION OF RECEIVER SENSITIVITY[4−10]

The sensitivity of a GPS receiver depends on two processing procedures: the acquisition and the tracking. In acquisition, long data can be used to improve the sensitivity. The tracking program, however, cannot use long data to improve sensitivity because of the finite length of the navigation data. The receiver must determine the navigation data every 20 ms. A maximum of 40 ms of data can be used to determine a navigation phase transition: 20 ms before the navigation data transition and 20 ms after.

The probability of error P_e for a BPSK (bi-phase shift keyed) signal, where the navigation data belong, is[4]

$$P_e = \frac{1}{2} erfc \left(\sqrt{\frac{E_b}{N_0}} \right) \quad (10.1)$$

where the *erfc* is the complementary error function, E_b is the energy in a bit, and N_0 is the noise power density, which can be considered as noise power per Hertz. The error function and complementary function will be discussed in

Section 10.7. The E_b/N_0 is related to the C/N_0 by the length of the navigation data t_b, which is 20 ms (20×10^{-3}). The relation can be written as

$$\frac{E_b}{N_0} = \frac{Ct_b}{N_0} = \frac{C}{N_0} \times 20 \times 10^{-3} \quad \text{and}$$

$$P_e = erfc\left(\sqrt{\frac{C}{N_0} \times 20 \times 10^{-3}}\right) \tag{10.2}$$

This probability of error P_e represents the error of 20 ms of data. However, 40 ms of data are needed to determine a phase transition. If a mistake is made in either the first or the second 20 ms of data, the detection of the phase transition will be wrong. Thus the probability of making a wrong decision on the phase transition will be $2P_e$.

Figure 10.1 shows the results of $2P_e$ versus C/N_0. In this figure the data points are obtained by the Monte Carlo method. For every data point there are 1000 or 2000 runs, and the probability of error is the total error divided by the total number of runs. The three points with the highest S/N have 2000 runs because the error rate is low. When the probability is low, a large number of runs are needed to confirm the result. For example, at $C/N_0 = 24$ dB the probability of error is about 1.5×10^{-3}, calculated from Equation (10.2). The

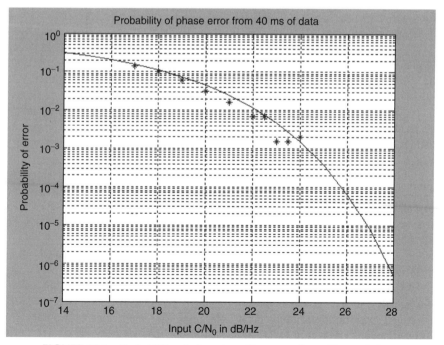

FIGURE 10.1   Probability of phase transition error from 40 ms of data.

number of runs can be found through the law of large numbers[6]. The law can be written as

$$P\{|x_n - p| < \varepsilon\} \geq 1 - \frac{1}{4n\varepsilon^2} \tag{10.3}$$

In words, Equation (10.3) can be described as follows: x_n is the measured probability of occurrence (in this case it is the probability of error) and p is the expected probability (or the expected probability of error). When $C/N_0 = 24$ dB, $p = 1.5 \times 10^{-3}$. The symbol ε represents the expected error. For example, if the expected error is 100% of p, then $\varepsilon = 100\% \times p = 1.5 \times 10^{-3}$. P can be considered as the confidence level, which is also a probability of occurrence, and n is the number of runs required.

The next step is to find the total number of runs required. The discussion above can be summarized as follows: To obtain a confidence of 80% so that the measured probability of error x_n is within 100% of p (or $\varepsilon = 1.5 \times 10^{-3}$), Equation (10.3) is written as

$$P\{|x_n - p| < \varepsilon\} = 0.8 = 1 - \frac{1}{4n\varepsilon^2} = 1 - \frac{1}{4n \times 2.25 \times 10^{-6}} \quad \text{or}$$

$$n = \frac{10^6}{9 \times 0.2} = 5.56 \times 10^5 \tag{10.4}$$

It is impractical to run the simulation this many times. By this equation, the 1000 runs only provide a confidence of 0.8 with $\varepsilon = 1.2p$ and $p = 2.5 \times 10^{-2}$, which is about the fifth point from the top in Figure 10.1. The remaining five points have an even smaller probability of error, and therefore should provide less confidence. However, because the data points and the curve are close, Equation (10.2) will be used to find the probability of error. The actual method of determining the navigation phase transition is presented in Section 10.17.

Now let us find the average time it takes for a mistake in navigation data to be produced. Let us refer this time as the false alarm time. The false alarm time can be obtained every 20 ms from $20 \times 10^{-3}/2P_e$ because $2P_e$ is the probability, which is used to obtain Figure 10.1. Figure 10.2 shows the results.

As mentioned in Chapter 5, it takes 30 seconds to collect five subframes and 18 seconds to collect the first three subframes that contain the ephemeris data to calculate the user position. It appears that it requires about 18 seconds of data without navigation data error, which in turn requires that $C/N_0 = 24.3$ dB, as shown in Figure 10.2. When $C/N_0 = 23$ dB, the average false alarm time is about 4.2 seconds, and when $C/N_0 = 25$ dB, the average false alarm time is about 53.2 seconds. The navigation data do not change rapidly from page (containing 5 subframes; Section 5.5) to page. If a navigation data error is made, the parity-check will find the bit error, and the data will be ignored. The mistake may be corrected from multiple numbers of subframes (the same subframe number on different pages.) For simplicity, let us choose $C/N_0 = 24$ dB ($S/N = -39$ dB referenced to 2 MHz bandwidth) as the sensitivity goal where it corresponds to

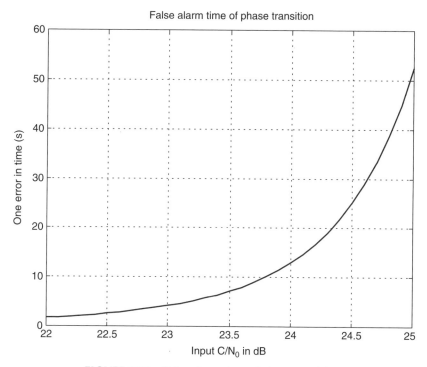

FIGURE 10.2   False alarm time of phase transition.

about 13 seconds false alarm time. Under this condition the input signal strength is at -150 dBm. Because the nominal signal strength of a GPS signal is -130 dBm, an additional 20 dB improvement in processing gain is required above normal processing. The simulated data show that a signal with $C/N_0 = 23$ dB can still be tracked for about at least 20 seconds[8]. By the aided approach introduced in Section 10.1, $C/N_0 = 16$ dB can be achieved[9].

10.4   PROBABILITY OF DETECTION AND FALSE ALARM RATE[1–3]

The probability of detection P_d and the probability of false alarm P_{fa} are important parameters for acquisition. They are functions of input S/N and threshold. For a fixed input S/N, a lower threshold produces a higher probability of detection and a higher false alarm rate. A higher threshold produces a lower probability of detection and a lower false alarm. Depending on the threshold setting, there are numerous combinations of P_d and P_{fa}. A commonly used S/N in radar detection of a single pulse is about 14 dB. Under this condition, two sets of P_d and P_{fa} will be given as an example. The probability of detection can be 90% ($P_d = 0.9$) with a probability of false alarm about 10^{-7} ($P_{fa} = 10^{-7}$)[1,2]. With the same S/N a different threshold can provide approximately $P_d = 0.5$ with

$P_{fa} = 10^{-11}$. Let us use $P_d = 0.9$ with $P_{fa} = 10^{-7}$ in the following discussion. The false alarm time T_{fa} defined in the previous section is

$$T_{fa} = \frac{1}{P_{fa}B} \tag{10.5}$$

where B is the bandwidth of the receiver. For the C/A code receiver the knoll-to-knoll bandwidth is 2.046 MHz. If the receiver has a bandwidth of 2.5 MHz the corresponding $T_{fa} = 4$ sec ($1/(2.5 \times 10^6 \times 10^{-7})$).

Let us discuss the probability of detection and false alarm time from a slightly different approach by using the detection of a GPS signal as an example. Suppose that the input signal is digitized at 5 MHz; thus in each ms there are 5000 data points. Each chip time of the C/A code is about 0.9775 μs (1 ms/1023). There are about 5 data points in a chip time. Suppose that in the acquisition method the correlation is calculated 2500 times. Each time the C/A code moves slightly less than half the chip time or 2 data points. The result is 2500 correlation outputs per ms. The chance to have one false alarm in one ms is $2.5 \times 10^{-4}(2500 \times 10^{-7})$. In other words, there is one false alarm in about every 4 sec or $T_{fa} = 4$ sec, which has the same value of the previous estimation. It should be noted that $T_{fa} = 4$ sec is for this special case. It there are 5000 correlation outputs per ms, the chance of having one false alarm in one ms is $5 \times 10^{-4}(5000 \times 10^{-7})$. Under this condition, the false alarm time is about $T_{fa} = 2$ sec. However, the probability of detect will also change. For this discussion, let us use $T_{fa} = 4$ sec.

For 1 ms acquisition, the search covers 2500 points in the time domain and 20 components in the frequency domain. The 20 frequency components are required to cover 20 kHz in 1 kHz steps. Thus each ms of data generates 50,000 (20×2500) correlation outputs. With a probability of false alarm of 10^{-7} the false alarm time is 200 ms ($1/(50,000 \times 10^{-7})$). Statistically this result means that for every 200 ms of data processed, there is one mistake.

The overall situation can be summarized as follows: With a $S/N = 14$ dB and performing acquisition on 1 ms of data, a certain threshold can provide 90% probability of detection and 1 mistake in every 200 ms. This is an acceptable result. From this discussion one can conclude that with $S/N = 14$ dB, a satisfactory result should be obtained. In later sections of this chapter, the goal is to achieve $S/N = 14$ dB from the combination of coherent and noncoherent integrations.

10.5 COHERENT INTEGRATION GAIN

By stripping the C/A code from the input data, the remaining signal becomes a continuous wave (cw). Once the input signal becomes a cw signal, fast Fourier transform (FFT) can be used to find the frequency, and this operation is referred to as coherent integration. Coherent integration is the first step in any acquisition method to find the GPS signal. The coherent signal processing gain can be found from the corresponding bandwidth, which is related to the data length.

If 1 ms of data is processed coherently, the equivalent bandwidth is 1 kHz ($1/10^{-3}$ sec). Because the cw signal is narrow band, the amplitude of the signal is not affected by narrowing the bandwidth. However, the noise is limited by the bandwidth, and the equivalent noise power is at -144 dBm at 1 kHz bandwidth. Thus 1 ms coherent integration can produce a S/N of 14 dB ($-130 + 144$) for an input of -130 dBm. From past acquisition experience this approach generates satisfactory results. Therefore it further confirms that $S/N = 14$ dB is a reasonable value to use for signal acquisition.

If 2 ms of data are processed coherently, the corresponding bandwidth is 500 Hz, and the noise floor is at -147 dBm. This method produces an S/N of 17 dB ($-130 + 147$). In general, the additional coherent integration gain G_c of n ms over 1 ms is

$$G_c = 10 \ \log(n) \qquad (10.6)$$

If $n = 2$ and 10 ms, the additional gains are $G_c = 3$ and 10 dB, respectively.

The length of coherent integration is limited by the navigation data, and the navigation data occur every 20 ms. One common maximum length used for weak signal acquisition is 10 ms because, in two consecutive 10 ms of input data, at least one set of data is without a navigation data phase transition. The coherent gain of 10 ms is 10 dB. Additional gain must be obtained from noncoherent integration. The following discussion is based on this assumption.

Sometimes the acquisition can use longer than 10 ms of data for coherent integration. For example, one can use 16 ms for coherent integration and ignore the navigation phase transition. The output will be degraded if navigation data transition is included in the data set. Other methods can be used to accommodate the phase transition. For example, performing 20 ms coherent integration can generate 1 set of correlation outputs. This acquisition can be performed 20 times by sliding the input data 1 ms each time to generate 20 sets of outputs. If there is no data transition in the input data, then all the 20 sets of outputs should have similar amplitude. If there is a data transition, then the highest output in the 20 sets should represent the 20 ms of input data without data transition. However, when the signal is weak, the highest output cannot be easily identified.

10.6 NONCOHERENT INTEGRATION

Noncoherent integration uses the outputs generated from coherent integrations to achieve additional S/N gain. Typically a set of long input data is divided into many blocks and coherent integration is performed on all the blocks. After the coherent integration every frequency output is complex and can be put into amplitude form. The amplitudes from the all coherent integration of the same frequency are summed. As a result, the otherwise weak signal will be enhanced, leading to a higher S/N.

For example, from 1 ms coherent acquisition, 2500 amplitudes in time domain will be generated. If 20 frequency bins each separated by 1 kHz are covered, the overall outputs from 1 ms of data can be represented by a 20 × 2500 matrix. If the input signal is not strong enough to be detected, a subsequent 1 ms of data will be processed using the same process and another 20 × 2500 matrix will be generated. These two metrics can be summed, and the result is a new 20 × 2500 matrix. This addition process is the noncoherent integration, which sums the output of the same frequency. Since in the adjacent ms of data the initial C/A codes are at about the same location, this process will enhance the output in the time domain. The number of noncoherent integration can be increased by summing subsequent 1 ms outputs.

The same process can be applied to 10 ms coherent integration instead of 1 ms. Ten ms of data can produce 2500 time domain outputs and 200 frequency domain outputs, separated by 100 Hz. Thus the output is a 200 × 2500 matrix. Subsequent 10 ms of data can be coherently processed and summed together to perform the noncoherent integration.

When adding the time domain outputs together, one must take into consideration the Doppler frequency. During acquisition the locally generated C/A code does not take the Doppler frequency into consideration. Thus the initial phase of the C/A code in the input signal may change position slightly with respect to the locally generated C/A code. If the Doppler is zero, then there is no initial point shift. A higher Doppler frequency causes more initial phase shift. When the results obtained in time domain are summed together, a proper time shift needs to be added according to the frequency component. This problem will be further discussed in Section 10.12.

In obtaining the amplitude of the output, a squaring operation is required, which is equivalent to the video detector in radar. Based on this similarity, the concept of noncoherent integration, discussed in radar books, is used to process GPS signal.

10.7   NONCOHERENT INTEGRATION LOSS AND GAIN[1,3,11]

In this section the gain of the noncoherent integration will be obtained. This discussion is based on Barton[3]. The procedure is to find two factors: the coherent integration gain and the noncoherent integration loss. The noncoherent integration gain equals the coherent integration gain minus the noncoherent integration loss. The coherent integration gain is given by Equation (10.6). Therefore the main goal of this section is to find the noncoherent integration loss.

If the probability of detection and probability of false alarm are given, the integration loss factor can be found from curves in Skolnik[1] and Barton[3]. Here a Matlab program will be used to find the integration loss.

The ideal detectability factor $D_c(1)$ is a function of probability of detection P_d and probability of false alarm P_{fa}, which can be written as

$$D_c(1) = [erfc^{-1}(2P_{fa}) - erfc^{-1}(2P_d)]^2 \qquad (10.7)$$

where $erfc^{-1}(y)$ is the inverse complimentary error function. The inverse error function is available in the Matlab program, but the complimentary inverse error function is not currently available. In order to use the Matlab program, the $erfc^{-1}(y)$ must be expressed in terms of $erf^{-1}(y)$.

The error function is defined as

$$y = erf(x) = \frac{2}{\sqrt{\pi}} \int_0^x e^{-t^2} dt \quad \text{or}$$

$$x = erf^{-1}(y) \tag{10.8}$$

The complimentary error function is defined as

$$z = erfc(w) = 1 - erf(w) \quad \text{or}$$

$$erf(w) = 1 - z \tag{10.9}$$

The inverse of these two functions can be written as

$$w = erfc^{-1}(z)$$

$$w = erf^{-1}(1 - z) \tag{10.10}$$

Thus,

$$erfc^{-1}(z) = erf^{-1}(1 - z) \tag{10.11}$$

Substituting this relation into Equation (10.7) yields

$$D_c(1) = [erf^{-1}(1 - 2P_{fa}) - erf^{-1}(1 - 2P_d)]^2 \tag{10.12}$$

The corresponding noncoherent integration loss is

$$L(n) = 10 \times \log\left[\frac{1 + \sqrt{1 + 9.2n/D_c(1)}}{1 + \sqrt{1 + 9.2/D_c(1)}}\right] \text{dB} \tag{10.13}$$

where n is number of noncoherent integrations. Given $P_d = 0.9$ and $P_{fa} = 10^{-7}$ as discussed in Section 10.4, the value of $D_c(1) \approx 21$ can be obtained from Equation (10.12). If n is given the noncoherent integration loss, $L(n)$ can be obtained from Equation (10.13). When n is a large value, the integration loss is approaching $L(n) = 5 \times \log(n)$.

The noncoherent integration gain G_i can be written as

$$G_i(n) = G_c(n) - L(n) = 10 \ \log(n) - L(n) \tag{10.14}$$

Figures 10.3 and 10.4 show the noncoherent integration loss and the noncoherent integration gain, respectively. It shows that when the noncoherent integration number n is small, the gain is relatively large. For example, when the noncoherent integration number is 2, the noncoherent gain $G_i(2) \approx 2.7$ dB, which is only 0.3 dB from the coherent gain of 3 dB. When $n = 100$, $G_i(100) \approx 14.6$ dB, which is about 5.4 dB lower than the coherent gain of 20 dB.

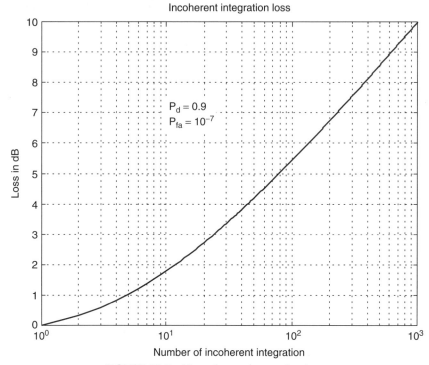

FIGURE 10.3   Noncoherent integration loss.

The discussion above can be summarized as follows: For a desired S/N of 14 dB (or $P_d = 0.9$ and $P_{fa} = 10^{-7}$) the $D_c(1) = 21$. From Equation (10.13) the integration loss can be found. Then Equation (10.14) can be used to find the noncoherent integration gain.

Let us use two examples to illustrate the combination of coherent and noncoherent integrations. Suppose that the input signal strength is at -150 dBm (or 20 dB below the normal power level.) One approach is to use 1 ms of data for coherent integration and the rest of the gain is obtained from noncoherent integration. For 1 ms coherent processing, the corresponding bandwidth is 1 kHz, and the noise floor is at -144 dBm; thus the S/N is at -6 dB ($-150 + 144$). In order to achieve $S/N = 14$ dB, the noncoherent integration should provide 20 dB gain. The result obtained from Figure 10.4 is that $n \approx 1000$. This approach requires approximately one second of noncoherent integrations. Every 20 ms, however, there is a possibility of a navigation data transition. If one takes this into consideration, the required noncoherent integration could be slightly longer because a data transition in a certain ms will decrease its coherent integration.

The other approach uses 10 ms coherent integration. With 10 ms coherent integration, the equivalent bandwidth is 100 Hz, and the corresponding noise floor is at -154 dBm. The S/N is 4 dB ($-150 + 154$.) In order to achieve a S/N of 14 dB, an noncoherent gain of 10 dB is required. This requires approximately 19

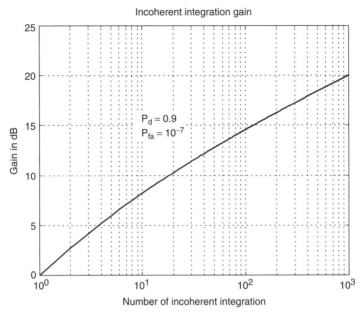

FIGURE 10.4   Noncoherent integration gain.

noncoherent integrations. In order to sum the results from data without navigation data phase transition, a maximum of 38 data blocks of 10 ms will be processed coherently. The coherent results from blocks 1, 3, 5, etc., will be summed and the results from blocks 2, 4, 6, etc., will be summed. One of these two summations should provide the desired results.

However, in actual acquisition all the results obtained from coherent integrations will be summed together without separation into odd and even groups. In actual processing, the peak obtained in the time domain will be shifted because of Doppler frequency. This shift must be considered when performing noncoherent integration.

An article by Psiaki[11] states that by processing 4 seconds of data through the 10 ms coherent integration, a C/N_0 of 18 dB can be achieved. Four seconds of data mean that $n = 200$ because the odd and the even 10 ms of data are added separately. From Figure 10.4 the noncoherent integration gain is slightly over 16 dB. Thus the overall gain is about 26 dB: 10 dB from the coherent integration and 16 dB from noncoherent gain. The sensitivity can be improved by 26 dB from $C/N_0 = 18$ dB to the desired value of 44 dB $(18 + 26)$. This can be considered as another example to illustrate the noncoherent integration gain.

10.8   ACQUISITION CONSIDERATIONS OF WEAK SIGNALS[12]

The actual acquisition method will be discussed in the following sections. The method discussed is without a priori information, such as knowing the range of the carrier frequency. In general, the acquisition method generates two important parameters: the carrier frequency and the initial phase of the C/A code. From this information the signal can be tracked. For weak signals the acquisition is also required to provide the initial position of navigation data. In every 20 ms there is a data point where a data transition may occur. For a GPS signal of normal strength the tracking program will find the navigation data transition. Detecting the existence of a data transition for a weak signal, however, is difficult. To detect the exact location of a data transition in 1 ms is even more difficult. The acquisition should provide this information by using long data record such as 1 second. In 1 second there are usually many data transitions separated by multiple of 20 ms. One can use many data transitions to find the exact location. This discussion should become clearer when the actual approach is discussed in Section 10.17. The tracking program must determine whether there is a data transition at every 20 ms after the acquisition. Once the exact data transition location is obtained from the acquisition, the tracking program can determine the existence of a data transition more effectively. This decision on whether there is a data transition every 20 ms, as discussed in Section 10.3, determines the receiver sensitivity limit.

Acquisition performed on long data through coherent integration increases the receiver sensitivity. The calculation, however, is also more complicated and time-consuming. Recall the circular correlation in frequency domain discussed in Section 7.7. The number of operations can be treated as proportional to $N \times \log_2 N$, where N is the number of data samples. If the data length is increased from 5000 points (1 ms) to 50,000 (10 ms), the number of operations is increased by about 12.7 times. Usually the acquisition complexity also limits the total number of data points.

The acquisition through coherent integration is also limited by the navigation data phase transition. In any length of data there is the possibility of having one data transition. If there is a data transition in the input data, the output energy from coherent integration will be spread into neighboring bins in the frequency domain. Using two consecutive two 10 ms of data for acquisition, it is guaranteed that in one set of the 10 ms data there is no data transition. As mentioned in the previous section, in actual acquisition the noncoherent integration is summing all the sequential outputs from the coherent integrations rather than summing the odd and even blocks separately. By this approach the effect of data transition in the input data is neglected. Akopian[12] suggests using 16 ms of data for coherent integration, which is then followed by noncoherent integration. From limited experiments it was found that the difference from 10 and 16 ms acquisition (combining coherent and noncoherent integrations) was not significant. The following discussion is based on 10 ms coherent integration followed by noncoherent integration.

Before discussing the actual acquisition method, the concept of output sampling rate will be presented next because it is used in the acquisition.

10.9 OUTPUT SAMPLING RATE

The concept of output sampling rate will be discussed here. The overall Doppler frequency is 10 kHz (from -5 to 5 kHz) caused by the satellite motion, as discussed in Section 3.6. This is the bandwidth the acquisition must search. In order to cover this bandwidth, the sampling speed must be 20 kHz for real input and 10 kHz for complex input. However, the input signal is sampled at 5 MHz to accommodate the 2.046 MHz input bandwidth. If the C/A code is stripped off from the input signal, the input becomes a cw signal with a frequency range of 10 kHz caused by the Doppler effect.

The input signal can be down converted into baseband and stripping the C/A code by the following operation:

$$s_b(t) = s(t)C(t) \exp(j2\pi f_0 t) \tag{10.15}$$

while $s_b(t)$, $s(t)$, and $C(t)$ are the baseband, the input and the C/A code signals, respectively, and $f_0 = 1.25$ MHz is the center frequency of the input signal after down conversion, as discussed in Section 7.1. From this equation one can see that the baseband signal is a complex quantity because of the operator $\exp(j2\pi f_0 t)$. The $s_b(t)$ has two output frequencies: the sum and difference between the input frequency and f_0. The sum frequency can be referred to as the high-output frequency. The difference frequency is often called the baseband frequency with a bandwidth of 10 kHz. For a complex signal with 10 kHz bandwidth, one can sample at 10 kHz.

In order to differentiate the two sampling frequencies, the 5 MHz sampling is referred to as the input sampling rate, and the 10 kHz sampling as the output sampling rate. There can be different input and output sampling rates. To convert from the input to the output sampling rate, one can sum many input points into one output point. For example, to change from 5 MHz to 10 kHz, 500 input data points can be summed into one output point because $5 \times 10^6/10 \times 10^3 = 500$. The summation process is equivalent to a low-pass filter, which can filter out the high-output frequency in $s_b(t)$.

10.10 COHERENT INTEGRATION FOR PERIODIC SIGNAL

This acquisition method only applies to periodic signal; therefore it applies only to C/A code, not to the P code signal. A more general acquisition that can process both the C/A and P code signals will be discussed in Section 12.4. Since the acquisition presented in this section is simpler than the general acquisition method, it is actually used in finding weak signals.

Only the coherent integration will be discussed in this section. The noncoherent integration operation will be discussed in the next two sections. The coherent method is similar to the modified circular correlation in Section 7.9. It can be considered an extension of the modified circular correlation acquisition. Ten ms of input data will be used in an example to illustrate the operation. The input is down-converted to baseband and then divided into 10 groups. Each group contains 1 ms of data (5000 points). Since the C/A code is periodic, each ms of data can be correlated with the same specific C/A code. The detail of this correlation operation is the same as described in Section 7.9. Each ms of data generates 2,500-point outputs. This operation can be further explained through Figure 7.4, which is redrawn as Figure 10.5. In this figure, each input data point and the corresponding point of the local code is multiplied and summed together to generate one output point. The procedure is equivalent to multiplying the input and C/A and summing 5000 points into one output point. The local code rotates two data points instead of one and performs the multiplication and summation again. Thus there are a total of 2500 output points to cover all the possible positions of the local C/A code.

Ten ms of input data generate ten 2500-point outputs. The ten outputs are the correlation results in 10 ms time. These results can be converted into the frequency domain. To convert to the frequency domain, a FFT is performed on every 10 corresponding points in the time domain. This operation is performed 2500 times. The final result is a 10 by 2500 point matrix with 2500 points in the time domain and 10 points in the frequency domain with a resolution of 100 Hz. This procedure can be illustrated in Figure 10.6. The index for the maximum amplitude of the matrix represents the initial phase of the C/A code and the carrier frequency at 100 Hz resolution. Since in the 1-ms coherent integration, 5000 points are summed into one point, the output sample time is 1 ms; thus this entire operation covers a frequency range of only 1 kHz. Alternatively, one can consider that there are 10 frequency components with a resolution of 100 Hz. The overall frequency coverage is 1 kHz.

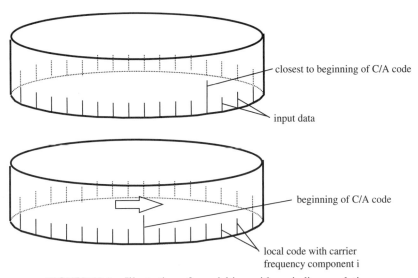

FIGURE 10.5 Illustration of acquisition with periodic correlation.

FIGURE 10.6   Illustration of 10-point FFT operations.

In order to cover more bandwidth than 1 kHz, the process above must be performed several times. For example, if another kHz should be searched, the input data must be converted to a baseband using a different frequency, f_0. To cover 10 kHz frequency range, the input must be down-converted 11 times, each separated by 1 kHz. This operation can be achieved by changing the frequency, f_0, in Equation (10.15) from $f_0 = 1245$ kHz to 1255 kHz in 1 kHz steps. The overall output is a matrix of 100 by 2500 points. Or there are 100 frequency bins separated by 100 Hz and 2500 time bins separated by 400 ns.

In order to save computation time, the 2500-point operation can, at times, be reduced to 2048 points. Since 2048 is base 2 number, the operation is slightly faster. Although the outputs are not evaluated in detail, it appears that this results in very little noted degradation.

10.11   RECOVER LOSS ON IN-BETWEEN FREQUENCIES[13]

In the above-discussed acquisition method, FFT is used to find the frequency. Whenever FFT is used with a rectangular time window, the frequency response is a sinc function ($\sin x/x$). There are always the possibility that an input frequency will fall exactly between two frequency bins, where $x = \pi/2$. When this situation occurs, the amplitude of the signal will drop due to

$$20\log\left(\frac{\sin x}{x}\right)\bigg|_{x=\pi/2} = 20\log(0.6366) = -3.92 \text{ dB} \qquad (10.16)$$

In searching a weak signal, this amount of loss cannot accepted. The goal of this section is to recover this loss.

One approach to remedy this loss is to down-convert the input frequency every 500 Hz, rather than 1 kHz. Under this condition the coarse frequency separation will be decreased from 1000 to 500 Hz, and the fine frequency resolution will be decreased from 100 to 50 Hz. A signal falling exactly between two frequency bins can be calculated from $x = \pi/4$ of the sinc function, which can be expressed as

$$20 \log \left(\frac{\sin x}{x} \right) \Bigg|_{x=\pi/4} = 20 \log(0.9003) = 0.91 \text{ dB} \qquad (10.17)$$

This is an acceptable value. Although this operation is very effective, it is usually computationally intensive.

A more efficient way is to operate on two adjacent frequency bins to generate a frequency component at the center between the bins. This method can be applied to any FFT-related operations. The following detailed discussion will continue the example from the previous section.

Before the frequency domain is shown, the filters in conventional and FFT outputs are illustrated because there might be a slight difference in appearance. A conventional filter bank response is shown in Figure 10.7a, whereas a filter bank produced through FFT is shown in Figure 10.7b.

An input signal at frequency f_1 is shown in Figure 10.7a. It is easy to understand the outputs from A and B shown in the figure. The output signal appears at the same frequency as the input signal.

In a FFT filter bank the outputs are only available on frequency bins. If the same input signal at f_1 occurs in Figure 10.7b, the outputs are shown in the two adjacent bins. Both contain the information of the input signal. For example, in Figure 10.7b, the amplitudes represent the attenuation of the filter. If the phases of successive outputs from either bin are also used to find a finer frequency resolution, the frequency of the input signal can be obtained. Thus one can conclude that the output from a frequency bin carries the frequency information of the input signal. The phase difference between the outputs at A and B is very close to $180°$ for a signal in between bins, which can be illustrated from the following equation:

$$x(n) = e^{\frac{j2\pi(k+0.5)n}{N}}$$

$$X(k) = \sum_{n=0}^{N-1} e^{\frac{j2\pi(k+0.5)n}{N}} e^{\frac{-j2\pi kn}{N}} = \sum_{n=0}^{N-1} e^{\frac{j\pi n}{N}} = \frac{1 - e^{j\pi}}{1 - e^{j\pi/N}}$$

$$= \frac{2}{1 - e^{j\pi/N}} \approx \frac{2}{-j\pi/N} = \frac{j2N}{\pi}$$

$$X(k+1) = \sum_{n=0}^{N-1} e^{\frac{j2\pi(k+0.5)n}{N}} e^{\frac{-j2\pi(k+1)n}{N}} = \sum_{n=0}^{N-1} e^{\frac{-j\pi n}{N}} = \frac{1 - e^{-j\pi}}{1 - e^{-j\pi/N}}$$

$$= \frac{2}{1 + e^{j\pi/N}} \approx \frac{2}{j\pi/N} = \frac{-j2N}{\pi} \qquad (10.18)$$

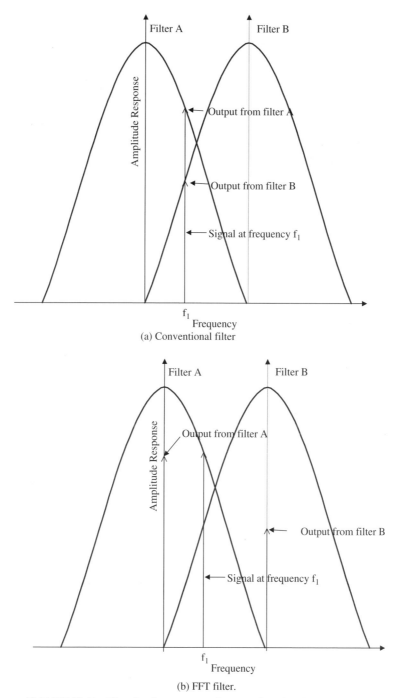

FIGURE 10.7   Filter bank response of conventional and FFT outputs.

where $x(n)$ is a complex input signal. The input frequency is between frequency bin k and $k + 1$ and is represented by $k + 0.5$. The $X(k)$ and $X(k + 1)$ are the two desired frequency bins, and they are 180° apart. In the equation above an approximation can be made if the value of N is assumed to be large, such as $1/N \gg 1/N^2$. For a 5000-point FFT the error is about 9^{-16}, which is negligible.

Figure 10.8 shows the frequency domain result of the periodic method discussed in the previous section. In this figure only 2 kHz of frequency is shown. One frequency range is from 0 to 1 kHz, labeled filter bank A with 100 Hz resolution, but not including the 1 kHz frequency bin. The other one is from 1 to 2 kHz, labeled filter bank B with 100 Hz resolution, but not including the 2 kHz frequency bin. The frequency bins in both filters A and B are labeled from 0 to 9. There are a total of 20 frequency bins in Figure 10.8. In order to cover ±5 kHz, 100 total frequency bins are required. In this figure the output signal amplitude is affected by two filters: one filter has a bandwidth of 1 kHz and the other one has a bandwidth of 100 Hz. The amplitude change between filters A and B is shown in this figure.

In this discussion, both the wideband and narrow band filters are included, but the discussion will be concentrated on the wideband filter. Usually the wideband filter A is considered to only cover ±500 Hz because the adjacent filters cover frequencies beyond this range. In this discussion the frequency cover is extended to ±1,000 Hz. Since in the FFT output there are only positive frequency components, $A+$ and $A-$ are used to represent the positive and negative frequencies of filter A. Similar representations $B+$ and $B-$ are used for filter B. The circles on

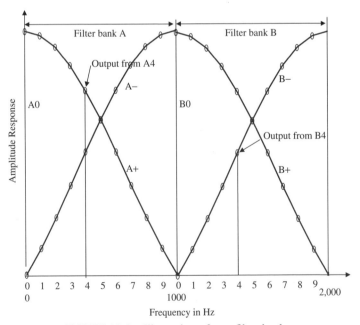

FIGURE 10.8   Illustration of two filter banks.

the lines $A+$, $A-$, $B+$, and $B-$ represent the outputs of the narrow band filters. The shape of the narrow band filter is not shown. For example, a signal at $A4$, as shown in the figure, without the narrow band filters, the outputs will appear at $A0$ and $B0$. Since there are narrow band filters, the outputs will appear at $A4$ and $B4$. The signal appears at $B4$ because the negative frequency is represented by $B-$. The amplitude difference is caused by the shape of the wideband filters A and B. The phase difference of these two outputs is close to $180°$. When the input is at filter 5 of A (500 Hz from 0 Hz), it has the same amplitude in filter 5 of B (-500 Hz from 1000 Hz.) The difference of these two outputs is equivalent to their summation because of they are $180°$ out of phase. This operation doubles the amplitude of the original outputs. Therefore the signal in the center portion between of filter banks A and B can be obtained by taking the difference between outputs A and B to increase the overall amplitude. For example, the outputs of filters 4 to 6 of A and the outputs of 4 to 6 of B can be replaced by their differences.

In this operation, although the signal amplitude is increased through subtraction of two corresponding outputs from wideband filters A and B, the noise also increases. In order to keep the noise uniform across all channels, the outputs of the subtraction channels are divided by $\sqrt{2}$ because noise is increased by $\sqrt{2}$. When the signal is 500 Hz away from a channel, its amplitude is 0.6366 (Equation (10.16)). After we double this value and divide by $\sqrt{2}$, the amplitude becomes 0.9. The corresponding degradation is 0.91 dB, which agrees with the result of Equation (10.17). Therefore this approach recovers about 3-dB loss. Figure 10.9 shows the outputs of 4 to 6 obtained from the difference. The amplitudes of points 4 to 6 are divided by $\sqrt{2}$ to compensate for the noise increase. The best improvement by this method is also shown in this figure.

This same idea applies to the narrow band filters designated as 0's to 9's, as shown in Figures 10.8 and 10.9. They have a frequency bandwidth of 100 Hz. To cover a 10 kHz frequency range there are 100 narrow band filters. If a signal falls at the center between two frequency bins, the amplitude decreases. In order to improve detection of signals between frequency bins, the outputs from adjacent bins are subtracted to increase the output amplitude. The amplitude obtained from an adjacent bin combination is also divided by $\sqrt{2}$ to compensate for the noise increase. Through this operation there are a total of 199 frequency bins: 100 from the circular correlation and 99 from the difference between two frequency bins. The frequency resolution is 50 Hz, and the time resolution is 400 ns. A certain threshold can be used to determine whether a signal is detected.

A slightly different approach can be used. That is the amplitudes of the adjacent channels can be added together rather than subtracting in complex quantity. Since this operation is equivalent to noncoherent integration, as discussed in Section 10.7, the improvement is about 2.7 dB rather than 3 dB. To accommodate this gain, the summation will be divided by 1.466 rather than $\sqrt{2}$ because $20\log(2/1.466) = 2.7$ dB.

Figure 10.10 shows the filter bank response from the periodic acquisition through simulation[13]. In these figures, each individual filter has a bandwidth

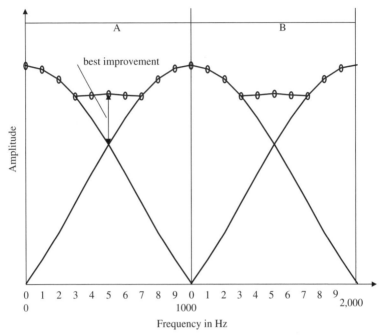

FIGURE 10.9 Filter bank outputs after compensation.

of 100 Hz. There are three wideband filters separated by 1 kHz, and in each of them there are 10 narrow band filters. Figure 10.10a shows the filter response without remediation through the loss recovery method. The maximum output difference is about 6 dB, which is caused by the wideband and narrow band filter responses. In Figure 10.10b the wideband filter responses are remediated. The maximum output difference is reduced to about 4.8 dB. In Figure 10.10c the outputs are remediated for both the wideband and narrow band filters. The maximum output difference is reduced to about 2.8 dB. From these results one can conclude that the remediation can recover about 3.2 dB of loss from 6 to 2.8 dB for the worst case. It is expected that acquisition through this method will provide a frequency accuracy of about 100 Hz. The sensitivity can vary about 2.8 dB.

10.12 TIME FREQUENCY ADJUSTMENT IN NONCOHERENT INTEGRATION[14]

In the previous sections the coherent integration performed on 10 ms of data is discussed. Successive 10 ms of data will be coherently processed and their output amplitude will be obtained as the final results. If the search frequency range is ± 5 kHz, there are eleven 1-kHz coarse frequency bins from -5 to 5 kHz. The total outputs from 10 ms of data can be represented by a 100 by 2500 matrix with 100 frequency bins separated by 100 Hz and 2500 time bins separated by 400 ns. Through recovering loss of frequency in between bins discussed in the last section, additional 99 by 2500 outputs are obtained. Thus the overall output matrix is 199 by 2500.

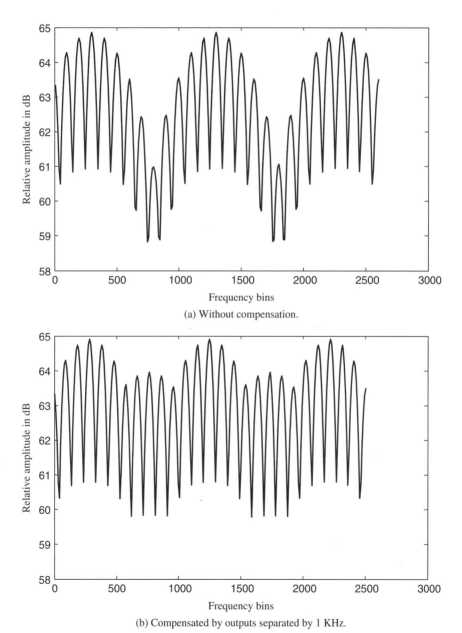

(a) Without compensation.

(b) Compensated by outputs separated by 1 KHz.

FIGURE 10.10   Simulated outputs from periodic acquisition.

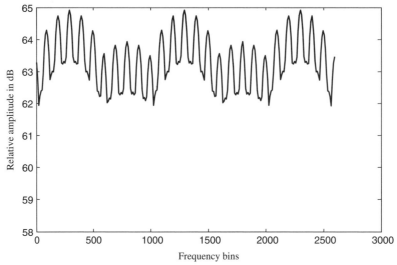

Relative amplitude in dB

Frequency bins

(c) Compensated by outputs separated by 1 KHz and 100 Hz.

FIGURE 10.10 (*continued*)

If ten 10 ms of data are processed, the overall data length is 100 ms. The outputs are ten 199 by 2500 matrices. Noncoherent integration adds all of these matrices together to form one 199 by 2500 matrix. If there is Doppler frequency shift in the input signal, this approach will misalign some data, and the result will be degraded. The goal of this section is to remedy this degradation by aligning the time domain with different Doppler frequencies.

The Doppler phenomenon can be explained as follows. If a satellite sends a 1 ms signal at the L1 frequency of 1575.42 MHz, it contains 1575.42×10^3 cycles of RF wave. The corresponding wavelength λ is

$$\lambda = \frac{c}{f_{L1}} \tag{10.19}$$

where c is the speed of light and $f_{L1} = 1575.42$ MHz. If an observer receives this signal with a positive Doppler frequency shift of f_d, the wavelength received by the observer changes to λ_d, which can be expressed as

$$\lambda_d = \frac{c}{f_{L1} + f_d} \tag{10.20}$$

The observer will receive the same number of cycle in less than 1 ms because of the positive Doppler frequency shift. The time required to receive the same number of cycles will be

$$t = \frac{10^{-3}\lambda_d}{\lambda} = \frac{10^{-3}f_{L1}}{f_{L1} + f_d} \approx 10^{-3}\left(1 - \frac{f_d}{f_{L1}}\right) \tag{10.21}$$

where 10^{-3} represents 1 ms. Since usually $f_d \ll f_{L1}$, the approximation can be used. This equation applies to both positive and negative Doppler frequencies. For a 5 kHz Doppler shift the time change for 1 ms is $10^{-3} - t = 3.17 \times 10^{-9}$ sec or 3.17 ns, where t can be obtained from Equation (10.21). This time change is smaller than the sampling time of 200 ns; thus the same C/A code can be used for acquisition without regenerating it. The digitized C/A code, however, changes slightly for different Doppler frequencies. For example, if two sets of 5000 points data are generated with Doppler frequencies of 0 and 5 kHz for satellite 1, only 6 out of 5000 points have different values. Thus use of the same C/A for acquisition can be justified.

For 10 ms coherent integration, the corresponding time change due to 5 kHz Doppler frequency shift is 31.7 ns (10×3.17). If 20 noncoherent integrations are performed, the total length of time is 200 ms. A 5 kHz Doppler frequency shift can cause a time change of 634 ns (200×3.17), which corresponds to three sampling intervals of 600 ns. If this time is not adjusted for the noncoherent integration, the peaks will no longer line up. Figure 10.11 shows that the Doppler frequency is uncompensated. In this set of data, satellite 5 has a Doppler frequency of -2900 Hz. Ten ms coherent integration is performed. In order to exaggerate this effect, a long data record (2 seconds) is processed. The amplitude from Figure 10.11a is about 8.7×10^6. Figure 10.11b shows a close-up version, where the peak is spread out. Figure 10.12 shows the compensated results. The amplitude from Figure 10.12a is about 2.3×10^7. Figure 10.12b shows a close-up, where the peak is sharp. For this long time noncoherent integration, the peaks can be improved by about 8.4 dB ($20 \times \log(23/8.7)$) with this method.

The time shift can be incorporated in the acquisition as follows[14]. The time shift for each block of data can be computed as

$$T_d = \frac{f_d T}{f_{L1}} \quad \text{and}$$
$$S_d = \exp(-j2\pi k_n T_d) \tag{10.22}$$

where T_d is the required time shift, T is the time delay between the current data block and the first data block for the noncoherent integration, S_d is the frequency domain equivalent of T_d, and k_n is the frequency component of the FFT kernel function. In obtaining the correlation results, the delay S_d is used in the frequency component of the locally generated signals.

10.13 THRESHOLD DETERMINATION FOR GAUSSIAN NOISE[15-17]

Once the procedures above are accomplished, the acquisition program can produce a reasonable peak at the correct initial phase of the C/A code and the carrier frequency. At this stage, a decision must be made to determine whether the peak represents a true signal. To accomplish this goal, a threshold must be set. If the peak is higher than the threshold, it is a signal. Otherwise, it is not.

(a) Overall output.

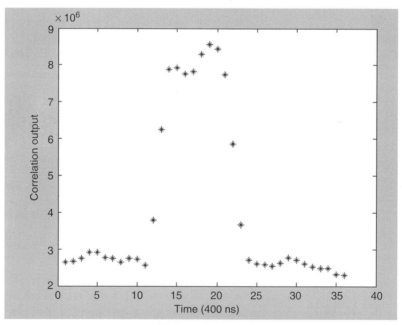

(b) Expanded output.

FIGURE 10.11 Noncoherent integration from uncompensated Doppler effect.

(a) Overall output.

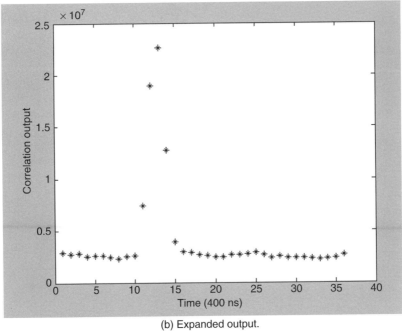

(b) Expanded output.

FIGURE 10.12   Noncoherent integration from compensated Doppler effect.

Two approaches are used to determine the threshold. The first approach assumes that the noise is Gaussian and uses simulated results to obtain thresholds. This method can be referred to as threshold by simulation. The advantage of this approach is that the noise of the collected data needs to be calculated only once, and the result will be used to determine the threshold. The output peak can be compared with the threshold. The second method uses real data and determines the threshold experimentally.

In order to determine the sensitivity of a receiver, two parameters must be determined: the probability of detection and the probability of false alarm. With a given S/N, only one of the parameters needs to be chosen, and the other is obtained accordingly. For example, as discussed in Section 10.4, assume $S/N = 14$ dB. If the probability of false alarm is chosen to be 10^{-11}, the probability of detection is about 50%. If the probability of false alarm is chosen to be 10^{-7}, the probability of detection is about 90%. In the following discussion the probability of false alarm is chosen, which is the equivalent of choosing a certain threshold. The false alarm is determined by the number of acquisitions performed. The objective is to have less than one false alarm in every 100 acquisitions.

The threshold determination is based on noise distribution. Once the noise distribution is known, the probability of false alarm can be determined. If the input noise is Gaussian, the amplitude of Gaussian noise is Rayleigh distribution, which can be written as[15]

$$p(r) = \frac{r}{\sigma^2} e^{-r^2/2\sigma^2} \tag{10.23}$$

Figure 10.13 shows this distribution. Simulated Gaussian noise is used to evaluate the noise distribution through coherent integration process. Gaussian noise with standard deviation of 1 and 0 mean is used as input for 10-ms coherent processing. The output is noise and the result shown in Figure 10.14. Figures 10.14a show the histogram with amplitude versus the probability of occurrence. Figure 10.14b shows the time domain output. The output from Figure 10.14a is close to a Rayleigh distribution as expected.

However, the output from the noncoherent integration is no longer Rayleigh distribution. Figures 10.15a and 10.15b show the noise distribution from noncoherent integration 2 and 20 times, respectively. As expected when the number of noncoherent integrations is large, the distribution approaches Gaussian by the central-limit theory[6].

The noise distribution through noncoherent processing is discussed in reference[16]. The threshold can be determined from a given probability of false alarm, but the solution is not trivial. For this reason the threshold is determined experimentally. The method used is the acquisition method discussed in the previous section with Gaussian noise as input. During each run a new set of noises is generated randomly. The acquisition is performed 100 times and the highest output is selected. In order to generate a more dependable result, the acquisition is performed 1000 times. Every 100 times the highest output is selected. The 10 highest outputs are averaged into one value. The threshold

FIGURE 10.13 Rayleigh distribution.

FIGURE 10.14 Simulated Gaussian noise through coherent integration.

FIGURE 10.15   Noise distribution after noncoherent integration.

is set equal to the averaged value, which represents one false alarm of every 100 acquisitions performed. If the noncoherent integration number is 10, then 10 outputs obtained from coherent integration will be summed together. In order to generate 1,000 runs, the acquisition must be performed 10,000 times because each time 10 acquisitions will be performed. Thus, to generate the threshold is time-consuming, and this is why only one data point (19 summations) was obtained above 10 summations. The result is shown in Figure 10.16. An extrapolated line is drawn through all the points. The threshold values are obtained from the extrapolated line at each summation value from summation equal to 1 to 20. The values are also listed in Table 10.2 as well as in Figure 10.16.

These results are obtained from a noise with standard deviation of 1 and 0 mean. In actual collected data the noise usually still has a mean close to zero, but the standard deviation can be any value. In order to use the values in Figure 10.16, the threshold must be a function of the standard deviation of the noise. The relationship between the threshold values and the standard deviation of the noise can be found through the following method. The standard deviation of the noise is varied and the 10 ms acquisitions without summation are used to determine the threshold. This operation is performed 1000 times to obtain the desired threshold. The results are shown in Figure 10.17. Similar tests are performed with summations. Figures 10.18a and 10.18b show the results of summing 2 and 5 times.

TABLE 10.2 Measured and Extrapolated Thresholds for 1 to 20 Noncoherent Integrations for 1 False Alarm in Every 100 Acquisitions

Number of Summations	Measured Threshold	Extrapolated Threshold
1	1814	1998
2	2754	2712
3	3518	3412
4	4214	4097
5	5486	4768
6	6600	5424
7	7826	6066
8		6693
9		7306
10		7904
11		8488
12		9057
13		9612
14		10153
15		10678
16		11190
17		11686
18		12168
19	12,660	12636
20		13090

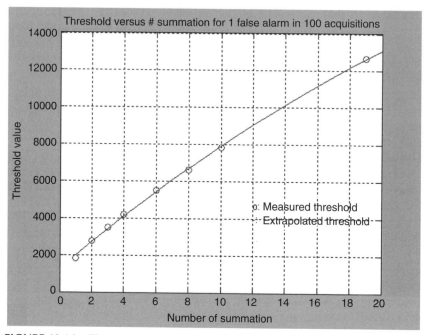

FIGURE 10.16 Threshold versus number of summations for 10-ms data acquisition.

FIGURE 10.17 Threshold versus standard deviation of noise without summation.

(a) 2 summations.

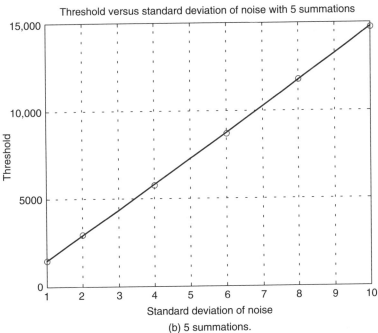

(b) 5 summations.

FIGURE 10.18   Threshold versus standard deviation of noise with two summations.

In these three figures it is shown that the threshold obtained is a linear function of the standard deviation of the noise, which is reasonable. If actual data are used for acquisition, the first step is to find the standard deviation of the noise. Since in a GPS system with 2 MHz bandwidth data collection arrangement the signal is about -19 dB below the noise, the signal strength will not affect much of the standard deviation of the noise. The standard deviation of the noise can be measured in the time domain using 1 ms of input data. Once the standard deviation of noise is determined, the threshold in Figure 10.16 can be multiplied by the standard deviation and used as the actual threshold.

With 1 false alarm over 100 runs a threshold is determined. This threshold is used to determine the probability of detection. In the next section we show some results obtained through simulated signals.

10.14   PROBABILITY OF DETECTION OF SIMULATED SIGNALS

Simulated data are used to evaluate the probability of detection. Different signal conditions can be selected to perform the test. The best signal conditions are defined as ones whose input frequency is coincident with one of the frequency bin and the initial phase of the C/A code is on a digitized data point. Under this condition the locally generated signal can be perfectly matched to the input signal. This condition is used for testing the probability of detection. In order to add minor uncertainty, the carrier frequency is arbitrarily set randomly ±50 Hz within a frequency bin. From these results the worst signal condition can be predicted.

The noise used has a 0 mean and standard deviation of 1 and the S/N is varied by adjusting the signal strength. For every S/N value the test is performed 100 times. For each run different noise is generated with 0 mean and standard deviation of 1. The number of detections is recorded. Figure 10.19 shows the S/N versus number of noncoherent summations for detecting the signal about 90 times out of 100 runs. This curve is limited to about $C/N_0 = 25.5$ dB because it is difficult to obtain 90 detections when the input signal is weak. A separate test has been carried out on the weak signals around $C/N_0 = 24$ dB. This curve confirms the noncoherent integration gain discussed in Section 10.7. A smaller number of summations makes more improvement.

For the lower S/N value test, the initial phase of the C/A code of the input signal is set to zero, but the Doppler frequency is random within ±5 Hz to provide a lower uncertainty. The results are listed in Table 10.3. In this table each data point is obtained from only 100 runs.

TABLE 10.3   Signal Detection at C/N_0 around 24 dB

C/N_0 (dB)	25	24.5	24	23.5	23
P_d	82	77	74	60	50
False alarm	2	11	4	11	9

S/N versus number of incoherent integration
to generate 90% detection

FIGURE 10.19   *S/N* versus number of summations for 90% detections.

The probability of detection is defined as that the correction initial phase of the
C/A code is obtained within 19 integrations. When the number of integrations is
beyond 19, it is considered that the signal is not detected. When an erroneous
result is obtained within 19 integrations, it is considered as a false alarm. For
example, at $C/N_0 = 24$ dB, the signal is correctly detected 74 times; a wrong
information is obtained 4 times, and there are 22 runs in which the outputs do not
cross the threshold. It appears that the false alarm is higher than the desired value
of 1 per 100 runs. Since both the methods of determining the threshold and the
probability of detections use very limited data sets because of long calculation
time, the error can be large.

Another way to detect a weak signal is based on the assumption that the signal
is in the input data. Under this condition one can detect the signal by not using
a threshold, but perform noncoherent integration 19 times. After the integration
the highest value will be used as the desired results. This approach yields only
correct and incorrect results. The detection results are listed in Table 10.4. These
results have slightly higher probability of detection (correct results). It takes
more time to process the signal because 19 integrations must be performed in
each operation. In both tables the results are obtained from 100 runs, and they
should not be considered as typical cases.

The worst operating condition is that the frequency of the input signal is
at the center of two frequency bins. After adjacent frequency bin operations
are performed as discussed in Section 10.11, the loss is about 2.8 dB. The worst
digitizing error is 100 ns (half the sampling time of 200 ns) from the initial phase
of the C/A code, and the sensitivity loss is about 0.92 dB. Therefore the overall
degradation should be about 3.72 dB. The overall noncoherent integration gain

TABLE 10.4   Signal Detection at C/N_0 around 24 dB without Threshold

C/N_0 (dB)	25	24	23
Correct results	94	78	61
Incorrect results	6	22	39

required is 13.72 dB instead of 10 dB, which is the most favorable condition. In order to obtain 13.72 dB of gain, the noncoherent integration must be summed about 72 times. As mentioned previously when every other 10 ms of data is added together to accommodate the navigation data, a total of 1440 ms of data will be processed. From the discussion above, it is clear that to acquire a signal at $C/N_0 = 24$ dB, a fair amount of processing is needed. For simplicity, however, it is assumed that with 19 summations a $C/N_0 = 24$ dB can be achieved. The last point in Figure 10.19 is based on this assumption.

10.15   THRESHOLD DETERMINATION FROM REAL DATA

Weak GPS signals have been collected to evaluate the acquisition method discussed in the previous sections. Data were collected inside a building, also collected outdoor under a shelter, and under trees. There were 16 data sets and the correct detection, false alarms, and missed signals were recorded. Some results show that a signal at times can be clearly observed from the visual output display but not detected. To understand this problem, the acquisition results (from 20 noncoherent summations of 10 ms coherent integration) were studied. The purpose was to find the noise distribution and the threshold. In order to find the noise distribution, a certain signal that was not in the input data had to be acquired. In other words, the acquisition was performed on a specific satellite (i.e., satellite 6), but this satellite was not in the input data. Such acquisition results can be considered as the noise floor, and they are shown in Figure 10.20. The results are displayed in the time domain as well as in a histogram, and they are from different data sets. After 20 summations the results should be close to a Gaussian distribution, as discussed in Section 10.13 and Figure 10.15.

The results in Figure 10.20 were generated from 2048 point FFT operation as discussed in Section 10.10. Figure 10.20a shows the distribution to be close to a Gaussian distribution. The distributions – Figures 10.20b and 10.20c are not close to Gaussian. In Figures 10.20d through 10.20f the noise floors have amplitude variations. The reason for these deviations is not clear, since the data were not collected under controlled environments. However, from the different distributions it is clear that a fixed threshold based on a Gaussian distribution will not be the optimum selection, especially for the distributions of Figures 10.20d through 10.20f.

A different approach was tried to detect the signals. In this method, the coherent integration uses 10 ms of data. After every noncoherent integration, the result is obtained in the time domain shown in Figure 10.20. From this result two

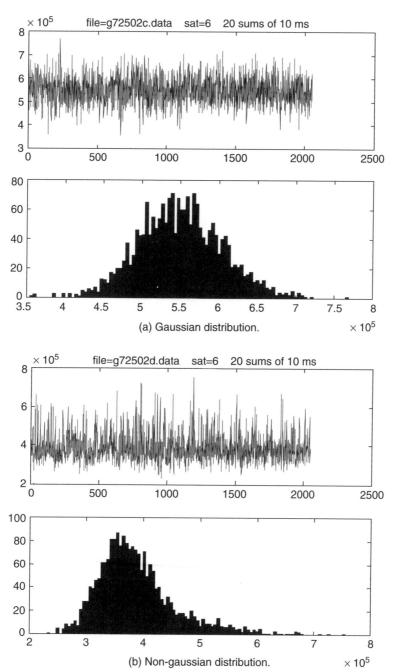

FIGURE 10.20   Noise floor from 20 noncoherent integrations of 10-ms coherent integrations.

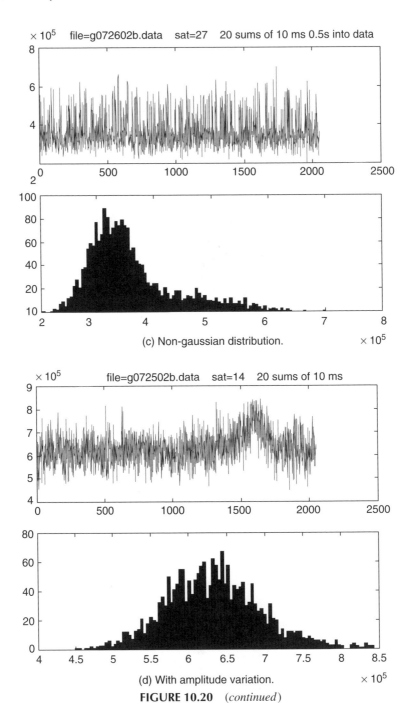

(c) Non-gaussian distribution.

(d) With amplitude variation.

FIGURE 10.20   (*continued*)

(e) With amplitude variation.

(f) With amplitude variation.

quantities – detection signal power and detection noise power – are defined as follows: First, the highest output is considered as the desired signal. Second, the remaining outputs, excluding 9 outputs around the signal, are considered to be noise. Since the signal output may spread into its neighborhood bins, ± 4 outputs around the highest output (9 total outputs) are arbitrarily chosen and are excluded in the detection noise power calculation. Third, the detection signal power P_{ds} and detection noise power P_{dn} are defined as

$$P_{ds} \equiv (S - b)^2$$

$$P_{dn} \equiv \frac{\sum_N (n - b)^2}{N} \tag{10.24}$$

where b is the dc bias obtained from the noise components, S is the amplitude of the highest output, n is the individual noise component, and N is the total noise components. Since there are 2500 time domain outputs from the acquisition, $N = 2491$ because 9 components are excluded from the detection noise power calculation. From experimental results it appears that when

$$\left(\frac{P_{ds}}{P_{dn}} \right)^{1/2} \geq 7.5 \quad \text{or}$$

$$\left(\frac{P_{ds}}{P_{dn}} \right)_{dB} \equiv 10 \log \left(\frac{P_{ds}}{P_{dn}} \right) \geq 17.5 \text{ dB} \tag{10.25}$$

a good combination of probability of detection and false alarm rate can be achieved. The quantity in this equation can be considered as the postdetection signal-to-noise ratio (S/N), which is different from the S/N used in Section 10.2. In the postdetection S/N, the bias is subtracted from both the signal and the noise. Comparing with the threshold derived from the previous section, this method provides the following advantages:

1. It usually takes less noncoherent integrations to detect a true signal.
2. It yields a high signal detection probability. Some signals that can not be detected with the previous threshold can be detected with this new threshold.
3. It generates less false detections.

In this detection criterion, the output distribution is not considered.

It should be noted that in calculating S/N conventionally, the signal and the noise power contains the dc components. The quantity $(P_{ds}/P_{dn})_{dB}$ is different from the conventional S/N, where

$$\frac{S}{N} = \frac{P_s}{P_n} = \frac{S^2}{\sum n^2 / N} \tag{10.26}$$

Comparing with Equation (10.24), the dc bias b is included in the signal and the noise power. The conventional S/N ratio has also been tried for signal detection. The result is inferior to this method. As a result this newly defined post detection threshold is used for signal detection.

In later sections, when a simulated signal is used, the conventional S/N will be used because the noise is well behaved. In the study using real signals, the post-detection S/N will be used because the noise floor distribution is not predicable.

10.16 FINE FREQUENCY CALCULATION[18]

For processing a strong signal, a frequency accuracy of ± 50 Hz is adequate to find the navigation data phase transition. For weak signals this frequency accuracy is inadequate because a higher correlation peak is needed. The closer the measured frequency is to the true frequency, the higher is the correlation peak. The correlation is used to determine the phase transition. It is desirable to have a frequency resolution of around a few Hz (e.g., 2 Hz). In order to achieve this resolution, long data must be used. The navigation data, however, limits the usage of long data for coherent processing. The following approach is used to achieve high-frequency resolution.

The approach to measure fine frequency is through squaring of the processed input signal in units of 1 ms. The acquisition can find the initial phase of the C/A code, which can be used to strip off the C/A code. After the C/A code is removed from the input signal, the phase transition caused by the navigation data still exists. It is well known that if a bi-phase coded signal is squared, the phase transition can be eliminated and the output is a cw signal with a frequency equal to twice the input frequency. This method, however, decreases the signal-to-noise ratio because the noise is also squared.

An estimation of this approach with squaring will be briefly discussed[18]. If the signal amplitude is A and the noise is n, the received signal x is

$$x = A + n \tag{10.27}$$

The corresponding S/N in dB is

$$\left(\frac{S}{N}\right)_{dB} = 20 \log \left(\frac{A}{\sigma_n}\right) \tag{10.28}$$

where σ_n is the standard deviation of the noise. Let us define a signal y as the square of the x, the result is

$$y = x^2 = A^2 + 2An + n^2 \tag{10.29}$$

In this equation, A^2 is considered as signal amplitude and $2An + n^2$ is noise. The corresponding $(S/N)_{sq}$ in dB can be considered as the energy ratio as

$$\left(\frac{S}{N}\right)_{sqdB} = 10\log\left(\frac{A^4}{< (2An + n^2)^2 >}\right) = 10\log\left(\frac{A^4}{3\sigma_n^4 + 4A^2\sigma_n^2}\right)$$

$$= 10\log\left(\frac{A^2}{\sigma_n^2}\right) - 10\log\left(4 + 3 \times 10^{-10\log(A^2/\sigma_n^2)/10}\right) \quad (10.30)$$

$$= \left(\frac{S}{N}\right)_{db} - 10\log\left(4 + 3 \times 10^{-((S/N)_{db}/10)}\right)$$

In the operation above, $<>$ is the expectation value, and the noise distribution n is assumed to be Gaussian. The relation between the input $(S/N)_{dB}$ and the $(S/N)_{sqdB}$ is shown in Figure 10.21. This curve indicates that when the input S/N is low, squaring will more than double the input S/N, such as from -40 dB to approximate to -84.78 dB. When the input S/N is high, the output S/N is decreased by approximately 6 dB, such as from 60 to 53.98 dB.

The application of this relation can be illustrated with an example. If the $C/N_0 = 24$ dB, 1 ms of processed data has a $S/N = -6$ dB $(24-30)$ because 1 ms has a nominal bandwidth of 1 kHz. Two ms of processed data have a $S/N = -3$ dB $(24-27)$, and 4 ms of processed data have a $S/N = 0$ dB. The corresponding squared output S/N's from 1, 2, and 4 ms are approximately -18, -12.99, and -8.45 dB from Equation (10.30) or Figure 10.21.

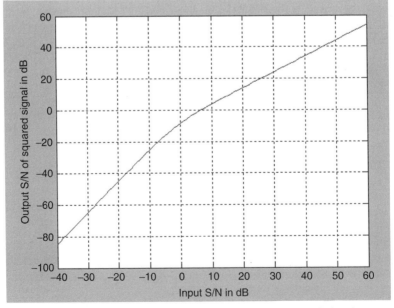

FIGURE 10.21   Input S/N versus output S/N of squared signal.

Six hundred ms of data through coherent processing can provide about 27.78 dB ($10 \times \log(600)$) of gain. In combining with the squaring operation of 1 ms, the overall S/N is about 9.78 dB ($27.78 - 18$). For the 2 ms summation, the coherent gain is 24.77 dB ($10 \times \log(600/2)$) because only 300 data blocks can be processed coherently. The overall S/N is about 11.1 dB ($24.77 - 13.67$). For the 4 ms summation, the coherent gain is about 21.76 dB ($10 \times \log(600/4)$), and the overall S/N is 12.21 dB ($21.76 - 9.55$). From this simple example, one can see that by increasing the number of ms of data summed, a higher overall S/N can be achieved. However, longer ms of data summed can increase the chance of including a navigation data transition, which is undesirable. For example, if 1 ms is used for squaring, the navigation has no effect. If 2 ms is summed, the possibility exists that in every 10 blocks there is one with navigation data. If 4 ms is summed, the possibility exists that in every 5 blocks there is one with navigation data. The blocks with navigation data will reduce the coherent processing gain. Of course, there is also the 50% chance that the navigation date transition has no effect on the 2 ms data summation and 25% chance on the 4 ms summation.

The actual processing is as follows: The input frequency is down-converted to a baseband of less than 1 kHz, such as near 100 Hz, because the input frequency can be determined about ± 25 Hz through the acquisition. After the down-conversion, the C/A code is stripped off, and every millisecond of data is averaged into one data point. This averaging is equivalent to sampling at 1 kHz. Thus, the total frequency coverage is 1 kHz because the outputs are complex. If 200 ms of data are used, the achievable frequency resolution is 5 Hz. Since the frequency is doubled, the frequency resolution is 2.5 Hz and the frequency accuracy can be ± 1.25 Hz.

Simulated data are used to test this method for signals with C/N_0 near 24 dB. In these simulations, the number of ms data summed is 1, 2, and 4. This input signal has a Doppler frequency of 55 Hz, which is arbitrarily chosen, but worse than the frequency accuracy of ± 25 Hz obtained from acquisition. It is also chosen that the measured frequency must be within ± 2 Hz of the Doppler frequency; otherwise, it is considered to be a failure. Since the tracking method for weak signal can achieve frequency accuracy of ± 0.5 Hz, ± 2 Hz is arbitrarily selected as the acquisition requirement. This test runs 100 times with different sequences of noise, and the number of successful results is recorded. In the tests, the actual data length is from 240 to 600 ms. The results are shown in Table 10.5.

In this table 600 simulated ms data are generated. The first 240 ms of data are used for the 240-ms test. Similar approaches are used for the 360- and 480-ms tests. From these results one can see that longer data and a larger number of ms summed combined will improve the frequency reading. For this special criterion of ± 2 Hz, from 240 to 360 ms the improvement is significant, but from 360 to 600 ms little improvement is realized. In these simulations the navigation data are not included. In using real data, if there is a navigation phase transition in the summed data, the final results from the 2-ms and 4-ms summations will be

TABLE 10.5   Correct Frequency Readings in 100 Runs

Data Length (ms)	ms Summed	C/N_0 (dB)		
		25	24	23
240	1	59	39	19.5
240	2	74.5	56	40.5
240	4	82	67	58
360	1	89.5	65.5	45
360	2	98.5	90	63
360	4	100	94	85.5
480	1	99	93	75.5
480	2	100	99.5	95.5
480	4	100	100	99
600	1	100	99	91
600	2	100	100	98
600	4	100	100	100

slightly degraded because blocks containing navigation data will reduce coherent processing gain.

It appears that with a data length of 480 ms and after summing 2 ms together, the result is satisfactory. There are a maximum of 24-phase transitions in 480 ms of data. If 2 ms summed together, there are 240 outputs. Twenty-four navigation phase transitions can only affect 10% of the data, and there is still enough data to obtain the correct frequency. If 10 ms are summed together, there are 48 outputs, and 24-phase transitions can affect 50% of the data. Thus summing of many ms of data may not be advisable.

10.17   FIRST NAVIGATION PHASE TRANSITION DETERMINATION

In a conventional GPS receiver the information, such as carrier frequency and initial C/A code phase obtained from the acquisition, can be used to track a signal. The navigation data phase transition can be found through the tracking program. For a weak signal it is difficult to determine the exact location of the navigation data transition from the tracking program. In order to process the data efficiently, the exact location of the navigation data phase transition is needed such that one can start to process the data every 20 ms. Thus in this section the determination of the location of the navigation phase transition is discussed. In general, it is easier to determine whether there is an existence of data transition than to determine the exact location of a data transition.

The basic approach to determine the location of the data transition is as follows: After finding the fine carrier frequency, as discussed in the previous section, the C/A code and carrier frequency can been stripped off from the input signal and 1 ms of data is summed into one point. Forty ms of reference data (40 points) can be generated to correlate with 40 ms of input data to find whether a data transition exists. There is probability, however, of false detection and missing detection. It appears that at least 20 positive identifications of data transitions are needed to find the exact location of the data transition. In actual collected data sometimes there is no data transition in 400 ms. From this observation it is difficult to use a predetermined amount of data to find the data transition location.

The following steps show the detailed procedure to find the location of the data transition. In some of the steps the approaches are obtained through experience.

Step 1. The first step is to perform correlations with the input signals. For each transition test two references are used: one with a transition at the middle and the other one without transition. Figure 10.22 shows two input signals, each containing 80 points of data. One input has a phase transition in the middle, and the other one does not, as shown in Figures 10.22a and 10.22b. The reference contains 40 points with a phase transition in the middle, as shown in Figure 10.22c. Figure 10.22d is obtained by linearly correlating the input in Figure 10.22a against the reference in Figure 10.22c. The correlation output has 41 points, and each point is obtained by 40 input data points. The correlation output is obtained by shifting the 0 point of the reference against the input data from 0 to 40 points. Figure 10.22e is obtained in a similar way by correlating the input in Figure 10.22b with the reference. When the input has a phase transition, the output has a peak with amplitude 40; otherwise, the output is zero, as shown in Figures 10.22d and 10.22e.

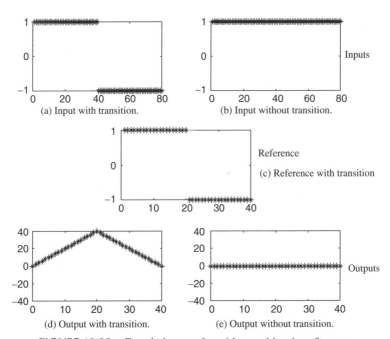

FIGURE 10.22   Correlation results with transition in reference.

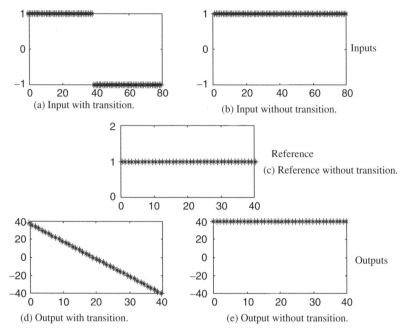

FIGURE 10.23 Correlation results with no transition in reference.

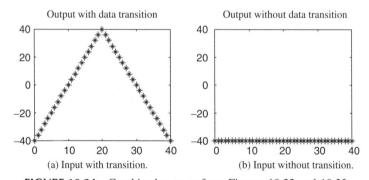

FIGURE 10.24 Combined outputs from Figures 10.22 and 10.23.

The difference of the two correlation peaks is 40. Similar results are shown in Figure 10.23. In this figure the reference does not contain a phase transition.

These outputs will be used with the outputs from Figure 10.22 to determine if a phase transition exists. Figure 10.24 is obtained by subtracting the absolute values of the Figure 10.23 outputs from the absolute values of Figure 10.22. When there is a transition in the data, the peak output is 40; otherwise, −40 and the difference is 80. Compared with the outputs from Figure 10.22 the difference is doubled. Thus the detection of a data transition can be improved.

This same method is used in the tracking program to find whether there is a navigation data transition every 20 ms.

Step 2. In determining the location of the data transition many data transitions are needed. In this step the existence of data transition must be detected. Perform correlation on 1 second of data points using the method mentioned in step 1 and there should be about 1000 outputs. It is assumed that the highest output is near a data transition. From this output every 20 points (20 ms) is checked to determine whether there is a data transition. At the desired point, three points are averaged. If the result is greater than zero, there is a data transition. For example, if the highest output in one second of data is at location 108, then the points 8, 28, 48, 68, 88, 128, 148, ..., will be checked for data transition. At every point, three points are averaged, such as at point 88 where the results at 87, 88, and 89 are averaged together. If the averaged value is greater than zero, a data transition is declared. If the averaged value is less than zero, no data transition is detected. A total of 20 data transitions are needed to find the exact location of the data transition. Once 20 possible transitions are found, the following steps are used to determine the exact location.

Step 3. Using the example above, if the data transitions are found at 28, 88, 108, 128, ..., the points around these points will be selected. The center point with ± 9 points is selected. Nineteen points are selected with center at the transition point. For example, points 19–37, 79–97, 99–117, 119–137, ..., are selected. There are 20 groups and each group contains 19 points. The points in the groups are added together, such as $19 + 79 + 99 + 119 + \ldots, 20 + 80 + 100 + 120 + \ldots,$ $\ldots 37 + 97 + 117 + 137 + \ldots,$ and the result has 19 values. The highest value is the location of the data transition.

The preceding approach was tested with simulated data. One second of data was generated and every 40 ms had a data transition. The method could detect 20 data transitions and use the result to obtain the location. It should be noted that some true data transition may be missed, and some false data transition may be detected. Three different C/N_0 values were evaluated. Each case was tested 1000 times with a different noise sequence. The results are shown in Table 10.6. It is interesting to note that when the C/N_0 changes by one dB, the result does not change rapidly. It might be that the results were obtained from averaging over a long record of data and the noise effect is reduced. Once the exact location

TABLE 10.6   Probability of First Data Transition Error

C/N_0 (dB)	25	24	23
Probability of error in percentage	0.5	1.2	3.3

of the navigation transition is found, the information can be passed on to the tracking program.

REFERENCES

1. Skolnik, M. I., *"Introduction to radar systems,"* McGraw-Hill, New York, 1962.
2. Tsui, J. B. Y., *"Microwave receivers with electronic warfare applications,"* Wiley, New York, 1986.
3. Barton, D. K., *"Modern radar system analysis,"* Artech House, Norwood, MA, 1988.
4. Ziemer, R. E., Peterson, R. L., *"Digital communications and spread spectrum systems,"* pp. 113–115 Macmillan Publishing Co. New York, 1985.
5. Lin, D., Tsui, J., Morton, J., "Sensitivity limit of a stand-alone GPS receiver and an acquisition method," ION GPS 2002, pp. 1663–1667, Portland, OR, September 24–27, 2002.
6. Papoulis, A., *"Probability, random variables, and stochastic processes,"* P. 263, McGraw-Hill, New York, 1991.
7. Simon, M. K., Hinedi, S. M., Lindsey, W. C., *"Digital communication techniques signal design and detection,"* p. 478, Prentice Hall, Englewood Cliffs, 1995.
8. Lin, David, AFRL/SNRP, private communication.
9. VanGraas, Frank, Professor Ohio University, private communication.
10. Psiaki, M. L., "Block acquisition of weak signals in a software receiver," ION GPS 2001, pp. 2838–2850, Salt Lake City, UT, September 11–14, 2001.
11. Akopian, D., Valio, H., Turunen, S., "Fine frequency resolving acquisition methods for GPS receivers," ION GPS 2002, pp. 2515–2523, Portland, OR, September 24–27, 2002.
12. Lin, D., AFRL/SNRP private communication.
13. Lin, D., Tsui, J., "Sensitivity improvement of a low cost commercial GPS receiver using software approach," ION 2004 National Technical Meeting, San Diego, CA, January 26–28, 2004.
14. Skolnik, M. I., *"Introduction to radar systems,"* p. 34, McGraw-Hill, New York, 1962.
15. Parkinson, B. W., Spilker, J. J., "Global positioning system: Theory and applications," vol. 1, pp. 403–403, American Institute of Aeronautics and Astronautics, Washington, DC, 1996.
16. Tsui, J., *"Digital techniques for wideband receivers,"* 2nd ed., p. 288, Artech House, Boston, 2001.
17. Liou, L., AFRL/SNRP private communication.

CHAPTER ELEVEN

Tracking Weak Signals

11.1 INTRODUCTION

In order to track weak GPS signals, the receiver must be in a low dynamic range environment. Although a high dynamic range case might be solved with additional aid, such as an inertial system, it will not be included in this book because it cannot be considered as an autonomous receiver. The method discussed in this chapter is originally designed for processing weak signals. It appears, however, that this method can also be used to process ordinary (relative strong) signals without significantly increasing the process time. The difference between tracking weak and strong signals is that longer data length is needed for tracking weak signals. The main purpose of tracking is to find three parameters: the carrier frequency, the navigation data transition, and the accurate C/A code initial phase. The carrier frequency is used to generate the local signal for continuing the tracking process. The local signal contains both the C/A and the carrier frequency. The navigation data transition is determined by the approach discussed in detail in Section 10.17. The accurate C/A code initial phase is needed to obtain the pseudorange. The method to obtain the accurate C/A code initial phase is also referred to as the fine time measurement. This measurement is used to determine the C/A code initial phase between two digitized points. The acquisition can only roughly determine the initial C/A code on a digitized data point. If the input signal is sampled at 5 MHz, the time between two adjacent points is 200 ns, which corresponds to 60 m, and is not accurate enough for determining the pseudorange.

There are several possible approaches used to track long input data, and each approach produces a slightly different result. One approach will be discussed

Fundamentals of Global Positioning System Receivers: A Software Approach, Second Edition,
by James Bao-Yen Tsui
ISBN 0-471-70647-7   Copyright © 2005 John Wiley & Sons, Inc.

in detail here. Before we proceed to the discussion of method, some trade-offs between performance and calculation speed will be pointed out.

11.2 FREQUENCY OF REGENERATING OF LOCAL C/A CODE

In Section 8.10 we saw that the local C/A code is only generated once in order to save operation time. Since the signal is sampled at 5 MHz, the corresponding sampling interval is 200 ns. Under the worst mismatch condition the locally generated C/A code and the input C/A code can be off by 100 ns, which is half of the sampling time. Since each chip of the C/A code is 977.5 ($10^6/1023$) ns, the 100 ns mismatch can cause approximately a 0.94 dB loss in the peak correlation output, as was calculated in Section 8.11. If the input signal is relatively strong, this loss does not greatly affect the operation. Under weak signal conditions it is desirable to maximize the operation gain. In other words, it is desirable to match in time the local C/A code and the C/A code of the input signal.

Two factors can cause a mismatch in the C/A codes between the input and the local signals. One is caused by the sampling frequency, which will be discussed in Section 11.13. The other cause is related to Doppler frequency, as was briefly discussed in Section 10.12. The mismatch caused by Doppler frequency shift will be discussed some more here.

The obvious way to match the C/A code of the input signal is to generate the local code every ms. For each ms the local code is digitized at a different point to match the input. The operation is time-consuming but may be applied later when a faster computer is available. The other approach is to generate the C/A code every few ms. The frequency of generating the local C/A can be understood by a brief example. Let us assume that the maximum Doppler frequency is about ±5 kHz, caused by the satellite motion. The rate change of the Doppler shift caused by the receiver motion will be neglected because the receiver must be in a low dynamic environment. As discussed in Section 3.6, it takes about 15.8 ms (approximated as 16 ms) to shift 100 ns in the C/A code if the Doppler frequency is at ±10 kHz. Thus it takes about 31.5 ms to shift 100 ns when the Doppler is at ±5 kHz.

Based on this time of 31.5 ms and the navigation data length of 20 ms, it is decided that the C/A code is regenerated every 20 ms because at 20 ms the initial C/A is shifted less than 100 ns. This means that the local C/A can match the input C/A code every 20 ms. At the end of 20 ms and under the worst case condition (at ±5 kHz Doppler) the mismatch is about 63.5 ns ($100 \times 20/31.5$). In order to achieve a better match, this mismatch time can be separated into 2 portions, each having 31.75 ns. One portion is placed at the beginning and the other at the end of the 20 ms. In the 20 ms duration the average mismatch is about 16 ns (half of 31.75 ns). Another reason to separate the mismatch into two equal portions is to better estimate the fine time. These points will be further discussed in Section 11.11. A simple estimation of correlation peak loss is about 0.14 dB, which is obtained from $20 \log[(977.5 - 16)/977.5]$. This can be considered as the worst loss because the Doppler frequency used is ± 5 kHz. It is decided that this loss is acceptable. Thus the C/A code is generated every 20 ms. Every time only 1 ms of C/A code is generated, it will be used for 20 ms.

However, the local C/A code can also be generated every second. Because the maximum mismatch can be 100 ns, this method can be applied to all signal conditions. Especially when the sampling frequency is not an integer, as discussed in Section 11.13, this method can still track the input signal. Under this condition the average mismatch time can be 50 ns, which corresponds to a loss of 0.46 dB.

If the computer speed is adequately improved, the C/A code might be generated every ms to obtain the maximum output. The improvement, however, in sensitivity is limited only about 0.14 or 0.46 dB as shown in the preceding paragraphs.

11.3   CARRIER FREQUENCY MEASUREMENT REQUIREMENT

It was discussed in the Section 10.3 that the sensitivity of a GPS receiver is limited by how accurately navigation data can be decoded. In order to determine the navigation data correctly a very accurate carrier frequency is needed. In addition the carrier frequency can also be used to help determine the pseudorange, which can be considered as carrier-aided processing, and the details will be discussed in Sections 11.9 to 11.11. Both arguments suggest that an accurate carrier frequency is needed. The question is how much accuracy is needed.

As discussed in Sections 3.7 and 3.8, the maximum Doppler frequency change is about 1 Hz per second. If one second of data is processed through the fast Fourier transform (FFT), the resolution is about 1 Hz, which matches the rate of change of the Doppler frequency. If longer data, such as 2 seconds, are used to measure the frequency, the resolution is about 0.5 Hz. But during this time the carrier frequency may change 2 Hz; thus this resolution and the frequency change is not matched. In other words, the frequency measured may cover several frequency bins, and the accuracy will not improve. Based on this argument 1 second of data is used to calculate the carrier frequency. Better than 1-Hz accuracy can be obtained by comparing amplitudes of adjacent frequency bins and extrapolating the result.

11.4   ONE MILLISECOND DATA PROCESSING AND INPUT DATA SELECTION

Before the discussion of input data selection, two terms will be defined here.

1. *Initial C/A code point (or initial point of a C/A code):* This refers to the first digital point of a C/A code, which is obtained from digitizing the first chip of the code. This chip is the first one of 1,023 chips generated, as discussed in Section 5.6.

2. *Initial C/A code phase (or initial phase of a C/A code):* Although this term has been used previously, it is restated here again to emphasize the relationship to the initial C/A code point. The initial C/A phase is a data point, which is closest to an initial C/A code point. Most likely in a data set there is no data point that matches the initial C/A code point. One can consider that the initial code point is between two input data points. For example, if there are 5000 data points in one ms and the initial C/A point is between the 1000 and 1001 data points but closer to 1000, the initial C/A code phase will be 1000. Sometimes the initial C/A code phase is also referred to as the offset time.

In both the tracking methods discussed in Chapter 8 and in this chapter, the input data are processed in 1 ms blocks. Since the data are sampled at 5 MHz, 5000 input data points will be grabbed and processed each time. There are several differences, however, between these two methods. The method in Chapter 8 will be briefly repeated here. The local C/A code is generated only once, and its initial phase must match a C/A code in the input signal. If the initial phases of the two C/A codes differ by more than 100 ns, the input data must shift by one point, as was illustrated in Section 9.2. In order to make sure that the input data contain 1 ms of data of every satellite with all possible C/A code initial phase, 2 ms of input data are taken every time. Each time the input data move only 1 ms, as shown in Figure 11.1. In that approach the locally generated C/A code always starts from the initial point.

In the present tracking method the locally generated C/A code does not start from the initial of the code. It is generated to match the first data point in the input signal. The initial point of the local C/A code can be anywhere in the code, and it changes every 20 ms because of the code regeneration. Under this condition two situations are different from the previous method. First, there is no need to shift the input data to match the local C/A code phase, as discussed in Section 10.2, because the worst mismatch after 20 ms is about 63.5 ns, which

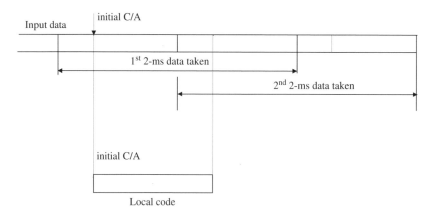

FIGURE 11.1   Input data taken for method discussed in Chapter 8.

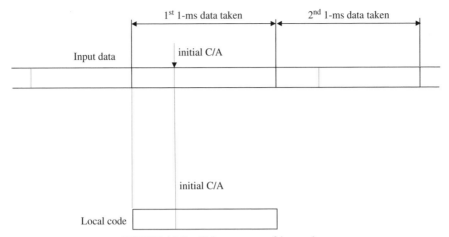

FIGURE 11.2   Take one ms of input data.

is less than 100 ns. Second, it requires only 1 ms of input data to contain an entire C/A code; thus only 1 ms of input data is taken each time. This situation is shown in Figure 11.2. Any 1 ms of input data (5000 points obtained by sampling at 5 MHz) contains all the satellites, but different satellites have different C/A codes with different initial points. In this figure only one satellite is shown.

In the acquisition of weak signals three parameters are obtained: the rough initial phase of the C/A code, the carrier frequency, and the beginning of the navigation data. The initial phase of the C/A code and the carrier frequency are used to generate the local signal. The beginning of the navigation data can be used to determine the beginning of the tracking process. It is discussed in Section 10.17 that 40 ms of data (20 ms before and 20 ms after the navigation phase transition) are used to determine whether a navigation phase transition exists. In every 20 ms there is one possible navigation data transition. The data transition is also an initial C/A code point. The tracking should start with a

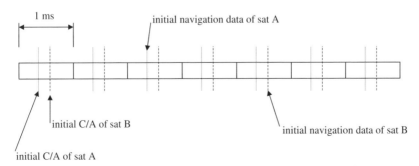

FIGURE 11.3   Tracking starts from a navigation data point.

certain ms of data, which is also the beginning of a navigation data. In other words, the selected initial C/A code is also a navigation data transition point. It should be noted that the navigation data transition point may or may not have an actual phase transition. For this reason, the tracking starts at a specific initial C/A code within the first 20 ms of data, as shown in Figure 11.3. In this figure the initial C/A codes of two satellites, A and B, are shown. The beginning of the navigation data of satellite A is in the third ms; thus, the tracking starts from the third ms. For satellite B the tracking starts from the fifth ms. Once the initial tracking starts, the tracking will run continuously.

11.5 GENERATION OF C/A CODE

In this section the generation of a digitized C/A will be presented because it is different from the program listed in the end of Chapter 7. Here the Doppler frequency and the initial code phase will be included in the code. By including the Doppler frequency in the C/A code, the peak correlation output will be improved slightly. First, the Doppler frequency affecting the sampling frequency will be discussed. Second, the Doppler frequency also affects the initial code phase of the C/A code generated in 20 ms time intervals.

The Doppler frequency effect was discussed in Section 10.12. However, there are some differences between the discussion of that section and here. In Section 10.12, the C/A code was only generated once. The Doppler frequency was used to adjust the time shifting for noncoherent processing. In this section, the C/A code is generated every 20 ms to better match the input data by taking the carrier frequency into consideration. In generating 1 ms C/A code, if the carrier frequency is taken into consideration, the output will be slightly different. For example, two sets of 5000 digitized points are generated from 1 ms of C/A code of satellite number 1: one set with zero Doppler frequency and one with a Doppler frequency of 5 kHz. Between the two sets of data there are about 6 points having different values that are dependent on the starting sample point. If the Doppler frequency is taken into consideration in generating the C/A code, the number of difference points can be reduced. This operation results in a slightly higher correlation peak. The following paragraph will derive the relation between the sampling frequency and the Doppler frequency.

If the Doppler frequency is positive, then the equivalent length of the C/A code shrinks in time at the receiver, as shown in Figure 11.4. Figure 11.4a shows the original C/A code and the sampling time. Figure 11.4b shows the shortened C/A code with the same sampling rate. Another way to express this effect is to decrease the equivalent sampling frequency but keep the length of the C/A code at a constant value, which is equivalent to extending the sampling time, as shown in Figure 11.4c. Note that the digitized values obtained from Figures 11.4b and 11.4c should be the same.

The decreased sampling frequency is used to digitize the C/A code. The equivalent sampling frequency f_{se} and the corresponding sampling time t_{se} can be

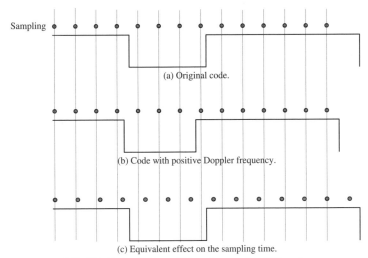

Sampling

(a) Original code.

(b) Code with positive Doppler frequency.

(c) Equivalent effect on the sampling time.

FIGURE 11.4   Effect of Doppler on the C/A code.

expressed as in Equation (10.18),

$$f_{se} = \frac{f_s f_{L1}}{f_{L1} + f_d} \approx f_s \left(1 - \frac{f_d}{f_{L1}} \right) \quad \text{or}$$

$$t_{se} = \frac{1}{f_{se}} = \frac{t_s (f_{L1} + f_d)}{f_{L1}} = t_s \left(1 + \frac{f_d}{f_{L1}} \right) \tag{11.1}$$

where $f_s = 5$ MHz is the original sampling frequency, $f_{L1} = 1575.24$ MHz is the $L1$ frequency, f_d is the Doppler frequency, $t_s = 1/f_s$ and t_{se} are the original and equivalent sampling times. It should be noted that this equation is different from Equations (10.17) and (10.18). The two equations in the previous chapter calculate the C/A code length change; thus, a positive Doppler will result in code length shrink. A one ms long code will have an effective length of smaller than 1 ms. In Equation (11.1) the equivalent sampling frequency f_{se} is decreased or the equivalent time t_{se} is increased. By this effective sampling frequency, the desired digitized data can be obtained.

Now let us discuss the generation of a C/A code with any initial code phase. The term offset time (t_{of}) is used for the initial C/A code phase. The offset time can only be a positive value in the actual program. If the offset time is shifted to the right, it is considered as a positive value, as shown in Figure 11.5. If the offset time is shifted to the left it is considered a negative value. The negative value can be represented by an equivalent relatively long positive offset as shown in the figure. The offset time t_{of} is also affected by the Doppler frequency, and the equivalent offset time t_{ofe} can be expressed as

$$t_{ofe} = \frac{t_{of} (f_{L1} + f_d)}{f_{L1}} \tag{11.2}$$

FIGURE 11.5   Offset time of the C/A code.

This equation is identical to Equation (11.1). The actual program for generating the digitized C/A code is listed at the end of this chapter. The inputs include the sampling frequency and the offset time.

The actual generation of the C/A code every 20 ms occurs as follows. We will use a numerical example to illustrate the operation. The important factor is the determination of the offset time change between the two C/A codes with Doppler frequency taken into consideration. If the Doppler frequency $f_d = 5$ kHz, the t_{se} can be obtained from Equation (11.1). The difference between t_s and t_{se} is -6.3475×10^{-13} sec. In 20 ms there are 100,000 samples. Therefore the total difference between two offset times is -6.3475×10^{-8} sec (or -63.475 ns). This value is less than half of the sample time of 100 ns. It implies that if the first C/A has zero offset, 20 ms later the C/A code has an offset time of -63.475 ns. Since only positive values can be used in the C/A code digitization program, this offset time must be changed to a positive value as 999.9365 μs $(1000 - 0.063475)$. This offset time and the correct Doppler frequency will be used to generate the next 1 ms of C/A code. As this process continues, the offset time will accumulate. For example, after another 20 ms the offset becomes $-(2 \times 63.475)$ ns if the

FIGURE 11.6   Generation of 1 ms of C/A code twice for 40 ms of operation.

Doppler frequency does not change. The generation of the C/A is illustrated in Figure 11.6. In this figure the initial phase offset of the first C/A code is zero, and this code is used for the first 20 ms. The initial phase offset of the second C/A code is −63.475 ns (or 999.9365 ns), and this code is used for the second 20 ms.

11.6   GENERATION OF LOCAL CODE AND CORRELATING ON 1 MS INPUT SIGNAL

The locally generated signal is calculated every ms, and it contains the C/A code, as well as the carrier frequency. The local signal can be written as

$$lc_k = C_k \exp[j2\pi f_k t + j\theta_k(n)] \tag{11.3}$$

where lc_k represents the local code, C_k is the C/A code, f_k is the carrier frequency and $\theta_k(n)$ is the initial phase of the carrier frequency for satellite k, respectively. This equation is used to generate 20 ms of data and each ms contains 5000 data points. It should be noted that C_k does not start from the initial point of the C/A code but from the point where the initial C/A code phase matches that of the input signal, as discussed in the previous section.

As in the equation above, the same C/A code (C_k) can be used to generate the local signal for the entire 20 ms time period. The carrier frequency phase angle $\theta_k(n)$, where $n = 1$ to 20, must be adjusted every ms in order to keep the carrier frequency continuous for the entire 20 ms without phase discontinuity. The carrier signal is a complex quantity, and its frequency f_k stays constant for one second of time. At the end of one second the frequency f_k is re-calculated, and a new value is generated. At the end of every ms, the carrier frequency phase angle is determined by the carrier frequency f_k, as well as the previous phase angle $\theta_k(n)$. Suppose that at the nth ms the carrier frequency phase angle is $\theta_k(n)$. After 1 ms the new value of the phase angle is $\theta_k(n+1)$, which can be calculated from the previous value $\theta_k(n)$ and the frequency f_k by the following equation:

$$\theta_k(n+1) = \text{angle}\{\exp[j2\pi f_k 5000 t_s + j\theta_k(n)]\} \tag{11.4}$$

where angle in this equation means to find the angle from the real and imaginary portions of the exponential function (exp). The value 5000 represents 5000 data samples in one ms of data and t_s is the sampling time. Every ms a new phase angle value $\theta_k(n)$ as the initial phase of the carrier is used in Equation (11.3) to generate the new local signal for the next ms. It is usually assumed that $\theta_k(1) = 0$.

Every 20 ms the C/A code is generated again and the time elapsed between the current 20 ms and the previous 20 ms C/A code is calculated using Equation (11.1). The actual calculation is illustrated in the last paragraph of the previous section. Both the C/A code and the carrier signal are digitized to 5000 points; thus the local signal has 5000 points. The local C/A code is shifted to generate the early and late codes, but the same carrier frequency is used in all

Carrier signal

1 ms

Generated every ms phase must be continuous
The same carrier frequency is used for 1 second

C/A code

One ms of C/A code used for 20 ms

FIGURE 11.7   Generation of carrier and C/A code signals.

three local codes: the early, the prompt, and the late code. This subject will be further discussed in Section 11.9.

The generation of the local signal from Equation (11.3) can be summarized as follows: The carrier phase $\theta_k(n)$ is updated every ms, and the phase must be continuous at 1 ms interval. The C/A code C_k is updated every 20 ms, and the carrier frequency f_k is updated every second. This operation is illustrated in Figure 11.7.

Once the prompt local code is generated, it is multiplied by 1 ms input signal point by point to obtain 5000 product values. Since the frequencies of the local and input signals are very close, the 5000 product values have approximately the same values. Figure 11.8 shows the operation on 3 ms of data, which are divided into segments A, B, and C. In this figure the product values from each ms are divided into two portions. The portion donated by a is the part before the initial C/A code phase, and the part after the initial C/A code phase is designated as b. Because the initial C/A code phase is used to determine the pseudorange, it is desirable to put 1 ms of data together, starting from an initial C/A code phase. To accomplish this goal, the portions from two adjacent ms are grouped together to form a new set of segments (shown in the figure as 1 and 2). For example, the b portion of A and the a portion of B are grouped together and designated as 1. The b portion of B and a portion of C are grouped together and designated as 2. Each segment contains 5000 points. These 5000 points are then summed into one point, which is the peak correlation output from one ms of data and the locally generated code. Thus, after these operations, 1 ms of input data (5000 points) only generates one output data point. Let us reference this output point as the compressed output. The compressed output is obtained from 5000 input data starting from an initial C/A code point. The compressed output is usually obtained from two consecutive

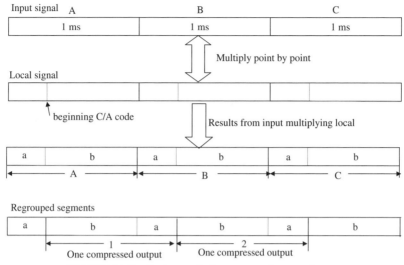

FIGURE 11.8   Peak correlation outputs from each ms.

sets of 5000 input data points by combining *a* and *b* portions, as shown in Figure 11.8.

One second of data only generates 1000 compressed outputs. Since one compressed data point is the result of processing 1 ms input data, the corresponding bandwidth is 1000 Hz. The operation speed requirement on compressed data is not stringent because of the low output data rate.

11.7   OBTAINING NAVIGATION DATA AND FINDING CARRIER FREQUENCY

Once 1 ms of data is compressed into one data point, the actual navigation data transition can be found. Since the location of the navigation data transition has been determined in the acquisition program and the tracking program starts from the beginning of a navigation data, as discussed in Section 11.4, the navigation data transition can be determined every 20 ms. The approach uses 40 ms of compressed data and two reference signals: one with phase transition at 20 ms and one without phase transition. The procedure has been discussed in detail in step 1 of Section 10.17. Once the navigation data transition is determined, the navigation data of 1 sec (1000 points) can be generated. These navigation data are multiplied with the compressed data, and this operation will strip off the navigation data from one second of input (1000 points). The result is a complex continuous wave (cw). An FFT is performed on these data putting the results in the frequency domain, which covers a bandwidth of 1 kHz with a resolution of 1 Hz. The amplitude of a typical frequency domain is shown in Figure 11.9.

(a) All frequency bins.

(b) 100 Frequency bins around peak.

FIGURE 11.9 Frequency output.

Since the input has been down converted to base band, the peak output is usually very close to the direct current (dc) component. Figure 11.9a shows the output of the 1000 frequency components. Figure 11.9b shows the 100 components near dc, which is obtained by plotting points from 951 to 1000 and 1 to 50 of Figure 11.9a. This operation shifts the outputs from near zero in Figure 11.9a to near the center in Figure 11.9b.

In general, the input frequency does not fall on a frequency bin. Under this condition, the frequency is spread into several frequency bins, as shown in Figure 11.9b. One can improve the frequency reading from the two adjacent frequency bins with the highest outputs as [1]

$$f_{del} = \frac{A(1)}{A(0) + A(1)} \tag{11.5}$$

where f_{del} is a shift frequency with resolution less than 1 Hz from the peak output, $A(0)$ is the highest amplitude in Figure 11.9b, and $A(1)$ is the higher one of the two adjacent frequency bins of $A(0)$. By this equation, the frequency can be adjusted to less than 1 Hz. The direction of the true frequency lies toward the higher neighboring bin. With this process, a very accurate carrier frequency can be obtained.

11.8   CALCULATING SIGNAL-TO-NOISE RATIO (*S/N*)

This operation is not essential to the tracking program; however, it is nice to have a *S/N* estimation every second. Once the frequency bins are calculated, the power of the signal can be calculated. If the noise power and the signal power are obtained, the *S/N* can be calculated. Theoretically the noise of the signal can be obtained from either the frequency or the time domain. From Figure 11.9a one can take a portion of the spectrum that does not include the signal to represent the noise power. The noise measured this way, however, varies from second to second by about 1 dB based on real data. This may be caused by the nonuniform distribution. A closer examination of the noise in Figure 11.9a reveals that the noise is slightly higher around bins 200 and 900. The noise of additional real data will be discussed in Section 10.15.

The noise power obtained from 1 ms of data in time domain every second varies much less than 1 dB. Therefore the noise power is measured in the time domain with 1 ms of data. In this operation the signal is included in the noise calculation. Because the nominal signal is about -19 dB under the noise in a 2 MHz bandwidth system, the error caused by including the signal can be neglected. The total noise power in one second can be obtained by multiplying the measured noise by 1000 because there are 1000 frequency bins. The noise power per second P_n is

$$P_n = \mathrm{var}(x_i) \times 1000 \tag{11.6}$$

where var represents variance and x_i ($i = 1 : 5000$) is 1 ms of input signal in time domain.

The signal power is obtained from the frequency domain. As shown in Figure 11.9b, the signal may disperse into several frequency bins. In practice, it is difficult then to determine exactly the noise power. Thus, it is arbitrarily decided that only 5 frequency bins, two on each side of the highest output, will be used to measure the signal power. An additional requirement is that the outputs from the four neighboring frequency bins must be 3.5 times greater than the noise amplitude (square root of noise power) to qualify as signal power. The constant 3.5 is arbitrarily chosen. The reason for having this requirement is to take into consideration the case where the signal is centered on one frequency bin and the neighboring four frequency bins contain most of the noise. Under this condition the neighboring bins should not contribute to the signal power. The signal power P_s can be calculated as

$$P_s = \sum_{k=-2}^{2} A(k)^2 \quad \text{where}$$

$$A(k) > 3.5\sqrt{P_n}$$

(11.7)

$A(0)$ is the highest output in Figure 11.9b, and $A(k)$, $k = -2, -1, 1,$ and 2 are the four neighboring bins. P_n is noise power obtained from the previous equation.

Because the signal is measured in 1 Hz resolution, the S/N calculated through this method is equal to C/N_0. For normal signal strength, the S/N calculated from second to second has a standard deviation of about 0.1 dB, based on processing results of experimental data.

11.9 BASIC IDEA OF FINDING THE PSEUDORANGE[2]

The idea of finding the fine time, which is related to pseudorange, is based on the correlation output, as discussed in Section 8.11. The only difference is that in Section 8.11 the fine time is calculated every ms, and the results from every 10 ms are averaged into one point. It appears that even for signals with normal strength, the correlation output is rather noisy. For a weak signal it is practically impossible to obtain a meaningful fine time from 1 ms of data. For this reason, one second of data is used to find the fine time.

One ms of input data (5000 points) is used to generate three output points. One is the compressed output, which is obtained by multiplying and summing the input with the prompt code. The other two points are generated by multiplying and summing the input with the early and late local codes. The generation of the compressed output from the prompt code was discussed in Section 11.6. The local early and late codes can be generated by shifting the 5000-point prompt C/A code by one or two data points. In Section 8.11 the early and late codes were obtained by shifting two data points; here they are obtained by shifting by one data point. By shifting one data point, the peak correlations from the early and late gates are slightly higher, but the tracking range decreases. Since the weak signal is tracked under low dynamic environment, the tracking range is not of primary concern. Higher peak correlation values can improve the S/N; therefore, the one data point shift is used to generate the early and late codes for weak signal tracking. The early and late local codes can be obtained in two steps: generation of the C/A code and the carrier frequency. The early C/A code is obtained by shifting the last point of the prompt C/A code to the front. The late C/A code is obtained by shifting the first point of the prompt local code to the end. The same carrier frequency generated in Section 11.6 is used for the early and late local codes.

After multiplying with 1 ms of input signal with both the early and late codes, each will have 5000 outputs. The 5000 outputs are divided into two portions at the initial C/A point and jointed with the outputs from two adjacent ms to form the compressed early and late outputs. The approach of generating the early and late compressed outputs is identical to the generation of the prompt point, as discussed in Section 11.6 and Figure 11.8. The only difference is the input data are multiplied with the local early and late C/A codes.

In this operation it is desirable to adjust the prompt correlation peak close to the true peak. To achieve this goal, locally generated C/A must match closely to the C/A code of the input signal to improve the S/N. When the prompt peak matches the true peak, the early and late correlation peak values are about 813.7 relative to the ideal peak value of 1023. These outputs are about 2 dB (20 log(813.7/1023)) below the prompt correlation peak. When the early and late codes are generated by shifting two samples, the early and late correlation peaks are about 4.6 dB below the prompt correlation peak. When weak signals are processed, higher early and late correlation peaks can provide a more accurate result.

To find the fine time x, the time difference between the measured peak and the true peak position, the same relation in Equation (8.45) can be used, which can be rewritten as

$$r \equiv \frac{y_{lsa}}{y_{esa}} = \frac{1 - x - d}{1 + x - d} \quad \text{or}$$

$$x = \frac{(1 - r)(1 - d)}{1 + r}$$

(11.8)

In this equation y_{esa} and y_{lsa} replace y_e and y_l in Equation (8.45). The values of y_{esa} and y_{lsa} are the amplitude of the summed early and late peak correlation outputs, respectively, and they will be further discussed in the next section. In this equation, d is the time between the early and prompt or the late and the prompt peaks. Both x and d are measured in unit of chip time of 0.9775 ms. Its value is 0.2046 instead of 0.4092 used in Equation (8.45) because this choice has slightly better noise performance but less tracking range.

In one second, 1000 compressed early outputs are generated. The 1000 compressed early outputs are summed together in vector form to produce one point;

thus the result is a complex quantity and represented by y_{es}. Similarly the compressed late outputs are summed to form one point and represented by y_{ls}. The amplitudes of y_{es} and y_{ls} are referenced as y_{esa} and y_{lsa}, respectively. These are the quantities shown in Equation (11.8). The actual summation method will be explained in the next section.

11.10   OBTAINING THE SUMMED EARLY AND LATE PEAK CORRELATION OUTPUTS (Y_{es} AND Y_{ls})

First, the approach to obtain y_{es} and y_{ls} will be presented. Then their meaning will be explained. Since the explanations for both y_{es} and y_{ls} are identical, only y_{es} will be used for illustration.

Every ms an early peak correlation result y_e is generated. Because the carrier frequency is a complex quantity, after frequency conversion the output is also complex. As a result the early peak correlation y_e is also complex. In one second, $1000y_e$ values are generated. These values are summed together in complex form rather than in absolute value, and the result is referred to as y_{es}. Summing in complex form is equivalent to coherent processing. The next step is to discuss the meaning of y_{es}.

In order to simplify the discussion, a numerical example will be used. If the Doppler is 5 kHz, it causes the time shift in every digitized input data sample by -6.3475×10^{-13} sec, which has been discussed in Section 11.5 and Equation (11.2). Every ms the time shift is -3.174×10^{-9} sec ($-6.3475 \times 10^{-13} \times 5,000$). Let us assume that the initial prompt C/A code matches the input signal exactly. Then the amplitude of the prompt peak correlation output will be 1023. The early and late C/A code will mismatch the input signal by one sample with correlation peaks of 813.7. Because the sampling interval is 200 ns, the next ms the early C/A code will mismatch the input by 196.826 ns ($200 - 3.174$ ns), and the corresponding peak correlation output will be slightly higher than 813.7. After 20 ms the early code and the input signal will be mismatched by about 136.53 ns ($200 - 3.17375 \times 20 = 200 - 63.475$). Thus the early correlation peak at 20 ms is even higher than that at the beginning of the input. The amplitude of the early output y_e is shown in Figure 11.10.

The initial output shown in the figure is represented by the solid line and the final output (20 ms later) by the dotted line. The outputs from the prompt and the late are decreased because the mismatch increases by 63.475 ns after 20 ms. The amplitude of y_e from each ms is shown in Figure 11.11. This figure can be considered a magnified version of y_e shown in Figure 11.10. After 20 ms the prompt local C/A code is regenerated and it matches with the input again. Thus, the amplitude of y_e decreases to the same initial value. As time goes on, this mismatch decreases again. This process repeats every 20 ms, or in 1 second it occurs 50 times. The results of the prompt peak correlation y_p and the late peak correlation y_l can be explained the same way.

The summation of these y_e will be referred to as y_{es}. From Figure 11.11 one can see that the result is the average value between the beginning and the

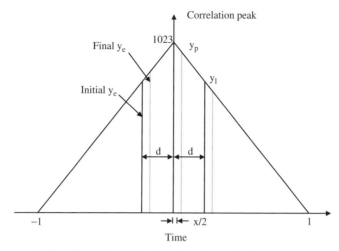

FIGURE 11.10 Variation of peak correlation outputs.

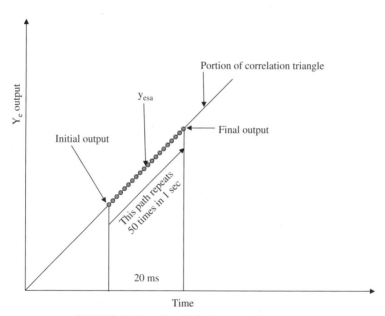

FIGURE 11.11 Detailed changes of y_e per ms.

ending points, or the mean of the $20y_e$ outputs. Since the reference signals used to generate y_e, y_p, and y_l contains the C/A code and the carrier frequency, the summation of the results can be considered coherent processing, and the results are labeled y_{es}, y_{ps}, and y_{ls}. If the carrier frequency is exactly right, y_e, y_p, and y_l should be added in phase and achieve the maximum value. Otherwise, the results might be slightly off. The amplitudes of three quantities, y_{es}, y_{ps}, and y_{ls}, are referred to as y_{esa}, y_{psa}, and y_{lsa}, respectively and they can be obtained by the following equation:

$$y_{es} = \sum_{n=1}^{1,000} y_e(n) \quad y_{ps} = \sum_{n=1}^{1,000} y_p(n) \quad y_{ls} = \sum_{n=1}^{1,000} y_l(n)$$

$$y_{esa} = |y_{es}| \qquad\qquad y_{psa} = |y_{ps}| \qquad\qquad y_{lsa} = |y_{ls}| \tag{11.9}$$

where $y_e(n)$, $y_p(n)$, and $y_l(n)$ are the 1 ms early, prompt, and late correlation peaks, respectively; y_{es}, y_{ps}, and y_{ls} are the 1 second early, prompt, and late correlation peaks, and y_{esa}, y_{psa}, and y_{lsa} are their corresponding amplitudes, respectively.

The y_{esa}, y_{psa}, and y_{lsa} are three points approximately at the middle between the initial and final points of y_e, y_p, and y_l, respectively. Figure 11.11 also shows the position of y_{esa}. From this discussion it is obvious that the summation does not change the meaning of y_e, y_p, and y_l, as presented in Section 8.11. The advantage is that the S/N is enhanced, and Equation (11.8) provides the desired fine time. This fine time is different from the fine time calculated in Chapter 8. The fine time calculated in Chapter 8 is from 1 ms of data, and the result represents the fine time of that 1 ms. The fine time obtained through Equation (11.8) is from 1 second of data, and the fine time is referred to the first C/A code initial phase in one second of data. This quantity can be related to the fine time at every ms, and the fine time of a certain ms can be calculated from the time delay caused by the Doppler frequency. The following example shows how to find the fine time of the first ms of data.

In this example the local C/A code and the C/A code in the input data are assumed exactly matched for the first ms. However, there might be a slight error. The fine time obtained from 1 second of data is used to find this error because 1 second of data should provide better accuracy. For example, at the beginning of the second ms the mismatch is 3.174 ns for a Doppler frequency of 5 kHz. At the twentieth ms the mismatch is 60.306 ns (3.174 × 19), and we let this quantity be represented by t. Thus the average fine time calculated should be 30.153 ns, which is half of x (or 60.306) ns, marked as $x/2$ in Figure 11.10. If measured at 30.153 ns, the fine time implies that the local C/A code and the C/A code of first ms are perfectly matched, as assumed. If the measured fine time is not equal to 30.153 ns, the difference between measured $x/2$ and 30.153 represents the misalignment between the local C/A and the C/A codes in the first ms. This misalignment can be used to determine the true fine time of the first ms. The rest fine times can be obtained from the first ms fine time and time shift (3.174 ns) between adjacent ms.

It should be noted that in Equation (11.8) the value of y_{psa} is not used in determining the fine time. The values of y_{psa} are the power level used to determine the navigation data and carrier frequency, as discussed in Section 11.7. Thus a high value of y_{psa} is desirable.

If the three local codes are generated every ms, then theoretically the locally generated prompt C/A code and the C/A code in the input data can be matched equally well at every ms because they are realigned at every ms. The early and late codes are off by one sample from the input data. Under this condition, the average fine time obtained from one second of data and the fine time of every ms are the same. Once the average fine time is found, no further operation is required to find the fine time of every ms. In addition the value of y_{psa} is always at the maximum, which can help to achieve a slightly better sensitivity in determining the navigation data.

11.11   ACTUAL TIME SHIFTING IN TRACKING

The example given in the previous section is used to help explain the calculation of the fine time of the first C/A code initial phase through one second of data. In the actual tracking, however, the operation is slightly different. At the beginning of the tracking the fine time of the first ms is not known. It can be any value x_0. The result of the very first second in tracking is illustrated in Figure 11.12. This is basically the same as Figure 11.10, and the only difference is that the y_p is off from the true peak by an amount x_0. At the end of the first second, the average fine time will be $x_0 + 30.153$ ns if the Doppler frequency is 5 kHz. In order to find the fine time of the first ms, the 30.153 ns must be subtracted from the measured average fine time. Starting at the next second, the fine time can be properly adjusted.

The adjustment is based on the idea of keeping the prompt peak correlation output y_p at maximum. Under this requirement the locally generated prompt C/A code and the C/A in the first ms of the input signal must not be exactly matched. From the assumed Doppler frequency of 5 kHz, the time difference after 19 ms of data is 60.306 ns, which is the x value. If in the first ms, the two C/A codes,

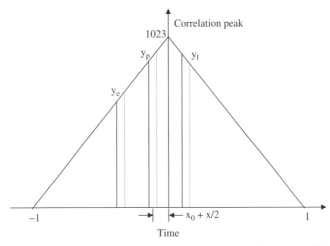

FIGURE 11.12   Resulting from a first second tracking. The initial time off for y_p can be any value.

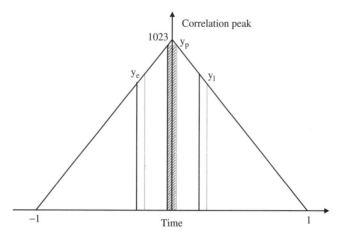

FIGURE 11.13   Tracking result from second 2 on.

one from local and one from input data, are aligned, at the 20th ms they will be off by x. If the two C/A codes in the first ms is off by $-x/2$, at the 20th ms they will be off by $x/2$. This arrangement will produce a slightly higher output y_p.

Based on this idea, after the first second, the mismatch time x_0 can be found. Since x_0 is known, in the next second the offset time can be chosen to any desired value. In the present approach the next second instead of matching the locally generated C/A code to the C/A code of the input signal, it is offset by $-x/2$ (or -30.153 ns). The result is shown in Figure 11.13. The shaded area represents the prompt peak correlation output. Comparing with the result in Figure 11.10, where the C/A codes are aligned in the first ms, the prompt peak correlation output from the present is slightly higher. With this arrangement, if x is measured correctly, the beginning position of y_p is off by -30.153 ns and the ending position is off by 30.153 ns. Under this condition the average time off is about 0 ns. The corresponding average amplitude loss is about 0.14 dB, which can be calculated by assuming the average amplitude is 15.77 ns (30.153/2) away from the correlation peak. The outputs from y_{esa} and y_{lsa} have the same amplitude. Once this relation is obtained the rest of the tracking can follow the same pattern.

This arrangement can also be used to better calculate the fine time, which might be a more important factor. Equation (11.8) or (8.45) is based on the assumption that the correlation peak is a triangle. As discussed in Section (8.12), however, due to the limited bandwidth of the receiver, the correlation peak is not a triangle. The exact shape of the correlation output is unknown. Therefore the result calculated from Equation (11.8) can only be considered an approximation. However, even if the correlation output shape is unknown, the fine time can be calculated very accurate, if the values of y_{esa} and y_{lsa} are equal. If the multipath effect can be mitigated, the correlation output shape can be considered symmetrical. Under this condition, if y_{esa} equals y_{lsa}, the fine time will be zero, which

TABLE 11.1   Fine Time Measured by Simulated GPS Signals (One C/A Code used 20 ms)

C/N_0	Mean Error(ns)	Standard Deviation(ns)
25	1.3	20
24	1.3	15
23	1.5	23
22	4	23
21	12	28
20	−4	30
19	−7	43
18	4	40

can also be found in Equation (11.8). The approach in Figure 11.13 makes y_{esa} approximately equal y_{lsa}; thus the error calculated will be small.

With this arrangement the fine time obtained from 1 second of data should be zero (or $x/2 = 0$). The difference between the two C/A codes in the first ms should be $-x/2$. If the fine time obtained from 1 second of data is not equal to zero, the difference between the two C/A codes in the input and the local signal should be adjusted by the fine time measured.

Simulated data are used to measure the fine time through this approach. Two sets of 8-second simulated GPS signals with Gaussian noise are used as input signals[2]. The results are listed in Table 11.1. Note that under the weak signal condition, the mean error is rather small, only a few ns, but the standard deviation increases. The tracking program usually can track a signal less than $C/N_0 = 24$ dB, as indicated in the table.

11.12   TRACKING PROGRAM WITH REGENERATION OF C/A CODE EVERY SECOND[3]

Since the tracking program can track signals below $C/N_0 = 24$ dB, the program must have enough sensitivity; therefore, the locally generated C/A code and C/A code of input data would have to match exactly. For this reason a tracking program without regenerating C/A code is studied. This type of tracking saves the generation of C/A code every 20 ms, but the tracking of the fine time of every ms can be cumbersome. In addition the locally generated C/A code often needs to be shifted a point. After every second the local C/A code and the carrier frequency are generated again.

Let us use a numerical example to help explain this approach. If the Doppler is 5 kHz, as discussed in Section 11.10, every ms the C/A code has a time shift of -3.174 ns. In one second the total time shift is -3174 ns. This time shift must be compensated; otherwise, the local C/A code and the C/A code in the input signal can no longer be aligned, and the input signal can no longer be tracked. It is desirable to make a data shift whenever the local and input C/A codes are off by half the sample time or 100 ns. For this example the local C/A code needs a shift after 32 ms of data. In 32 ms the total time mismatch is -101.568 ns (-3.174×32). The time shift step can only be 200 ns. After a shift of 200 ns the mismatch becomes 98.432 ns ($200 - 101.568$); thus, the next shift will take about 63 ms because the fine time will change from 98.432 ns to about -100 ns, which is close to 200 ns. The time required for this time shift can also be calculated as $200/3.174 \approx 63$ ms. In this approach, the mismatch of every ms must be accumulated. If the mismatch is over ± 100 ns, a data point shift will be needed. Figure 11.14 shows three lines of the ms y_e values.

In Figure 11.14 the first line has 32 y_e values, and the rest of the lines have 63 y_e values. All three lines of y_e's should be one on top of the other, but to show the delay time between them, they are drawn separately. There should be about 17 lines in one second of time. The starting points of the lines are not at the same time, as shown in Figure 11.11. The first line contains 32 y_e's. The next 15 lines each contains 63 y_e's. The last contains 23 y_e's. In order to obtain the fine time through averaging, all the 1,000 individual fine times must be tracked. The tracking of the fine time at every ms will be discussed in the next section. As will become clear in the next section, the tracking of a signal with inaccurate sampling frequency is a more general case than this one. Similarly the y_l's will be shifted in identical time with the y_e's. Once the summations of y_e's and y_{l}'s are obtained, the averaged fine time can be obtained in the same way, as discussed in Section 11.10.

Simulated data were used to evaluate the results. For each S/N, 100 runs with different noise sequence are performed. The results are listed in Table 11.2[3]

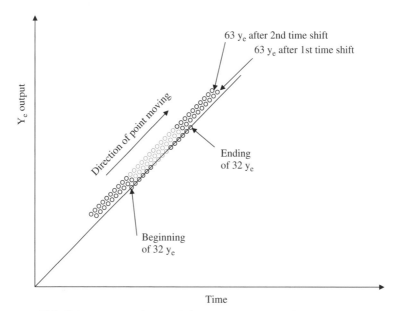

FIGURE 11.14   Early correlation peak y_e values with time shift.

TABLE 11.2 Fine Time Measured by Simulated Tracking Signals (One C/A Code used/sec)

C/N_0	Mean Error (ns)	Standard Deviation (ns)
25	−1.1	14
24	−2.1	16
23	−2.9	18
22	−7.5	20
21	−6.5	22
20	−3.2	27
19	1.3	30
18	4.3	35

Because the number of runs used in obtaining Tables 11.1 and 11.2 is rather small, it is difficult to draw a firm conclusion. From these limited data, however, one can see that the results are comparable. Theoretically the approach in Section 11.11 should have a smaller mean error because in this method the y_{esa} and y_{lsa} may not be equal. It is possible, however, to generate the C/A code every second and make y_{esa} and y_{lsa} of about the same amplitude. Under this condition the results from these two approaches should be very close.

11.13 TRACKING OF SIGNALS WITH NON-INTEGER SAMPLING FREQUENCY[4]

Non-integer sampling frequency means that in 1 ms, the C/A code sample generated is not an integer number. Strictly speaking, the number of digitized data per ms can hardly be an integer. For example, in the preceding discussion the sampling frequency was assumed to be 5 MHz, and the frequency was rather accurate. In integer signal tracking, one can assume that each ms of data contains 5000 digitized points and the round-off error is small. However, if the sampling frequency is off by 500 Hz, the sampling frequency becomes 5,000,500 Hz. With this sampling rate, 1 ms of data would thus contain 5000.5 samples. This is referred to as non-integer sampling frequency. In Section 11.11, it was assumed that the locally generated C/A code and the input C/A are matched within 100 ns (half the sampling time) at the end of 20 ms. If the mismatch, after a few ms, is larger than half the sampling time, the sampling rate can also be considered non-integer. When a non-integer sampling frequency is used, it has an impact on the tracking program, and this is why the subject is discussed here.

In the preceding example the number of data points in 1 ms used in the tracking program had to be truncated to an integer. Either 5000 or 5001 points could be chosen. With either selection in the second ms, the locally generated C/A code and the input signal are mismatched by half a sample time, or 100 ns. In the third ms the mismatch will be one sample. It is obvious that this situation does not fit the requirement for less than 100 ns of time shift after 20 ms.

This phenomenon actually occurs in a Motorola front end (Model M12 Oncore), as will be discussed further in Section 12.5. In this Motorola receiver the nominal sampling frequency is 5456 MHz and the corresponding sampling time is about 183 ns. This condition should call for 1 ms data to have 5456 points. The measured sampling frequency, however, is about 5,455,656.85 Hz. One ms contains about 5455.66 points of data. Since this value must be an integer, 5456 points are selected. Thus after 1 ms the mismatch between the input and the local code is 0.34 sample, if the Doppler frequency is zero.

One obvious way of dealing with the problem is to generate the local C/A code for a whole one second. The local C/A code can be generated every ms by calculating the proper initial phase at the beginning of each ms. Then the locally generated C/A code and the input signal will be closely matched. At every 20 ms the navigation data transition will be determined. At the end of each second the navigation data bits are stripped and the carrier frequency calculated. The early and late codes can also be used to multiply with the input signal to find the fine time at the one second interval. This fine time is included in generating the next second of C/A code. The C/A code generated at the beginning of each second is based on the previous C/A code, the carrier frequency measured, and the fine time calculated in the last second.

Another approach uses a similar method, as discussed in the preceding section. The C/A code is only generated once every second. At the end of each ms the difference between the input and the local code is registered. For example, assume that the sampling frequency is 5,455,660 Hz, and for each ms 5456 C/A data points are generated. At the end of the first ms, the input and the local C/A codes are mismatched by a 0.34 (5456–5455.66) sample interval. At the end of the second ms, the mismatch becomes a 0.68 sample interval. Whenever the C/A code in the input and the local C/A codes are mismatched by equal or more than 0.5 sample interval, the local code will be shifted one data sample to align with the input. Therefore, at the end of the second ms, the mismatch becomes a −0.32 (0.68−1) sample interval with a one sample shift. At the end of the third ms, the mismatch becomes a 0.02 (0.34−0.32) sample interval.

As the operation continues, the shift of one data point between the input and the local code happens about once every 2 or 3 ms. At every ms the early and late codes are also multiplied with the input signal. When the data are shifted by one point, the peak correlations from the early and late codes are performed accordingly. In every ms the initial C/A code of a certain satellite in the input data is tracked by the data shifting. After 1 second the navigation data will be determined at 20 ms intervals and stripped off. Then FFT is used to find the carrier frequency, which has a resolution of about 1 Hz. The 1000 early and late peak correlations accumulated will be used to find the fine time. The fine time is obtained by accumulating the mismatches of all the 1000 peak correlations.

In actual operation 5456 points of input data are taken each time. These data may contain signals from several satellites. In the tracking program, there is a pointer and a mismatch accumulator associated with each satellite signal. The pointer marks the initial phase of the C/A code. If the initial C/A code of a satellite is at 1000 of the 5456 data points, the pointer is at 1000. If there is no Doppler shift, after the first ms the mismatch is 0.34 sample. After two ms, the mismatch is 0.68, which is greater than 0.5 sample. The pointer is subtracted by one and the value is 999, which indicates a data point shift. After the shift, the mismatch becomes −0.32. The mismatch accumulator sums the mismatch of every ms. In this special case the first three values of the mismatch will be 0.34, −0.32, and 0.02, and accumulator sums these values. In general, the Doppler frequency must be taken into consideration in obtaining the values of the mismatch.

The next procedure is to find the fine time. After one second of operation another C/A code is generated, again, according to the carrier frequency and the fine time. The C/A code generated after one second of operation must match the input data C/A code. However, often there is an error in time, and this error is represented by ε. To see this, in the example above, suppose that at the end of the first 3 ms, the four mismatches are ε, $\varepsilon + 0.34$, $\varepsilon − 0.32$, and $\varepsilon + 0.02$. The early gate of the first 3 ms is shown in Figure 11.15. Because the prompt gate is off by ε the early gate will be off by $\varepsilon − 1$. This is because it is one sample before the prompt gate. Since the three early peak correlations are on a straight line, as shown in Figure 11.16 the following relation holds:

$$\frac{y_{e1}}{P - (\varepsilon - 1)} = \frac{y_{e2}}{P - (\varepsilon - 1 + 0.34)} = \frac{y_{e3}}{P - (\varepsilon - 1 - 0.32)}$$
$$= \frac{y_{e1} + y_{e2} + y_{e3}}{P - (\varepsilon - 1) + P - (\varepsilon - 1 + 0.34) + P - (\varepsilon - 1 - 0.32)}$$
$$= \frac{y_{e1} + y_{e2} + y_{e3}}{3P - (3\varepsilon + 2.8)} \tag{11.10}$$

where P represents the time when y_e approaches zero, as shown in Figure 11.16. Let us use d_n to represent the mismatch and extend this idea to the whole

FIGURE 11.15   First 3 ms mismatch of the prompt gate.

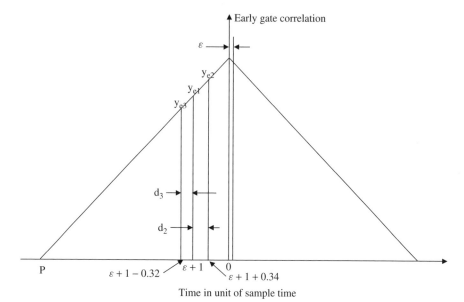

FIGURE 11.16   Peak correlations from three early gates.

one second. For the example above, $d_1 = 0$, $d_2 = 0.34$, and $d_3 = -0.32$. Equation (11.10) can be written with this notation as

$$\frac{y_{e1}}{P - (\varepsilon - 1)} = \frac{y_{e2}}{P - (\varepsilon - 1 + 0.34)} = \cdots = \frac{\sum_{n=1}^{N} y_{en}/N}{P - [\varepsilon - 1 + (\sum_{n=1}^{N} d_n/N)]} \quad (11.11)$$

where $N = 1000$ representing 1000 ms.

The average values of the early peak correlation y_{esa} and late peak correlation y_{lsa} are

$$y_{esa} = \frac{1}{N} \sum_{n=1}^{N} y_{en}$$

$$y_{lsa} = \frac{1}{N} \sum_{n=1}^{N} y_{ln} \quad (11.12)$$

The position of y_{esa} is at $-1 + \varepsilon + (\sum_{n=1}^{N} d_n/N)$, and the position of y_{lsa} is at $1 + \varepsilon + (\sum_{n=1}^{N} d_n/N)$ because the early and late gates are separated by two sampling units. The values of y_{esa} and y_{lsa} can be used to calculate the fine time x by way of Equation (11.8). It is desirable to make y_{esa} and y_{lsa} in Equation (11.12) equal, as discussed in Section 11.11 and shown in Figure 11.13. Then the position of y_{psa} can be found accurately. In order to accomplish this goal, the carrier frequency must be measured and the mismatch predicted after every ms in one second. At the end of each second the C/A code will be generated with the proper initial phase such that y_{esa} and y_{lsa} are approximately equal. The position (time) calculated from this equation is

$$x = \varepsilon + \frac{1}{N} \sum_{n=1}^{N} d_n \quad \text{or}$$

$$\varepsilon = x - \frac{1}{N} \sum_{n=1}^{N} d_n$$

(11.13)

This is almost the same as the tracking method discussed in the previous section. The only difference between this method and that discussed in the previous section is that the sample shifting can be very frequent. This is the tracking used on the Motorola front end, and the result is satisfactory.

11.14   CASES WHERE AN INITIAL C/A CODE POINT MATCHES THE 1 MS SELECTED DATA

For simplicity, the 5 MHz sampling frequency is used here. The input data are processed in 5000 point (1 ms) blocks. In general, in 5000 points of data the initial phase of the C/A code is somewhere in the middle, as illustrated in Figure 11.8. If a certain satellite has a positive Doppler frequency shift, one C/A code length is slightly less than 5000 points. In other words, the time between two consecutive initial phases of the C/A code is slightly shorter than 1 ms. As a result each time the 5000 input data points are grabbed, the numerical value of the initial C/A code point decreases, or the initial C/A code point is moves toward the beginning of the data, as shown in Figure 11.17. In this figure, in order to illustrate the shifting of the initial C/A code point, the effect is exaggerated. Recall that in Figure 11.4b, which shows a positive Doppler frequency, the length of the C/A code is shortened. For a 5 kHz positive Doppler frequency shift, in 1 second of input data, the length of the C/A code is shortened by about 3,174 ns, as discussed in Section 11.12. It takes about 315 seconds ($10^6/3174$) to shorten it by 1 ms. In 315 seconds, one additional ms of data is collected. This requires processing the 1 ms of data twice. In a different explanation, it takes 315 seconds to move a initial C/A code point from the end to the beginning of the input data set.

Although this occurrence is rare, eventually, the initial C/A code point starts moving to the front of the input data and becomes closely aligned with the first digitized data point. When this phenomenon occurs, the next C/A code phase also becomes closely aligned to the beginning point of the next 5000 points, as shown in Figure 11.17, blocks A and B. For simplicity, one can consider that in input data block A there are two initial C/A code points: one at the beginning and one at the end of the data. When this phenomenon occurs, the initial point

Initial point of C/A code

FIGURE 11.17   Initial C/A code moving toward the beginning of the data due to positive Doppler shifting.

of the local C/A code must be shifted from the beginning to the end in two consecutive ms. The input data in block *A* can be considered as the additional ms of data received, and it must be processed twice. The first time it is processed with a local initial C/A code point at the beginning, and the second time with a local C/A code initial point at the end. When block *B* is processed, the local C/A code initial point is already at the end. This operation processes one additional set of 5000-point data.

When the Doppler frequency is −5 kHz, the initial C/A code point in the data moves toward the end of the grabbed data, as shown in Figure 11.18. After some time the initial C/A code point will line up with the data selected. Because the time interval between two adjacent initial C/A code points is slightly longer than 5000 data points in one set of data, it may not contain an initial C/A code point, as shown in block *B* in Figure 11.18. In this case the local C/A code initial point must switch from the end to the beginning in two adjacent ms. The local code will multiply with input data in blocks *A* and *C*, but skip *B* because 1 ms of data is missed in every 315 seconds.

The discussion above can be summarized as follows: When the Doppler frequency is positive, one C/A code length is less than 5000 points of input data. In one set of 5000 input data, it may contain two initial C/A code points. This data set should be processed twice. When the Doppler frequency is negative, one C/A

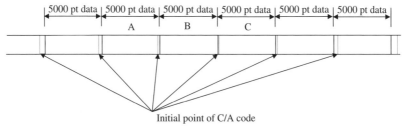

Initial point of C/A code

(a) *xy* Plane. (b) Altitude variation versus time.

FIGURE 11.18   Initial C/A code moving toward the end of the data due to negative Doppler shifting.

(a) *xy* plane.

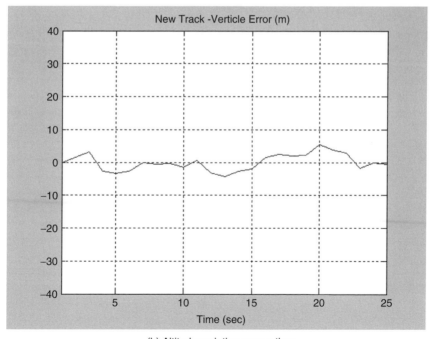

(b) Altitude variation versus time.

FIGURE 11.19   User position calculated through weak signal method.

(a)

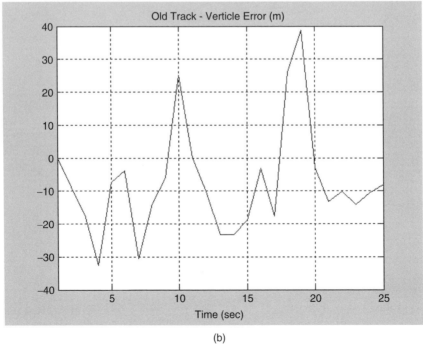

(b)

FIGURE 11.20 User position calculated through BASS tracking method.

code length is more than 5000 points of input data. In one set of 5000 input data, it may not contain an initial C/A code point. This data set should be skipped. This phenomenon also occurs when the sampling frequency is not an integer and a similar operation can be applied.

It was found that without this operation of processing one set of 5000 data points twice or skipping it, every so often the fine time obtained was off by 1 ms. This affected the pseudorange, and the calculated user location was found to have large errors. After modifications, the pseudorange was corrected and the user position also corrected. This method is also used in the tracking program discussed in Chapter 8, extended to 2 ms of input data taken each time.

11.15 EXPERIMENTAL RESULTS OF THE WEAK SIGNAL TRACKING

Experimental data collected using a 5 MHz sampling rate were tracked by two methods: the original BASS method discussed in Chapter 8 and the weak signal method presented in this chapter. The data were processed off line. The final result was the user position. The user position was calculated and displayed every second. Figure 11.19 shows the user position calculated from the weak signal method. Figure 11.19a shows the user position in xy plane, and Figure 11.19b shows the altitude variation versus time. Note that the maximum variation in xy plane is about ±3 meters and the maximum altitude variation is about ±5 meters.

Figure 11.20 shows the location calculated from the BASS tracking method. Figure 11.20a shows the location in the xy plane, and Figure 11.20b shows altitude variations plotted against time. Compared with the results in Figure 11.19, the improvements from the weak signal method are obvious. These improvements are expected because in the weak signal tracking one second of data is used to calculate the fine time, whereas in the BASS method, the fine time is calculated from 10 ms of data. The noise is reduced, which affects the calculated pseudorange.

REFERENCES

1. Tsui, J. *Digital Techniques for Wideband Receivers*, 2nd ed, p. 122, Artech House, Norwood, MA, 2001.
2. Lin, D., Tsui, J., "A weak signal tracking technique for a stand-alone software GPS receiver," ION GPS 2002, pp. 2534–2538, Portland, OR, September 24–27, 2002.
3. Lin, D., AFRL/SNRP engineer, private communication.
4. Lin, D., Tsui, J., "Sensitivity improvement of a low cost commercial GPS receiver using software approach," ION 2004 National Technical Meeting, San Diego, CA, January 26–28, 2004.

```
% digold.m call cacode
function code2 = digold(n,fs,fd,offset,sat);

% n       - number of samples
% fs      - sample frequency in Hz;
% offset - delay time in ns second must be equal to or
greater than -2/fs.
% sat     - satellite number;
% fd    - Doppler

gold_rate = 1.023e6;       %gold code clock rate in Hz.
f1=1575.42e6;
fs=fs*(f1-fd)/f1;
ts=1/fs;
tc=1/gold_rate;

code_in = cacode(sat);

offset=offset*f1/(f1-fd);
b = 1:n;
tmp=(ts*b+offset);
if(tmp(1)<0)
    tmp(1)=1023*tc;
    tmp(2)=ts;
    end
c = ceil(tmp/tc);
c=rem((c-1),1023)+1;
code2 =code_in(c);
```

```
% cacode.m
function [ss_ca]=cacode(svnum);

% modified by JT May 8 02 to generate satellite ca this
program contains WASP

%function ca=cacode2(svnum,fs,numsamp);
%   function to generate any of the 32 GPS C/A codes at a user
% specified sampling frequency.
%
% Input Arguments:
% svnum - the Satellite's PRN number
%         1-32 is traditional GPS PRN numbers
%         120 (38) is PRN for WAAS INMARSAT AOR-EAST (refer
to WAAS MOPS for shift)
%         122 (40) is PRN for WAAS INMARSAT AOR-WEST
%         ?? is PRN for WAAS INMARSAT Reserved
%         ?? is PRN for WAAS INMARSAT IOR
%         134 (52) is PRN for WAAS INMARSAT POR
%
% fs - desired sampling frequency
% numsamp - number of samples to generate
%           (sequence can extend for longer than a single
code period (1 ms)
%
% Output Argument
% ca - a vector containing the desired output sequence
%
% D. Akos - WPAFB AAWP-1

%this start the sequence one sample into the first chip (not
at leading edge of first chip)!
%dma comment 17-5-00
%
%this has been fixed as of 9-Dec-2000 but not fully
completely tested, but the first
%sample in the code should correspond to the first chip
now!!!  Also added the
%ability to find the GEO for WAAS by using PRN numbers above
32
%dma comment 9-Dec-2000
%modified by JT Oct 24, 02

% Constants
coderate=1.023e6;  %no consider given to the neglible Doppler
effect on code

% the g2s vector holds the appropriate shift of the g2 code
to generate
```

```
% the C/A code (ex. for SV#19 - use a G2 shift of
g2shift(19,1)=471)
g2s = [5;6;7;8;17;18;139;140;141;251;252;254;255;256;257;258;
469;470;471; ...
     472;473;474;509;512;513;514;515;516;859;860;861;862;863;
     950;947;948;950; ...
     145;175;52;21;237;235;886;657;634;762;355;1012;176;603;
     130;359;595;68;386];
% The last line gives 19 WAAS satellites their PRNs are from
120 to 138 thus,
if 119<svnum & svnum<139
  svnum=svnum+38-120;
end

g2shift=g2s(svnum,1);

% Generate G1 code
    %    load shift register
          reg = -1*ones(1,10);
    %
    for i = 1:1023,
        g1(i) = reg(10);
        save1 = reg(3)*reg(10);
        reg(1,2:10) = reg(1:1:9);
        reg(1) = save1;
    end,
%
% Generate G2 code
%
    %    load shift register
          reg = -1*ones(1,10);
    %
    for i = 1:1023,
        g2(i) = reg(10);
        save2 = reg(2)*reg(3)*reg(6)*reg(8)*reg(9)*reg(10);
        reg(1,2:10) = reg(1:1:9);
        reg(1) = save2;
    end,
    %
    %    Shift G2 code
    %
    g2tmp(1,1:g2shift)=g2(1,1023-g2shift+1:1023);
    g2tmp(1,g2shift+1:1023)=g2(1,1:1023-g2shift);
    %
    g2 = g2tmp;
```

```
%
%  Form single sample C/A code by multiplying G1 and G2
%
ss_ca = -g1.*g2;
%
```

CHAPTER TWELVE

GPS Receiver-Related Subjects

12.1 INTRODUCTION

In this chapter some subjects related to GPS receivers will be discussed. For the most part, the almanac data used can be found on the Internet. When the approximate time and the user location are known, the almanac data can be used to estimate the number of visible satellites and their approximate Doppler frequencies. This information can narrow down the satellite acquisition procedure.

A different means of acquisition will be presented that can be applied to periodic as well as nonperiodic signals, such as the P code. This method also consists of performing circular correlation in the frequency domain.

Other subjects include sampling frequency correction through the wide area augmentation system (WAAS) signal, strong and weak signal conditions in acquisition, simulation of GPS signals, impact on acquisition of filter bandwidth in front of ADC, real time operation of a software receiver, passive altimeter, the number of ADC bit required under jamming conditions, and calculating the satellite position through almanac data. An aided GPS system will be only briefly discussed.

12.2 INFORMATION FROM ALMANAC DATA[1]

From the acquisition method discussed in Section 10.10, it is evident that finding the initial C/A code and the carrier frequency is a time-consuming process. It is possible to obtain the almanac data (referred to as the YUMA data) from the internet (*http://www.navcen.uscg.gov/ftp/GPS/almanacs/yuma/*). These data are usually updated about once a week. The satellite constellation can be obtained from

Fundamentals of Global Positioning System Receivers: A Software Approach, Second Edition, by James Bao-Yen Tsui
ISBN 0-471-70647-7   Copyright © 2005 John Wiley & Sons, Inc.

these data through a Matlab program[1]. A discussion on finding the available satellite positions from almanac data is presented in Section 12.12. Programs can be also found on the Internet that provide satellite information. For example, one of these internet locations is *http://sirius.chinalake.navy.mil/satpred/*.

The number of available satellites and their carrier frequencies are of most interest here. An example will be used to illustrate the use of the YUMA data. In this example, an arbitrary time of year 2002, April 25, 8 : 04 AM, is selected, and the user position is Dayton, Ohio. The purpose is to find the available satellites and their Doppler frequencies. This particular time is associated with YUMA data 139, where 139 represents the week number. The number of satellites and their Doppler frequencies can be calculated from several YUMA data sets besides YUMA data 139. Four YUMA data sets 127, 138, 139, and 140 are used to

TABLE 12.1   Doppler Frequencies Calculated with Different YUMA Data

Reference:	Time: 2002 April 25, 8 o'clock AM 04 min 00 sec						
	Location: Latitude: 39.7772 Longitude: −84.0821 Altitude: 179.2						
1	2	3	4	5	6	7	8
YUMA Data	139	138		140		127	
	Doppler	Doppler	Frequency Off	Doppler	Frequency Off	Doppler	Frequency Off
Satellite	Hz	Hz	139–138	Hz	139–140	Hz	139–127
2						−2181	
4	3928	3930	−2	3925	3	3924	4
7	1525	1547	−22	1503	22	1826	−301
8	−2647	−2613	−34	−2683	36	−2250	−397
11	−1507	−1469	−38	−1546	39	−1045	−462
20	2698	2705	−7	2689	9	2736	−38
26	−1528	−1491	−37	−1567	39	−1015	−513
27	−3686	−3670	−16	−3702	16	−3466	−220
28	803	826	−23	780	23	1066	−263
29				283			
31	−3274	−3279	5	−3270	−4	−3319	45

find discrepancies in the results. As discussed in Sections 5.12 and 9.5, the week number has 10 bits. It started on January 5/6, 1980, and rolled over on August 21/22, 1999. The week numbers listed here are the values after the roll over.

Each sidereal day is 23 hours 56 minutes and 4.0954 seconds. It is 3 minutes and 55.9046 seconds shorter than a solar day. When a different set of YUMA data is used, this time difference must be taken into consideration. Since the YUMA data are separated by 1 week, the time should be corrected for 7 days. Some example results are shown in Table 12.1. In this table the satellite number is listed in column1. Columns 2, 3, 5, and 7 are the Doppler frequency values obtained from YUMA data 139, 138, 140, and 127, respectively. Columns 4, 6, and 8 list difference frequencies using the frequencies in column 2 as reference. The maximum difference frequency in columns 4 and 6 is less than 50 Hz, but one extra satellite (29) is obtained from data 140. When the YUMA data 127, which is 12 weeks away from YUMA data 139, is used, the Doppler frequency calculated can be off by over 500 Hz, as shown in column 8 for satellite number 26. Even with this frequency difference, the carrier frequency search can be

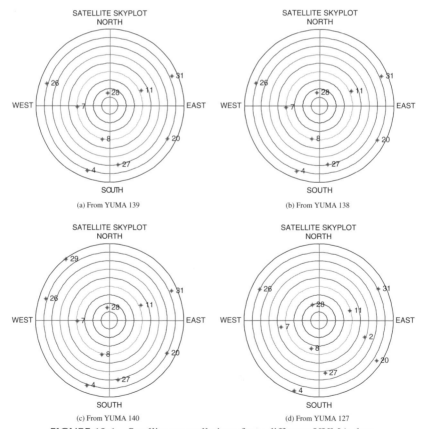

FIGURE 12.1 Satellite constellations from different YUMA data.

narrowed from ±5 kHz to about ±500 Hz. In order to keep a more accurate Doppler frequency, YUMA data close to the specific date should be used.

The satellite constellation for the given time and user position is shown in Figure 12.1. In these figures the constellations are obtained from YUMA data 139, 138, 140, and 127. It shows that the constellations are quite similar, even though YUMA 127 is more than 10 weeks away from the other three YUMA data sets.

If there are errors in the given desired time and user location, it is important to know the variation of Doppler frequency, which is used in acquisition. Again, for the time and user location selected earlier, but with errors deliberately introduced,

TABLE 12.2 Doppler Frequencies Calculated Using Different Time and Location

Reference:	Time: 2002 April 25, 8 o'clock AM 04 min 00 sec								
	Location: Latitude: 39.7772 Longitude: −84.0821 Altitude: 179.2								
1	2	3	4	5	6	7	8	9	10
Time off (min)	0	1		−1		1		−1	
Latitudinal distance off (Km)	0	100		100		−100		−100	
Longitudinal distance off (Km)	0	100		100		−100		−100	
Satellite	Doppler	Doppler	Frequency Off	Doppler	Frequency Off	Doppler	Frequency Off	Doppler	Frequency Off
	Hz	Hz	Hz	Hz	Hz	Hz	Hz	Hz	Hz
4	3928	3935	−7	3938	−10	3914	14	3920	8
7	1525	1588	−63	1620	−95	1427	98	1460	65
8	−2647	−2729	82	−2669	22	−2624	−23	−2563	−84
11	−1507	−1474	−33	−1406	−101	−1604	97	−1539	32
20	2698	2752	−54	2763	−65	2630	68	2642	56
26	−1528	−1625	97	−1575	47	−1481	−47	−1430	−98
27	−3686	−3712	26	−3692	6	−3680	−6	−3657	−29
28	803	784	19	829	−26	776	27	818	−15
31	−3274	−3286	12	−3290	16	−3257	−17	−3263	−11

if the time is offset by ± 1 minute, the distance is offset by ± 100 Km. The results here are obtained from YUMA data 139 and shown in Table 12.2.

In this table, column 1 lists the satellite number. The frequencies listed in column 2 are used as references. Note that for time, latitude, and longitude, the errors are all zero. Columns 3, 5, 7, and 9 are with different error combinations. For example, in column 5 the time is off by -1 minute, and the latitude and longitude are both off by 100 Km. Columns 4, 6, 8, and 10 give difference frequencies with the results in column 1 used as reference.

After the time error and location error are combined, the maximum difference frequency is about -101 Hz, for satellite 11 in column 6. This maximum difference frequency is a relatively small value compared to ± 5 kHz. Using the data in both Tables 12.1 and 12.2, one can estimate that the estimated Doppler frequencies are within ± 150 Hz. If one knows the Doppler frequency of a satellite in this range, the acquisition can be simplified. Instead of searching a frequency range of ± 5 kHz, a frequency range of ± 150 Hz should be adequate. The following section discusses acquisition with relative accurately carrier frequency information.

12.3   ACQUISITION WITH ACCURATE CARRIER FREQUENCY INFORMATION

If the carrier frequency of a certain satellite can be predicted rather accurately, such as within ± 150 Hz, the acquisition can be rather simple. Not only can the frequency range search be narrowed, but also the FFT length reduced. The method of acquisition for a coherent integration of 10 ms will be presented here.

The input signal will first be down-converted to a baseband frequency. Because the Doppler frequency is known from the YUMA data calculation, this frequency can be used in the down-converting operation. The down-converted frequency will be processed with a bandwidth of ± 50 Hz, because 10 ms of data have a corresponding frequency resolution of 100 Hz (1/10 ms). The 10 ms of data (50,000) can be divided into 10 sections. Each section contains 5,000 points, such as from $1 \sim 5000$, $5001 \sim 10,000$, ..., $49,5001 \sim 50,000$. The corresponding points in the 10 sections of data can be summed to the 5000 points shown in Figure 12.2.

From the resulting 5000 points, circular correlation is performed with 1 ms of reference C/A code data, and 2500 points are obtained, as shown in Section 7.9. This operation covers only ± 50 Hz, because the input data used is 10 ms long. In order to cover ± 150 Hz, the same operation must be performed two more times using appropriate down-converting frequencies. For example, if the input frequency estimated through the YUMA data is at f_0, the input signal can be down-converted by f_0, and the operation above can then be performed. To cover the additional frequency range, the input signal can be down-converted by $f_0 \pm 100$ Hz, and the same operation also performed two more times. From

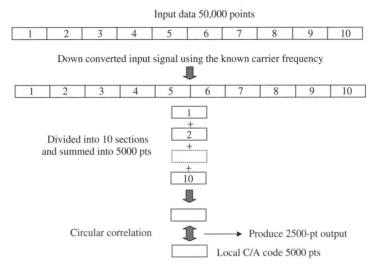

FIGURE 12.2   Folding 10 ms of data into 1 ms.

this operation, 10 ms of input data will generate output. This operation is much simpler than the method described in Section 10.10 because less bandwidth is covered.

Having less output data brings two advantages. One is that less searching time is involved in finding the maximum of the correlation outputs. Another is that the number of false alarm generated can be reduced. Although the probability of false alarm does not change, a smaller number of outputs can produce fewer false alarms.

In this operation when the input signal is ± 50 Hz from the desired frequency f_0, the output will be about 3.92 dB down. One can use the method of recovering loss between frequencies by operating on two adjacent outputs, separated by 100 Hz, and produce a 5 by 2500 matrix output. The undesirable loss, as discussed in Section 10.11, can then be reduced.

12.4   CIRCULAR CORRELATION BY PARTITION[(2)]

This approach can be applied to a periodic or a nonperiodic signal. The C/A code can be considered as periodic and the P code or P(Y) code as nonperiodic. The basic principle of this acquisition method will be explained first; then an example of actual use will be presented. The input signal must be down-converted into the baseband first. In order to discuss this method, the output sampling rate described in Section 10.9 will be used. From the output sampling rate, one can determine the overall processing bandwidth. Normally, it is desirable to cover a 10 kHz (± 5 kHz) bandwidth. When this bandwidth is used, the signal with Doppler

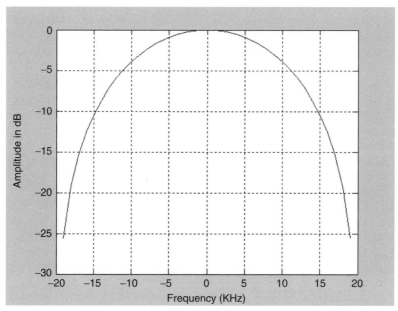

FIGURE 12.3   Insertion loss of a sinc function filter with 20 kHz resolution.

frequency at ±5 kHz will suffer a 3.92 dB processing loss, as discussed in Section 10.11. In order to reduce this loss, the processing bandwidth is increased to 20 kHz, as shown in Figure 12.3. With this bandwidth the signal at ±5 kHz has an acceptable loss of 0.91 dB, which is caused by the filter shape. If 1 ms of data is summed into one output point, it provides 1 kHz resolution. In order to have 20 kHz resolution, 50 μs (1 ms/20) of data can be summed into one output point.

Suppose that the data are collected by the system discussed in Section 6.8. The intermediate frequency (IF) is at 21.25 MHz and the sampling rate is 5 MHz. There is a total of 5000 input data points per ms, and 250 (5000/20) input data points can be summed into one output point.

First, a periodic signal will be used to illustrate the basic idea because this method can be applied to both periodic and nonperiodic signals. In this example 10 ms of input data with C/A code modulation are used for coherent integration. There are a total 50,000 points of input data in 10 ms. The signal is processed as follows:

1. The input signal is down-converted to baseband by multiplying it with a complex continuous wave (cw) signal. The frequency of the cw signal is at 21.25 MHz.

2. The locally generated 10-ms C/A code, containing 50,000 points, is multiplied with the input signal point by point, which results in a 50,000 point output.

FIGURE 12.4   Acquisition on 10 ms C/A code signal.

3. Summing 250 points into one point produces an output of 200 points.
4. The input frequency is found by performing FFT on the 200 data points.

Figure 12.4 shows one of the 2500 operations. If the local C/A matches the C/A code in the input signal, the output from the FFT has a strong single peak. If the local C/A code and the C/A code in the input do not match, the output is noiselike. Under this condition the reference C/A code is shifted two samples by moving the first two data points to the end. The same operations are performed from 2 to 4 until the correct initial C/A code is found. Because the C/A is repetitive over 1 ms time, the local signal is only needed to shift 2500 times instead of 25,000 times. The overall outputs from 10 ms of data are 50,000 points (200 in frequency domain and 2500 in time domain.) This operation is easy to understand but computationally intensive. The following method can accomplish the same goal with less calculation.

The circular correlation by partition method provides a simple way to perform the operation above. The down-converted input signal is divided into 250-point sections, and there are 200 sections in 10 ms. The objective is to locally generate 10 ms of C/A code and divide it into 200 250-point sections. The circular correlation of each section is illustrated in Figure 12.5. In this figure only the first two sections are shown. One section of local C/A code will circular correlated with two consecutive sections of input to guarantee that the portion of the C/A code is contained in the input. The 250-point C/A code must be padded with 250 zeros to match the length of the input signal. In this figure the shaded area represents the portion of C/A code in the input signal. The circular correlation on the two sets of 500 points of data can be performed the same as described in Sections 7.8 and 7.9. The details of the operation are as follows: First, FFTs are performed on both the input and the local C/A code to change them into frequency domain results. Since the down-converted input signal is complex, the spectrum will be like the one shown in Figure 7.5, where most of the information is in the first half of the spectrum. After the FFT the first 250 points are retained, since this is where the information in the input signal is concentrated. Next, multiply the input frequency domain data with the complex conjugate of the C/A code frequency domain data and take the inverse FFT of the result to generate the desired output.

FIGURE 12.5 Operation of 250-point C/A code and 500 input data.

The C/A code only has values in the first half, since the second half of the C/A code is padded with zero. Therefore the output only has information in the first half. Thus, in all, only 125 points of data are kept.

In effect, one uses 250 points of C/A code to generate 125 points of output data. This is equivalent to sliding 250 points of the C/A code over the input data in two-point steps. At each step the C/A multiplies the input data, which are summed into one point.

The same operation is performed on the rest of the input data (the remaining 199 sections), as shown in Figure 12.6. In this figure, section 1 of C/A code is correlated with sections 1 and 2 of the input signal. Section 2 of C/A code

FIGURE 12.6 Correlation of the first 250 points of the input and the C/A code.

is correlated with sections 2 and 3 of the input signal. Finally, section 200 of C/A code is correlated with sections 200 and 1 of the input signal. Overall, 200 125-point output data are obtained.

The C/A code is repeated every ms and there are only 20 (not 200) 250 point C/A codes in one ms. Since the correlation is only needed to perform in one ms, the preceding operation must be repeated 19 more times. These additional operations are shown in Figure 12.7. In this figure the operation is close to that shown in Figure 12.6. The only difference is that the C/A is shifted one section each time. In this figure the first two shifts and the last shift are shown. Because each operation generates 125 of the 200-point outputs, the total output is a 2500 (20 × 125) by 200-point matrix. Both the row and column of the matrix are in the time domain.

FFT is performed on the output data to obtain the frequency information. A 200-point FFT is performed 2500 times, and the results are 2500 by 200 outputs, which are the same results as shown in Figure 12.4. The amplitude of the outputs is obtained from the real and imaginary components. There are 200 frequency bins and 2500 time bins. These outputs are searched to find the highest value. If the highest output also crosses a certain threshold, the acquisition is complete. If the output does not cross the threshold, noncoherent integration is needed to improve the S/N.

From the 200-point FFT one can obtain 200 frequency bins, and each one is separated by 100 Hz. For signals whose frequency falls in the center of two

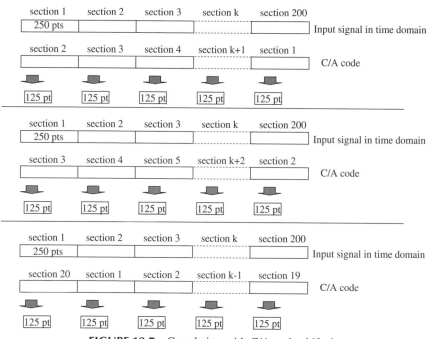

FIGURE 12.7 Correlation with C/A code shifted.

FIGURE 12.8   Acquisition on 10 ms P code signal.

100 Hz bins, the insertion loss is also 3.92 dB. In order to recover this loss, the outputs from two adjacent 100 Hz filters are combined, using the method similar to the one discussed in Section 10.11. Because there are two hundred 100 Hz filters, the overall coverage is ±10 kHz, as discussed in Figure 12.3. However, only the middle portion of the spectrum (±5 kHz) can be used. Therefore this method is not as efficient as the method discussed in Section 10.10.

The P code is nonperiodic and has been discussed in Section 5.4. If this method is applied to P code, the corresponding sampling frequency should be around 50 MHz to accommodate the 20-MHz bandwidth[2]. There are 500,000 digitized points in 10 ms of input data. In order to cover ±5 kHz, the filter bandwidth will still have to be 20 kHz. Under this condition 2500 input points are summed into one point, as shown in Figure 12.8. The operation is similar to the operation shown in Figure 12.4, but with two basic differences. First, the input and the reference both contain 500,000 points in 10 ms, and 2,500 points are summed into one output point in the time domain, which still contains 200 points. Second, to correlate with the input signal, the reference is shifted against the input signal instead of making itself wrap around, because the P code is nonperiodic. The searching time window can be any value and is not limited to 1 ms. In general, the requirement on the searching window is several ms. The actual acquisition will use circular correlation by the partition method, as presented in Figures 12.5, 12.6, and 12.7. The difference is that in Figures 12.5 and 12.6 the total number of data points increased by 10 times. In Figure 12.7 the reference is moving along the input signal rather than wrapping around it.

12.5   SAMPLING FREQUENCY CORRECTION THROUGH WIDE AREA AUGMENTATION SYSTEM (WAAS) SIGNAL[3]

The pseudorange calculated for a GPS receiver depends on the accuracy of the sampling frequency. If the sampling frequency is inaccurate, the pseudorange will have an error that reflects on the calculated user position. While an accurate oscillator is expensive, the accuracy of a low-cost oscillator is usually poor and the output frequency can be off from the specified value. Thus it is important

TABLE 12.3   Additional Satellites Generating Data

PRN	Code Delay Chips	First 10 Chips C/A Octal
120	145	0671
121	175	0536
122	52	1510
123	21	1545
124	237	0160
125	235	0701
126	886	0013
127	657	1060
128	634	0245
129	762	0527
130	355	1436
131	1012	1226
132	176	1257
133	603	0046
134	130	1071
135	359	0561
136	595	1037
137	68	0770
138	386	1327

to measure first the true sampling frequency of a receiver. One way to measure the sampling frequency is to use a frequency counter in the laboratory. However, this instrument may not be available to some users.

The WAAS uses geo-stationary satellites to receive data measured from many ground stations, and it transmits information to GPS users for position correction. Because the WAAS satellites are geo-stationary, the Doppler frequency caused by their motion is very small, typically on the order of a few tens of Hz. Thus the signal transmitted by the WAAS can be used to calibrate the sampling frequency in a GPS receiver. The WAAS signal frequency is at 1575.42 MHz. The sampling frequency of a C/A code GPS receiver is in the neighborhood of 5 MHz. The ratio of these two frequencies is about 300; thus 10 Hz inaccuracy in the WAAS frequency will be translated to about 0.03 Hz (10/300). The accuracy of the sampling frequency measured through this approach should be less than 1 Hz. However, the WAAS signal can be rather weak for users in certain areas. The noise in the received signal can be the dominant factor in determining the sampling frequency.

The C/A codes of the WAAS satellite can be generated through the same approach as discussed in Section 5.6. The only information needed here is the code delay chips in Table 5.3. The additional satellites are listed in Table 12.3[1]. These satellites are incorporated in the cacode.m program at the end of Chapter 11. The only satellite can be found in the received data (collected in Dayton, Ohio) is PRN 122 with a delay of 52.

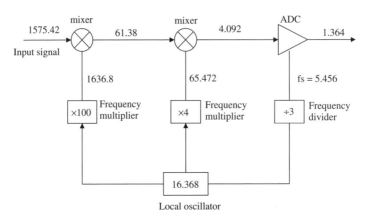

FIGURE 12.9   Frequency plan of a Motorola GPS receiver front end.

The C/A code in the WAAS is also periodic with 1 ms code length. However, the navigation data rate is 500 Hz, which is 10 times higher than the satellites in the GPS constellation. Because of this high data rate, long coherent integration will encounter many data transitions. Therefore, it is decided to perform 1-ms coherent integration and 20-ms noncoherent integration for acquisition. The data collection system used is a modified Motorola GPS receiver (Model M12 Oncore) front end. The frequency plan of the data collection system is shown in Figure 12.9. The numbers in this figure are frequencies in MHz, and they are considered nominal values because they are provided by the manufacture but are not accurate. The purpose of this discussion is to find the accurate sampling frequency (f_s.) The nominal sampling frequency $f_s = 5.456$ MHz.

The measurement is based on time domain analysis. The length of the C/A code transmitted is 1 ms. However, Doppler frequency can change this value. Since the signal from the WAAS satellite has near zero Doppler frequency, the time of each C/A code should be very close to 1 ms. For this operation it is assumed that Doppler frequency is zero and the received C/A code length is 1 ms. If the sampling frequency is correct, every C/A code will have 5456 points of digitized data. The acquisition will use 5456 points of data, which is 1 ms. With correct sampling, the initial C/A code phase will remain the same whether acquisition is performed on the 1st ms or the 1001th ms. If the assumed frequency is incorrect, the initial C/A code phase will change. This phenomenon is described in Figure 12.10. In Figure 12.10a the sampling frequency is at the correct value. The initial C/A code phase is a constant value from one ms to another. In Figure 12.10b the sampling frequency is higher than the nominal value; thus, in one ms there are more than 5456 sampled data points. The initial C/A code phase increases from one ms to another. In this example the initial C/A code measured will move toward the right. Based on the time difference, the true sampling frequency can be found.

FIGURE 12.10 Shift of initial phase of C/A code.

Theoretically this measurement is very simple and only two points are needed. For example, one can obtain the initial C/A codes at the 1st and 1001st ms through acquisition. If the initial point has shifted to the right, the sampling frequency is higher than the nominal value. If the initial C/A code measured in the 1st ms is at 2000 and the initial C/A code measured at the 1001st ms has shifted by p points, the result should be $2000 + p$. Since from 1st to 1001st ms covers 1 second of time, there are a total of $5,456,000 + p$ samples. Thus, the actual sampling time t_s is

$$t_s = \frac{1}{5456000 + p} \quad \text{or} \quad f_s = 5456000 + p \ Hz \quad (12.1)$$

This simple equation can be used to find the sampling frequency. Because the noise can cause an error in the measurement, this simple method maybe inadequate.

In the actual determination of the sampling frequency the process is as follows: One ms of data is used to perform coherent integration and 20 ms for noncoherent integration, and the initial point of the C/A code is determined every 20 ms. In 2 seconds this 20-ms operation is performed 100 times. The length of the data used is not critical but should depend on the signal strength. For a strong signal, less data can be used. The idea is to use 100 points of 20-ms data averaged to obtain a more dependable result. A typical initial C/A code measured over 2 seconds of time is shown in Figure 12.11, which contains 100 data points. Each data point is obtained from 20 ms of data. Note that most of the data can fit on a straight line, so we refer to this line as the "desired line," though there are many obviously erroneous points. In determining the initial C/A code sometimes the values wrap around. For example, if in the 1st 20 ms the initial C/A code phase measured is 4 and the second 20 ms the value is 5454, the first value can be adjusted by adding 5456. In this example, the first value will become 5460 (4 + 5456). In general, if there is a wrap around, either the value before the wrap around or the values after it must be adjusted by 5456.

The following operation is used to eliminate the erroneous points and to find the slope of the desired line. There might be many ways to accomplish this goal. One of the methods is presented here. Figure 12.12 shows the difference values

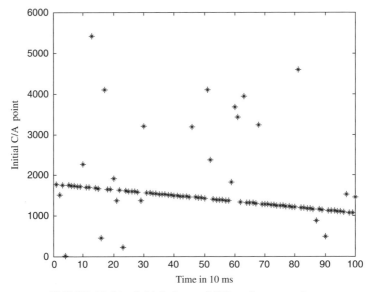

FIGURE 12.11   Initial phase of C/A code versus time.

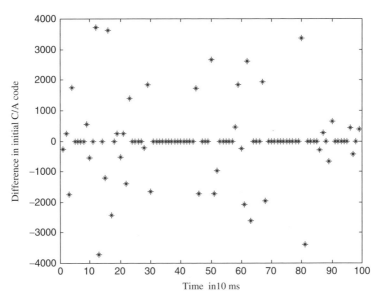

FIGURE 12.12   Difference of initial phase of C/A code versus time.

between two adjacent initial C/A codes. The majority differences are close to a constant value, and it is desirable to keep them. Figure 12.13 shows the sorted difference values in ascending order. The horizontal axis simply represents the index associated with the difference ranking order. The sorted results in the

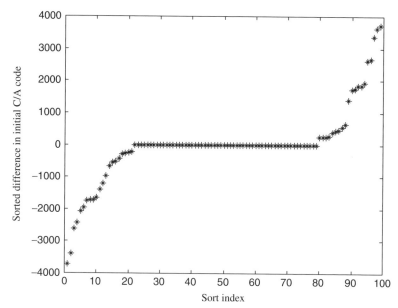

FIGURE 12.13   Sorted difference of initial phase of C/A code.

center portion are the desired ones, as shown in Figure 12.14. This figure can be considered a magnified version of Figure 12.13.

In Figure 12.14, because in the acquisition for the coherent integration the data are shifted 2 samples, the time resolution for the initial C/A code can be considered as 2 sampling points. Because the resolution is 2 samples, the measured initial C/A code change is either -6 samples or -8 samples every 20 ms. This is why the differences of initial C/A codes are divided into two groups: -8 and -6 in Figure 12.14. By arbitrarily using 25 points of the center portions of Figure 12.14 to avoid the erroneous difference values on the ends, an average value and a standard deviation can be found from these 25 points. The average value is close to the slope of the desired line. The standard deviation is then used as a limit to eliminate erroneous data points. One can draw a straight line passing through a "valid" initial C/A code data phase by using the average value as the slope. A valid initial C/A point is a point on the desired line in Figure 12.11. It can be determined from the difference between initial points. If the difference initial C/A code value obtained from two adjacent points n_k and n_{k+1} is very close to the average value, both n_k and n_{k+1} are valid points. Select ± 50 times of standard deviations as the threshold to eliminate the erroneous points. The value 50 is obtained from experimenting with the values shown in Figure 12.11. Figure 12.15 shows the two lines used to limit the "good" points. These good points are fitted into a straight line, as shown in Figure 12.16. From this line, one can find two points – the initial C/A codes corresponding to $x(1)$ and $x(101)$ – and call them $y(1)$ and $y(101)$. In this example, although there are

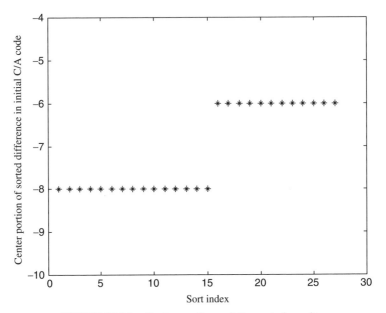

FIGURE 12.14 Center portions of the sorted results.

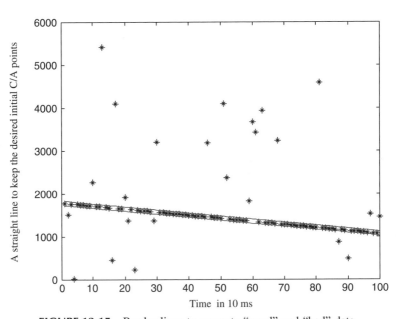

FIGURE 12.15 Border lines to separate "good" and "bad" data.

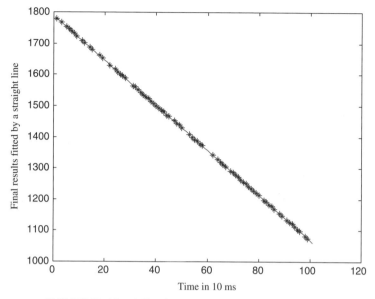

FIGURE 12.16   A line is extrapolated from "good" data.

only 100 raw data points obtained from the measurement, any values on the line can be used to find the total initial C/A code phase change. From $x(1)$ to $x(101)$ the span is 2 seconds of time. The number of points shifted per second can be calculated as

$$p = \frac{y(1) - y(101)}{2} \tag{12.2}$$

Once the value of p is found, the sampling frequency can be found from Equation (12.1).

Data obtained every 2 seconds are used to calculate the sampling frequency. The sampling frequency is calculated 54 times, which means the length of the data is 110 seconds. The same approach is used to determine the sampling frequencies of two Motorola front ends. The average sampling frequencies measured are 5,455,656.85 and 5,455,640.80 Hz with standard deviations of 0.22 and 0.26 Hz, respectively. A frequency counter is used to measure the front end with 5,455,640.80 Hz sampling frequency, and the reading is between 5,455,641 and 5,455,640 Hz, because the counter has a resolution of 1 Hz. A very stable signal generator is used to generate a signal at 5,455,640.80 Hz. The signal is measured by the frequency counter, which also displays 5,455,641 and 5,455,640 Hz. Our experiment indicates that the measurements are fairly accurate, and almost as close as those from the frequency counter.

Because the measured sampling frequency is about 5455.6 or 5456.7 kHz, which is off by more than 300 Hz from the designated sampling frequency, in one ms there will be 5455.6 data points. The tracking program discussed in Section 11.13 must be used to track the digitized output.

12.6 STRONG AND WEAK SIGNAL CONDITIONS[4-6]

Many satellite signals are blocked by tall buildings in an urban area crowded with high rise buildings. Only a few satellites can reach a GPS receiver directly. This is why signals at the receiver consist of both strong and weak signals. The strong signal referenced here is a signal at nominal power level and the weak signal is weaker than the nominal power level. Let us assume that S_1 is a strong signal and S_2 is a weak signal. During acquisition, the strong signal can be found easily. In order to find the weak signal, C/A code of S_2 must be used for acquisition. If the frequency of S_2 is not known, one must search within the anticipated frequency range. However, the C/A codes are not true orthogonal but near orthogonal with each other. The autocorrelation peak of a C/A code is 1023, and the largest cross-correlation peak is -65, as discussed in Section 5.7. The cross-correlation peak is about -23.94 dB ($20 \times \log(65/1023)$) below the autocorrelation peak. Thus, if the weak signal is about -20 dB below the strong one, the cross-correlation peak with the strong signal is comparable to the autocorrelation peak of the weak signal. One approach to locating the weak signal is to find all the strong signals and subtract them from the input data set.

Before the strong signal is removed, the effect of the strong signal on the weak signal acquisition will be shown. Suppose that the received GPS signal consists of a strong signal S_1, a weak signal S_2, and noise n, then

$$y = S_1 + S_2 + n \tag{12.3}$$

The noise ratios of the strong and weak signals are

$$\left(\frac{S}{N}\right)_1 = 10\log\left(\frac{P_1}{N_0}\right) \quad \left(\frac{S}{N}\right)_2 = 10\log\left(\frac{P_2}{N_0}\right) \tag{12.4}$$

where P_1 is the strong signal power, P_2 is the weak signal power, and N_0 (variance of n) is the noise power. The basic process used to find the weak signal is to correlate the input signal y with the locally generated C/A code (C_2) of the weak signal S_2. Since y also contains the strong signal, for the purpose of the discussion, the strong signal can be represented as

$$S_1 = A_1 C_1, \quad P_1 = A_1^2 \tag{12.5}$$

where A_1 and C_1 are the amplitude and the C/A code of the strong signal, respectively.

The basic idea of this approach is to treat the cross correlation between the C/A codes: C_1 of the strong signal and the locally generated C_2 as equivalent noise power. The equivalent noise power N_e can be then written as

$$N_e = E[(C_2 \otimes S_1)^2] = A_1^2 E[(C_2 \otimes C_1)^2] = P_1 K \quad \text{and}$$
$$K \equiv E[(C_2 \otimes C_1)^2] \tag{12.6}$$

where $E[]$ is the average operation and $\otimes$ represents correlation. For a sampling rate of 5 MHz, the value of K obtained is $K = 7.0654 \times 10^{-4}$. This value is obtained by the following procedure: Sum the square of the 1-ms cross-correlation outputs, and divide by 5000 to obtain N_e. Next, divide N_e by P_1 to obtain K. Since the autocorrelation peak is 5000, $P_1 = 5000^2$.

Since we have the equivalent noise power, we can rewrite the processed signal-to-noise ratio of the weak signal as

$$\left(\frac{S}{N}\right)_p = 10\log\left(\frac{P_2}{N_e + N_0/G}\right) \tag{12.7}$$

Substituting Equations (12.4) and (12.6) into Equation (12.7) yields

$$\left(\frac{S}{N}\right)_p = \left(\frac{S}{N}\right)_2 + G_{dB} - 10\log\left(KG10^{(S/N)_1/10} + 1\right) \tag{12.8}$$

where G is the acquisition processing gain in ratio and G_{dB} is gain in dB. In Equation (12.7) the thermal noise power N_0 can be decreased through acquisition, but the equivalent noise power N_e is not affected.

From Section 10.3 we know that the desired weakest signal that can be processed is -39 dB (reference to 2-MHz bandwidth and $C/N_0 = 24$ dB). In order to detect this signal, a S/N of 14 dB is required. Thus the required acquisition gain is 53 dB $(14 + 39)$ or $G = 2 \times 10^5$.

With the values G, G_{dB}, and K known, S/N_p can be calculated as a function of $(S/N)_2$. Figure 12.17 shows the relation between $(S/N)_2$ and $(S/N)_p$ for different strong signal levels. The dashed line at the left-hand corner of the figure corresponds to the case where no strong signal is present. This line shows a gain of 53 dB. For example, when the input $(S/N)_2$ is -39 dB, the processing output $(S/N)_p$ is 14 dB. The solid lines correspond to cases where a strong signal is present for $(S/N)_1 = -19$ to -13 dB in one dB step. The required output threshold of 14 dB is also indicated. The difficulties associated with a weak signal acquisition in the presence of a strong signal can be illustrated by a simple example. Suppose that the input is $S/N_2 = -35$ dB $(C/N_0 = 28$ dB), and the corresponding output is $(S/N)_p = 18$ dB without the presence of a strong signal. When a strong signal of $(S/N)_1 = -19$ dB is present the processed $(S/N)_p = 13.5$ dB, which is slightly lower than the desired value of 14 dB. When the strong signal is $(S/N)_1 = -13$ dB, the processed $(S/N)_p = 8$ dB.

From this analysis it is evident that in order to successfully acquire weak signals, it is necessary to remove the signals with nominal power levels. There is an obvious procedure to achieve this purpose. It works to remove the strong signals through direction subtraction. In its operation, the amplitude, the initial C/A code, the carrier frequency, and the phase of the carrier frequency must be obtained. To begin, write the strong signal as

$$S_1 = A_1 C_1 \sin(2\pi f t + \theta) \tag{12.9}$$

FIGURE 12.17 Processed weak signal $(S/N)_p$ versus input weak signal $(S/N)_2$ as a function of strong input signals.

where A_1 is the amplitude, C_1 is the C/A code with the correct initial phase, f is the carrier frequency, and θ is the carrier phase. This signal can be subtracted from the input signal in the time domain. If all these values are measured correctly, the strong signal can be removed almost thoroughly from the input. If all the quantities are not measured accurately, the strong signal can only be partially removed.

The other approach to remove the strong signal is referred to as the subspace projection. Although this idea applies to multiple-dimensional problems, a two-dimensional problem will be used to illustrate the basic idea. In the two-dimensional case, a signal y received by the receiver consists of two signals: a strong one S_1 and a weak one S_2. The strong signal S_1 can be obtained by acquisition on signal y. The strong signal can be either obtained correctly or with errors. Figure 12.18 shows the strong signal S_1, weak signals S_2, and the projection of y on the strong signal S_{1p}. Since S_1 and S_2 are near orthogonal, they can be considered nearly perpendicular to each other. Figure 12.18a shows that the signal S_1 is measured correctly and $S_1 = S_{1p}$. The vector sum represents the received signal y, and there is no noise. The weak signal S_2 can be found through

$$S_2 = y - S_1 \tag{12.10}$$

Figure 12.18b shows that the amplitude of the strong signal is measured wrongly from the received signal y. If the direct subtraction is used, the result

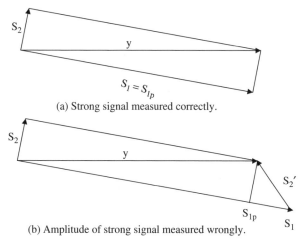

(a) Strong signal measured correctly.

(b) Amplitude of strong signal measured wrongly.

FIGURE 12.18 Strong and weak signals.

will be S_2', which is different from the correct signal S_2. The projection S_{1p}, however, still equals to the true signal S_1, even if the amplitude is off. Thus the weak signal obtained through the following relation can provide the correct answer:

$$S_2 = y - S_{1p} \tag{12.11}$$

This simple illustration demonstrates that the projection method is more robust than the direct subtraction.

In practice, calculating the projection involves more than two signals. To simplify, one divides the signals arbitrarily into two groups: the strong (S) and the weak (W) signals. First, the input signal is down-converted to baseband. Based on the vector space notion, the down-converted input signal can be written as[5]

$$y = a_w W + a_s S + n \tag{12.12}$$

where

y: $[y(0), y(1), y(2), \ldots, y(N-1)]^T$, with N being the number of samples in the down-converted input data and $[]^T$ the transpose of a matrix.

W: a matrix whose columns are unit vectors containing samples of the down-converted weak signals.

a_w: amplitude vector for the weak satellite signals.

S: a matrix whose columns are unit vectors containing samples of the down-converted satellite signals with nominal power.

a_s: amplitude vector for the nominal satellite signals.

n: white Gaussian noise vector with variance of σ_n^2.

As in the two-dimensional case, the goal is to find the projection $P_s y$ of y on S. This is obtained by the following matrix operation:

$$P_s y = S(S^T S)^{-1} S^T y \qquad (12.13)$$

Since the GPS satellite codes are nearly orthogonal, and also because of $a_w \ll a_s$,

$$a_w S^T W \ll a_s S^T \cdot S \qquad (12.14)$$

Therefore,

$$P_s y \approx S(S^T S)^{-1} S^T y = S(S^T S)^{-1} S^T (a_w W + a_s S + n) \approx a_s S + P_s n \quad (12.15)$$

where $P_s n$ represents the noise projection on the strong signal S. This means that the result from Equation (12.13) is the sum of the nominal signals and the projected noise. The weak signals can be obtained as

$$a_w W = y - P_s y \qquad (12.16)$$

In this operation the nominal signals are obtained from the input y through conventional acquisition, and they form the S matrix. Equations (12.13) and (12.16) are used to find the weak signals. Simulated data are used to compare the results from direct subtraction and the projection methods. The input signal contains three strong signals ($C/N_0 = 44 \sim 51$ dB,) one weak signal ($C/N_0 = 36$ dB), and the proper amount of noise. Random numbers within realistic ranges are used for each satellite C/A code initial phase offset, carrier frequency, and phase. The three strong signals are acquired first and written in the time domain according to Equation (12.3). The residues are found through both the direct and subspace projection methods. At each input power level from $C/N_0 = 44 \sim 51$ dB, 30 runs with different noise are performed. The purpose is to find the residue after the signals are subtracted. The results are shown in Figure 12.19. The dotted line represents the direct subtraction method, and the solid line represents the subspace projection method. The vertical bars show the standard deviations of the 30 runs. It appears that the subspace projection method produces small residue values, and the values do not increase with increasing of strong signal power.

Another set of simulated data are used to acquire the weak signal. The following results are obtained:

1. In the absence of any strong signal, the lowest input S/N that can be acquired is -40 dB ($C/N_0 = 23$), which is one dB below the minimum required S/N of -39 dB.
2. In the presence of strong signals, the weak signal that can be acquired depends on the strong signal level. Table 12.4 shows three sets of test results. The first column is the strong signal S/N level, and the second column is the weakest signal can be acquired. In this table the S/N is

FIGURE 12.19   Residue after the strong signal is subtracted from input.

TABLE 12.4   Weakest Signal Acquired with Strong Signal Presence

Strong Signal $(S/N)_1$	Weakest Signal Acquired $(S/N)_2$
-13 dB $(C/N_0 = 50$ dB)	-32 dB $(C/N_0 = 31$ dB)
-15 dB $(C/N_0 = 48$ dB)	-33 dB $(C/N_0 = 30$ dB)
-17 dB $(C/N_0 = 46$ dB)	-35 dB $(C/N_0 = 28$ dB)

referenced to a 2-MHz bandwidth system. The weak signals shown in this table are about 1 to 2 dB lower than the one shown in Figure 12.17. One needs to keep in mind, however, that the result in Table 12.3 is the lowest values that can be achieved in a number of simulation test runs. They are not statistical averages.

3. If the strong signal in the range of $(S/N)_1 = -13 \sim -9$ dB is removed using the projection method, the weakest signal that can be acquired is $(S/N)_2 = -39$ dB. Up to 3 strong signals were tested, and the weakest signal could be acquired is still $(S/N)_2 = -39$ dB. This represents an improvement of 4 to 7 dB from Table 12.4.

12.7   SIMULATION OF GPS SIGNALS

It is convenient to use simulated results to evaluate a certain signal processing method. Simulation can be performed easily, and the *S/N* results are accurate. In data collected from real satellites, the *S/N* ratio cannot be easily controlled. Collecting signals from a hardware simulator can be tedious, but sometimes a hardware simulator must be used to confirm simulated results.

One way to generate simulated data is to consider the noise first. The normal (Gaussian) distribution is usually used for the noise. The signal is generated based on the noise amplitude. The noise and signal are added together as the desired signal. In generating noise, the sampling frequency f_s is important. If the signal is sampling at 5 MHz, the input bandwidth is 2.5 MHz (5/2), which is determined by the Nyquist sampling theory. Considering the shape of a real filter with a bandwidth of 2 MHz, an input bandwidth of 2.5 MHz is a reasonable value. When the input bandwidth is 2 MHz, the thermal noise, as indicated in Section 10.2, is at about −111 dBm. If the input signal is at −130 dBm, the S/N is at −19 dB. The amplitude of the signal can be written as

$$\text{amp} = \sqrt{2} n_{st} (10^{dB/20}) \qquad (12.17)$$

where the $\sqrt{2}$ is for considering the power of the carrier signal and n_{st} is the standard deviation of the noise. In Matlab simulation, $n_{st} = 1$ is often used and the value dB is the desired S/N.

If the signal is sampled at a different frequency, the equivalent bandwidth will be different from 2 MHz, which will affect the noise floor and the S/N. For example, if the signal is sampled at 50 MHz instead of 5 MHz, the noise bandwidth is about 25 MHz, and the corresponding noise floor is at about −100 dBm. An input signal at −130 dBm will have a S/N of −30 dB instead of −19 dB.

If the noise is digitized at a higher frequency than 5 MHz, such as 50 MHz, and an input of 3 MHz bandwidth is desirable in the simulation, the noise should be filtered out by a 3 MHz filter before the signal is added. After the filter, the standard deviation of the noise can be obtained and the amplitude of the signal can be found using Equation (12.17).

The S/N problem does not exist with data collected through a hardware system, either from the satellites or from a simulator, because the filter in the receiver will limit the amount of noise.

12.8   ACQUISITION IMPACT OF FILTER BANDWIDTH IN FRONT OF ADC

Because the null-to-null bandwidth of the C/A code is 2.046 MHz, it is natural to assume that the filter bandwidth in front of the ADC will be around 2 MHz. It is interesting to question whether the filter bandwidth has an impact on the acquisition result. In this section, digital filters (rather than analog filters) with different bandwidth are used to evaluate the impact, since digital filters are easy to simulate.

The approach is to use real data collected from the GPS data collection system, described in Section 6.8. The output IF signal is sampled at 10 MHz rather than the 5 MHz, which is used in most examples in this book. The reason to use this sampling frequency is to avoid spectrum overlap. The analog filter at the RF front end of the ADC is about 3 MHz, but it does not have a sharp cutoff frequency response. Energy, including signals and noise, will extend over the 2.5-MHz frequency range. This energy will fold into the 2.5-MHz bandwidth and appear at the digitized output. If the sampling frequency is 10 MHz, the

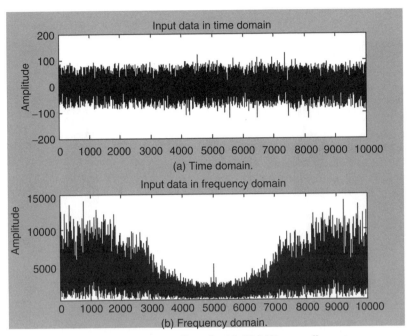

FIGURE 12.20   Data collected with 10 MHz sampling rate.

output bandwidth is 5 MHz. The 3-MHz analog filter can limit most of the energy in this 5 MHz bandwidth. The data collected are shown in Figures 12.20. Figures 12.20a and 12.20b show the time and frequency domain of the data, respectively. In Figure 12.20b the shape of the analog filter in the RF chain before the ADC is shown. As expected, the cutoff frequency of the filter is wider than 2.5 MHz. This data set contains five satellites that can be found using 1 ms of the data for acquisition.

First, the desired signals are acquired from the 10-MHz sampled data. After the desired signal is found the post detection S/N (defined in Equation (10.21)) is calculated. This result can be used as a reference. The postdetection S/N is used in this study because it can provide a better acquisition result, as discussed in Section 10.15. The input data are then digitally filtered, and the same acquisition method is applied. The digital filter used is a fourth-order Butterworth filter with different bandwidth. The bandwidth varies from 0.5 to 2 MHz in 0.1 MHz step. One set of the filtered data (1-MHz bandwidth) is shown in Figure 12.21.

The postdetection S/N-obtained from all five satellites, using 0.5 to 2 MHz bandwidth digital filters, are averaged. The plot is shown in Figure 12.22. The data point at bandwidth of 2.1 MHz represents the case where the input data are not digitally filtered. The highest S/N value achieved is 21.5 dB, when the data are not filtered (nf in the figure). When the filter bandwidth decreases to 1 MHz, the postdetection S/N also decreases to a value slightly above 20.5 dB.

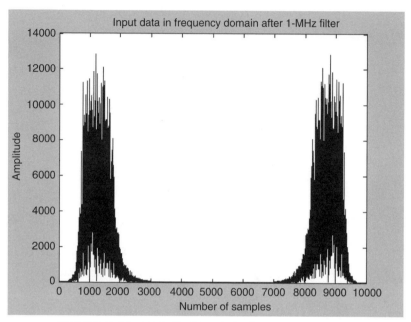

FIGURE 12.21 Digitally filtered (1-MHz bandwidth) data.

FIGURE 12.22 Detection S/N versus filter bandwidth with 10-MHz sampling rate.

The next test is using 5 MHz as the sampling rate to evaluate the bandwidth effect of digital filter. The approach is to decimate the digitally filtered data by 2. In other words, the 10-MHz sampled input data are decimated by 2, and then digitally filtered. The same acquisition method is used. The results are shown in Figure 12.23. When there is no filter (2.1 MHz), the detection S/N is less than 21 dB. This value is lower than 21.5 dB, the corresponding point in Figure 12.22. It is believed that this difference is caused by the noise distribution. The noise bandwidth is more than 2.5 MHz and folds into the receiver. Since 5 MHz is the regularly used sampling frequency in the software receiver, the results in Figure 12.23 can be used to evaluate the effect of the filter bandwidth.

Figure 12.23 shows that when the filter bandwidth is reduced to 1 MHz, the postdetection S/N is decreased by less than 1 dB, compared with a 2 MHz bandwidth. From this simple illustration, one can expect that a lower sampling frequency can be used and the receiver sensitivity will not decrease significantly. For example, if a 1-MHz analog filter is used instead of the present 2 MHz, the receiver might decrease by about 1 dB. With this filter bandwidth, however, the sampling frequency can be reduced to about 2.5 to 3 MHz. Although the low sampling rate can further decrease the receiver sensitivity slightly, it can reduce the computation burden. However, the receiver sensitivity is not the only concern. When a receiver has a narrow band, the correlation peak is rounded off, and this can affect the accuracy of the pseudorange measurements. If a narrow band filter and lower sampling frequency is desirable, further study is needed.

With the improvement in computer operation speed, it appears that 5 MHz sampling rate is a satisfactory value to operate in real time, and there is no need

FIGURE 12.23 Detection S/N versus filter bandwidth with 5-MHz sampling rate.

to reduce the sampling frequency. The next section will cover the number of bits required under strong interference.

12.9 NUMBER OF ADC BITS REQUIRED UNDER JAMMING[7]

In commercial GPS receivers, the ADC used is only one or two bits because all the signals have comparable strength under normal environment. Under intentional jamming, however, it is interesting to consider how many bits are needed to process the GPS signals. This problem can be very complicated because there are different types of jamming signals and many different approaches to mitigate the jamming signal. If the jamming signal cannot be completely removed, the GPS signal may not be found. In order to simplify this study, simulated data with a very simple jamming signal will be evaluated.

Only one jamming and one GPS signal will be used to study this problem. In order to further simplify the problem, the GPS signal strength considered will be at nominal power level, such that 1 ms of data can be used for acquisition. The correlation output then has 5000 points instead of 2500 points in the time domain to achieve the maximum correlation output. The following signal conditions are used:

1. The signals are sampled at 5 MHz; thus the equivalent bandwidth is 2.5 MHz.
2. The input GPS signal $S/N = -18$ dB. The noise bandwidth is 2.5 MHz because of the sampling rate used in this simulation. After 1-ms acquisition through coherent integration, the output $S/N \approx 14.4$ dB.
3. The jamming signal is a cw with a fixed frequency. The frequency matches one of the frequency bins after the FFT operation. Thus the jamming signal can be easily removed in the frequency domain.

The jamming-to-signal ratio (J/S) is the jamming power to signal power ratio, usually expressed in dB. The J/N is the jamming power to noise power ratio expressed in dB. The J/N is used as an input parameter in the simulation. Because the signal is 18 dB below the noise, the relation between J/S and J/N is

$$J/S = J/N - 18 \text{ dB} \tag{12.18}$$

For example, if the $J/S = 120$ dB, the J/N is 102 dB.

The input data are digitized by an ADC with a different number of bits to generate 5000 points. Before performing acquisition on the input data, an FFT operation is performed. Since the input data are real values, and the jamming signal is on a frequency bin, two spectral lines are produced that are symmetrical with respect to the 2500th frequency bin. This jamming signal is removed by making the two spectral lines equal to zero. The remaining results are converted back to time domain through an inverse fast Fourier transform (IFFT). A regular acquisition is performed on the data. The S/N in dB of the GPS signal is obtained after the acquisition. This operation is performed 500 times. Each time a different noise sequence with zero mean and the same standard deviation is used. The results are shown in Table 12.5.

As is shown in the table, if the input signals are digitized with an ADC having a high number of bits, the output S/N is close to 14.4 dB, which is the result without the cw jamming signal. If the number of bits decreases, the S/N also decreases, and this represents the bit reduction effect of the ADC. The S/N value of 14 dB is arbitrarily chosen as the desired value. The number of bits is

TABLE 12.5   Output S/N in dB of the GPS Signal after Processing

J/S in dB	120	100	80	60	40	20
Number of Bits						
20	14.35					
19	14.01					
18	12.88	14.52				
17		14.32				
16		14.20				
15		13.43	14.54			
14		11.11	14.40			
13			14.32			
12			13.71	14.46		
11			12.10	14.35		
10				14.35		
9				14.01		
8				12.60	14.42	
7					14.40	
6					14.08	14.4
5					13.02	14.06
4						13.98
3						12.77

obtained from this value. For example, when $J/S = 100$ dB, 15 bits produces S/N of 13.43 dB, and 16 produces S/N of 14.20 dB. Using a linear relation between these two points ($x = 15$, $y = 13.43$ and $x = 16$, $y = 14.20$), 15.74 bits will produce S/N of 14 dB. Similarly, other points are obtained, and a total of six data points are obtained, as listed in Table 12.6, with different J/S.

It is desirable to plot the number of bits versus the J/S value in Table 12.6. When the number of bits is small, the nonlinear effect of the ADC will occur, which will distort the input waveform. Then, the energy of the cw jamming signal may spread into other frequency bins. The jamming signal cannot be removed effectively. This effect will degrade the output S/N. In addition, when the ADC bit number is getting smaller, the sensitivity of the receiver may degrade slightly, as discussed in Section 6.12. Because of these effects, even at a lower J/S value, a relatively higher number of bits is needed than the expected value. Thus, the last data in this table are not used for the curve fitting.

Using only the first five points in Table 12.6, to draw a line through linear least square fit, the results are shown in Figure 12.24. Note that according to this curve, when the J/S is 20 dB, an ADC with approximately 2.6 bits is adequate. As discussed above, the actual required number of bits is 4.25 in order to produce an output S/N of 14 dB. Since only the high J/S cases are of interest, the curve

TABLE 12.6 J/S versus Number of Bits with Output $S/N = 14$ dB

J/S in dB	120	100	80	60	40	20
Number of bits	18.99	15.74	12.47	8.99	5.93	4.25

FIGURE 12.24 J/S versus number of ADC bits.

can be used to predict the required number of ADC bits. From this curve one can write an empirical equation to predicate the number of bits (N_b) required as

$$N_b \approx \frac{J/S(db)}{6} - 1 \quad \text{or}$$

$$N_b \approx \frac{J/N(dB)}{6} + 2 \tag{12.19}$$

For example, if the $J/S = 66$ dB, then $N_b = 9$ bits is the required value.

12.10   REAL TIME OPERATION OF A SOFTWARE RECEIVER[8,9]

With advancements in computer power, a software receiver could already operate in real time on a personal computer[8]. The computer operates at 2 GHz and actually tracks 8 satellites in real time. To show how this would work, a real time software receiver tracking only 4 satellites is demonstrated. The receiver performs acquisition, then tracking, and finally finding the user location.

The acquisition of the receiver takes some time. The acquisition covers 20 kHz bandwidth with frequency separation of 1 kHz. It takes about 0.15 second to find one satellite and about 0.6 second to find 4 satellites. Because the tracking speed is faster than the data input rate, the tracking will catch up with the input data. The transition from acquisition to tracking problem discussed in Section 9.15 will thus no longer be of concern.

If special attention is given to improving the operating speed of the software, a receiver should be able to track 12 satellites in real time[9]. For example, in this receiver the local frequency used for acquisition covers −10 to 10 kHz. Instead of generating the frequency every time, it generates all the required frequencies, separated by 175 Hz, and the results are stored. The receiver also uses fixed point computations. This operation will speed up the acquisition operation.

If a field programmable gate array (FPGA) is used instead of a personal computer, the receiver should be able to process wider bandwidth signals. For example, in order to study the multipath problem, a wider bandwidth, such as 20 MHz, is desirable to obtain a finer correlation peak. The sampling rate should increase to about 50 MHz. For such a receiver to operate in real time, a FPGA would be used.

It is anticipated that in the near future as software GPS receiver will become a reality because it uses a very simple front end. The performance should surpass the hardware receiver in higher sensitivity areas because more powerful acquisition can be implemented.

12.11   PASSIVE ALTIMETER[10,11]

This section will illustrate the flexibility of a software GPS receiver. A two-channel software GPS receiver with one antenna looking up and one looking

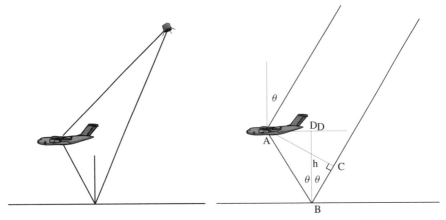

(a) Satellite and aircraft.   (b) Geometry to determine aircraft altitude.

FIGURE 12.25   Direct and reflected signals from a satellite to an aircraft.

down is used to determine the altitude of an aircraft. This altimeter is passive, and no transmitter is required.

The arrangement is to mount a two-channel receiver on the aircraft: channels 1 and 2. Channel 1 has an antenna facing upward. This antenna is right-hand polarized to receive signals directly from the satellites. Channel 2 has a left-hand polarized antenna facing downward to receive the reflected signals from ground. The schematic of each channel is the same as shown in Figure 6.5a. The local oscillators and the digitizing clocks in both channels are phase locked. The two channels are physically close. The same local oscillator and digitizing clock are used for both channels. Figure 12.25a shows the relation of the satellite and the aircraft. Figure 12.25b shows the geometric relationship and is used to determine the altitude of the aircraft. Because the satellite is much higher than the aircraft, the direct rays reaching the aircraft and the reflection point on the ground are parallel.

The height of the aircraft can be found as follows: The difference distance d between the direct and reflected paths is measured and expressed as

$$d \equiv AB + BC = AB(1 + \cos 2\theta) \quad \text{or}$$

$$AB = \frac{d}{1 + \cos 2\theta} \tag{12.20}$$

where AB and BC are distances and θ is the reflection angle, shown in Figure 12.25b. The altitude of the aircraft h (or DB) can be found as

$$h \equiv DB = AB \cos \theta = \frac{d \cos \theta}{1 + \cos 2\theta} = \frac{d \cos \theta}{1 + 2\cos^2 \theta - 1} = \frac{d}{2 \cos \theta} \tag{12.21}$$

If the distance d and the angle θ are known, the altitude of the aircraft can be found.

The next step is to find sensitivity of altitude calculated by different satellites. After taking the derivative of this equation with respect to θ, the result is

$$\frac{dh}{d\theta} = \frac{d[-(1+\cos 2\theta)\sin\theta + 2\cos\theta\sin 2\theta]}{(1+\cos 2\theta)^2} \tag{12.22}$$

In this equation, when θ is small, dh is small. This means that when the satellite is close to zenith, the error in altitude measured is small. In other words, a more accurate altitude measurement can be achieved by using satellites close to the zenith. In addition at the lower elevation satellites the effect is that the surface area, for which the reflection occurs, spreads out. Thus, there is a reduced resolution as well. Satellites near the zenith are selected for the actual altitude calculation from the reflected signals. The calculation is limited to satellites with $\theta < 60°$. The altitudes were calculated, from several satellites and the final value was obtained by the averaging the measurement results.

In the altitude measuring system here, channel 1 works like a regular GPS receiver. It receives and decodes the signals from the satellites. This channel produces two sets of information. First, it generates a reference signal that can be used to correlate with the signal in the second channel. The reference signal requires the initial C/A code, the carrier frequency, and the navigation data to perform a relatively long coherent integration. Since the direct signal is stronger than the reflected signal, this information can be obtained easily. Second, this channel finds the locations of all the satellites. The location of a satellite can be used to determine the angle θ in Figure 12.25.

Flight data were collected from a DC-3 aircraft over southeastern Ohio. The ground surface is mainly forest, with a rolling terrain. The maximum altitude is about 10,000 ft. The reflected signals proved to be relatively weak compared to the direct signals. In order to detect these signals, longer correlation peaks were required. The references were generated from the signals of the first channel. Theoretically, if the signal is known from the direct channel, long coherent integration can be used to obtain the correlation peak of the reflected signals. From experimental results, it appeared that the maximum coherent integration length was about 20 ms. No correlation peak of coherent integration longer than 20 ms increased. The reason is that the reflected signal could have had a slightly different Doppler than the direct signal. The Doppler frequency of the reflected signal could also have been affected by the altitude change, which represents the distance between the downward-looking antenna and the reflection point on the ground. In addition, reflected signal could have contained a distribution of delays, which can result in random phases.

After the 20-ms coherent integration, the results were further processed through 10 noncoherent integrations. The correlation outputs from 10,000 ft are shown in Figure 12.26. In these figures, the first peak is from a direct signal. Although the second antenna is facing downward and is left-hand polarized, it still can receive signals directly from a satellite. In Figure 12.26a, only the 20-ms coherent integration is displayed; the reflection cannot be easily identified. In Figure 12.26b,

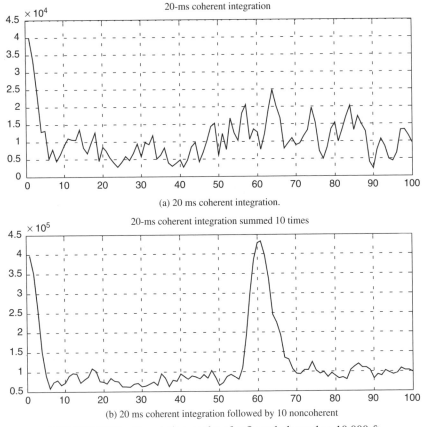

(a) 20 ms coherent integration.

(b) 20 ms coherent integration followed by 10 noncoherent

FIGURE 12.26 Correlation peaks of reflected channel at 10,000 ft.

the correlation peak of the reflected signal can be seen clearly after 10 noncoherent integrations. The time difference between these peaks can be related to the measured distance $d = AB + BC$ in Figure 12.25b. In the actual measurements, the positions of the peaks (direct and reflected signals) were calculated from the fine time resolution method, as discussed in Section 8.11.

Sometimes, the second channel cannot obtain a distinct correlation peak from the direct signal. Then, the correlation peak of the direct signal obtained from the first channel is used as the reference because it is prominent and easy to identify. If the correlation peak of channel 1 is used as reference, the distance between the upper and the lower antenna must be taken into consideration.

The amplitude of the correlation peak does not decrease rapidly with an altitude increase because the refection is from the earth surface rather than from a point source. The reflection from the earth surface does not have the R^4 relation, where R is path length from the source to the receiver. The R^4 relation is often used in a radar equation where the target is usually considered as a point source. This passive altimeter method allows one to find altitudes higher than 10,000 ft.

(a) From GPS altitude calculation. (b) From direction and reflected signals.

FIGURE 11.27   Altitude measured versus time.

Katzberg, Garrison, and Howell [11] were able to measure the reflected signal at about 8 Km over water by this approach. The lower limit of the altitude appears to be about 30 ft in this approach, because the fine time resolution method (Section 8.11) is used to calculate the peak. The lower altitude limit is not limited here by the sampling point, which is separated by 200 ns (approximate 200 ft).

A typical altitude result is shown in Figure 12.27. The upper curve is obtained from the altitude calculation of the software GPS receiver; therefore the height is referenced to sea level. The lower curve is obtained from the direct and reflected signals, and the height is referenced to the surface of the ground. These two curves match fairly well and show that the airplane is descending.

As mentioned earlier in this section, the altitude of the airplane is obtained from the averaged altitudes calculated from different satellites. In the averaging processing, a weighting is used on different altitude values. The weighting factor does not depend on the angle θ in Figure 12.25b but is a function of reflection signal strength. A stronger reflection signal has a large weighting factor because a better peak location can be determined.

12.12   SATELLITE POSITIONS AND DOPPLER FREQUENCIES FROM ALMANAC DATA [12]

This section shows the procedure to calculate the satellite positions from the almanac data. The information obtained from the almanac data can be considered

TABLE 12.7 Almanac Data from One Satellite

********* Week 240 almanac for PRN-01 *********		
ID	01	(Satellite number)
Health	000	(Health condition)
Eccentricity	0.5343914032E − 002	(e_s: Equation(3.23))
Time of applicability (s)	503 808.0000	(t_{yu}: Time of YUMA data)
Orbital inclination (rad)	0.9784364629	(i: Equation(4.45))
Rate of right ascen (r/s)	−0.8034620388E − 008	($\dot{\Omega}$: Equation(4.14))
SQRT(A) (m$^{1/2}$)	5153.525879	($\sqrt{a_s}$: Equation(3.17))
Right ascen at week (rad)	−0.2265141720E + 001	(Ω_e: Equation (4.14))
Argument of perigee (rad)	−1.671651819	(ω : Equation (4.12))
Mean anom (rad)	−0.1932000116E + 001	(M_0: Equation (3.27))
$Af0(s)$	0.3376007080E − 003	(a_{f0}: Satellite clock correction Section 5.12)
$Af1(s/s)$	0.0000000000E + 000	(a_{f1}: Satellite clock correction Section 5.12)
Week	240	(Week of YUMA data)

as a subset of ephemeris data. It does not contain most of the error correction parameters. Therefore, the satellite positions calculated will not be as accurate, but the information is adequate to simplify the acquisition program, as discussed in Section 12.2.

The basic approach to calculating the satellite position is similar to that discussed in Section 4.9. There are two main differences. First, in Section 4.9, the time is obtained from the ephemeris data, whereas in this approach of the time is an input parameter. Second, the user location is given. In Section 4.9, the user location was the final output of the receiver. In the present case, the user location can be used to determine the satellites visible in the sky. The velocities of the satellites can be found from the positions of the satellites as a function of time, such as find the satellite position every second.

An almanac file contains all the satellites in the sky. The data for each satellite are as listed in Table 12.7. The values in Table 12.7 were printed directly from YUMA satellite data, and the information in parentheses relates the data to the symbols used in this book. The following steps are required to find the satellites:

Step 1. Change from local time to GPS time. Suppose that the input time is the local time [given as year, month, date, hour, minute, second]. This time must be changed to GPS time first. The year, month, and date are used to determine the GPS week number (WN). The reference date of WN is 1999 August 21/22, although the GPS WN started on midnight January 5/morning January 6 1980. Its roll over happened in 1999. The number of dates passed since August 21, 1999, divided by 7 gives the WN and the remainder is the date of the week (DOW).

The WN is used to determine the week of YUMA data (WY) used. The DOW is used as part of the input time. The closer the WY to the WN, the more accurate the calculated results should be.

The hour, minute, and second should be changed to Greenwich time. For example, to the local time in Dayton, Ohio, there is added 4 hours for daylight saving time and 5 hours for winter time. The time is then changed to seconds. If the WN and the YUMA week numbers are different, a correction factor is needed. Every sidereal day and solar day is off by approximately $(4 \times 60 - 4.09)$ seconds, and every week is off by $7 \times (4 \times 60 - 4.09)$ seconds. Thus, the overall time in Dayton, Ohio, in the summer is obtained as

$$t_{GPS} = [\text{DOW} \times 24 + (h+4) \times 60 + \text{min}] \times 60 + \text{sec} + 7 \times (WN - WY)$$
$$\times (4 \times 60 - 4.09) \tag{12.23}$$

where h, min, and sec are input time representing local time in hours, minutes, and seconds, and $(h+4)$ represents the 4-hour adjustment from Dayton, Ohio, to Greenwich time in summer. The epoch time t_{oe} can be found as

$$t_{oe} = t_{yu} + a_{f0} \tag{12.24}$$

where t_{yu} and a_{f0} are from the YUMA data and t_{oe} is the time used in the next step of calculations.

Step 2.   Calculate satellite position. The equations used to calculate the satellite position are presented in Chapter 4. They are repeated here for convenience.

The nine quantities obtained from YUMA data are e_s, t_{yu}, i, $\dot{\Omega}$, $\sqrt{a_s}$, Ω, ω, M_0, and a_{f0}. They are used in the following equations repeated here from Chapter 4. The constants used in these calculations are $\mu = 3.986005 \times 10^{14}$ m^3/sec^2 and $\dot{\Omega}_{ie} = 7.2921151467 \times 10^{-5}$ rad/sec.

The mean motion is calculated as

$$n = \sqrt{\frac{\mu}{a_s^3}} \tag{4.33}$$

This equation is simplified because Δn is not available in the almanac data. The mean anomaly can be found as

$$M = M_0 + n(t_{GPS} - t_{oe}) \tag{4.34}$$

where time t_c is replaced by T_{GPS} and t_{oe} can be found in Equation (12.24) The eccentric anomaly E can be found through iteration as

$$E = M + e_s \sin E \tag{4.35}$$

The distance from satellite to the center of the earth r can be found as

$$r = a_s(1 - e_s \cos E) \tag{4.40}$$

The true anomaly v can be found from the following equations as

$$v_1 = \cos^{-1}\left(\frac{\cos E - e_s}{1 - e_s \cos E}\right)$$

$$v_2 = \sin^{-1}\left(\frac{\sqrt{1 - e_s^2}\, \sin E}{1 - e_s \cos E}\right) \tag{4.41}$$

$$v = v_1 sign(v_2)$$

The definition of ϕ is

$$\phi \equiv v + \omega \tag{4.42}$$

The Ω_{er} can be found as

$$\Omega_{er} = \Omega_e + \dot{\Omega}\,(t_{GPS} - t_{oe}) - \dot{\Omega}_{ie}\, t_{GPS} \tag{4.46}$$

where t is replaced by t_{GPS}.

The position of the satellite can be found as

$$\begin{bmatrix} x \\ y \\ z \end{bmatrix} = \begin{bmatrix} r \cos \Omega_{er} \cos \phi - r \sin \Omega_{er} \cos i \sin \phi \\ r \sin \Omega_{er} \cos \phi + r \cos \Omega_{er} \cos i \sin \phi \\ r \sin i \sin \phi \end{bmatrix} \tag{4.47}$$

The satellite position obtained is in earth-center, earth-fixed (ECEF) coordination. This approach is used to calculate the positions of all the satellites.

Step 3.   Coordinate transform. In this operation, three coordinate systems are included. The user position is usually given in latitude (L_c), longitude (l), and altitude (h). The satellite position is in ECEF system. The final satellite position is referenced to the user position in east, north, and height. The following coordinate transform can achieve this goal:

The user position is changed from L_c, l, and h into ECEF system to have the same coordinate with the satellite position through

$$x_0 = \frac{a_e \sin l}{\sqrt{1 + (1 - e_e^2)\tan L_c}} + h \times \cos l \times \cos L_c$$

$$y_0 = \frac{a_e \cos l}{\sqrt{1 + (1 - e_e^2)\tan L_c}} + h \times \sin l \times \sin L_c \tag{12.25}$$

$$z_0 = \frac{a_e(1 - e_e^2)\sin L_c}{\sqrt{1 - e_e^2 \sin^2 L_c}} + h \times \sin L_c$$

where x_0, y_0, and z_0 are the user position in ECEF system; a_e and e_e are the earth semimajor axis and eccentricity as discussed in Section 2.10. The satellite position referenced to the user position can be found as

$$e_0 = -\sin l(x - x_0) + \cos l(y - y_0)$$

$$n_0 = -\sin L_c \cos l(x - x_0) - \sin L_c \sin l(y - y_0) + \cos L_c(z - z_0) \quad (12.26)$$

$$u_0 = \cos L_c \cos l(x - x_0) + \cos L_c \sin l(y - y_0) + \sin L_c(z - z_0)$$

where e_0, n_0, and u_0 are the east, north, and up directions of the satellite with respect to the user position. One can select all the satellites above a certain elevation angle, such as $5°$, as the observable satellites in the sky.

The equations above provide the position of a satellite. If the input time is changed by, say, one second, the position of the satellite can be calculated again. The position difference between these two positions divided by the given time is the satellite velocity. Let us consider a unit vector u_{us}, which is from the user position to a certain satellite. The velocity component of that satellite along u_{us} can be found by projecting the satellite velocity on this unit vector. This velocity component can be used to determine the Doppler frequency of the satellite, which can be used for acquisition.

12.13 EMERGENCY GEOLOCATION WITH CELL PHONE OPERATIONS[13,14]

When a cell phone operator calls 911 to report emergency situation, it is desirable to locate the caller automatically. Often the caller is inside a building. If the building is equipped with a geolocation system, the caller's location inside the building can be identified. The location of the building must also be identified. This combination of information is sent to the emergency unit receiving the call. Another approach is to use the GPS signal to locate the caller. However, inside a building the GPS signal strength may be very weak. An autonomy receiver usually does not have the sensitivity to acquire and process the signals, as discussed in Chapter 10.

One of the approaches is to use a GPS base station in the neighborhood to receive signals at normal strength. The GPS signals of normal strength are acquired and tracked. In other words, the initial phase of the C/A code, the carrier frequency, and the navigation data can all be obtained from the signals of normal strength. This information can be transmitted through the cell system to the cell phone users in the vicinity. The cell phone unit will include a GPS receiver. This GPS receiver can perform acquisition based on the received GPS information. Since the carrier frequency of the GPS is known only a very narrow frequency range need to be searched. Since the navigation data are known, the sensitivity of the receiver is not limited by the navigation data length, as discussed in Section 10.3. The navigation data also provide the opportunity to perform

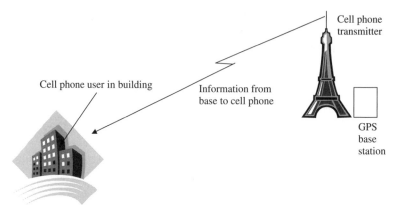

FIGURE 12.28   Concept of a cell phone GPS system.

coherent integration longer than 10 to 20 ms. This way, one can create a long local signal based on the cell phone received GPS information from the base station and perform correlation with received weak signals to acquire it. The satellite positions can also be obtained from the cell phone information. The GPS receiver with the cell phone unit is not required to decode the navigation data, and this limits the sensitivity of a GPS receiver. The only information required to find the cell phone operator position is the pseudorange. With the aid from the cell phone information, the sensitivity can be improved at least another 10 dB such that a $C/N_0 = 14$ dB can be achieved[13,14].

The concept of such a GPS cell phone system is illustrated in Figure 12.28. Under this arrangement, when the GPS signal at the cell phone is strong, it can determine the user position very rapidly, such as in a few ms because the receiver does not need to decode the signal. When the input signal is weak, the receiver can acquire the signal and determine the required pseudorange.

A software GPS receiver can have many other applications that are difficult for a hardware receiver to achieve. It is possible for a software GPS receiver to perform additional functions besides user position determination by using the same hardware but different software. As the processing speed improves, the bandwidth of a software receiver could be increased to process more wideband signals. It is likely that in the near future, there will be built several different types of receivers using different software but only one set of hardware.

REFERENCES

1. Braasch, M., Professor at Ohio University, has provided the Matlab program.
2. Lin, D., Tsui, J., "Direct Y(P)-code acquisition algorithm for software GPS receivers," ION GPS-99, pp. 363–367, Nashville, TN, September 14–17, 1999.
3. Kaplan, E. D., *Understanding GPS Principles and Applications*, Artech House, Norwood, MA, 1996.

4. Behrens, R. T., Scharf, L. L., "Signal processing applications of oblique projection operators," *IEEE Trans. Signal Proc.*, pp. 1413–1423, vol. 42, no. 6, June 1994.

5. Morton, Y. T., Air Force Summer Faculty Fellowship Program Final Report, 2002.

6. Morton, Y. T., Tsui, J., Lin, D., Miller, M., Zhou, Q., French, M., "Assessment and handling of CA code self-interference during weak GPS signal acquisition," ION GPS-03, pp. 646–653, Portland, OR, September 9–12, 2003.

7. Lin, David, AFRL/SNRP private communication.

8. Schamus, J., Tsui, J., Lin, D., "Real-time GPS software receiver," ION GPS 2002 pp. 2561–2565 Portland, OR, September 24–27, 2002.

9. Ledvina, B. M., Psiaki, M. L., Powell, S. P., Kintner, P. M., "A 12-channel real-time GPS L1 software receiver," *Proc. Institute of Navigation National Technical Meeting*, Anaheim, CA, January 22–24, 2003.

10. Liou, L., "Passive altimeter study using GPS flight data," ION GPS-03, pp. 1264–1269, Portland, OR, September 9–12, 2003.

11. Katzberg, S. J., Garrison, J. L., Howell, C. T., "Simple over-water altimeter using GPS reflections," ION GPS-99, pp. 1819–1827, Nashville, TN, September 14–17, 1999.

12. Leick, A., *GPS Satellite Surveying*, 2nd ed., Wiley, New York, 1995.

13. Sheynblat, L., "Satellite positioning system augmentation with wireless communication signals," Patent number 5,99,124 December 7, 1999.

14. Krasner, N. F., "Combined GPS positioning system and communications system utilizing shared circuitry," Patent number 6,002,363 December 14, 1999.

Index

Fundamentals of Global Positioning System Receivers: A Software Approach, Second Edition,
by James Bao-Yen Tsui
ISBN 0-471-70647-7   Copyright © 2005 John Wiley & Sons, Inc.